MATERIALS RESEARCH SOCIETY
SYMPOSIUM PROCEEDINGS VOLUME 1109

Transparent Conductors and Semiconductors for Optoelectronics

December 1-5, 2008
Boston, Massachusetts, USA

Printed from e-media with permission by:

Curran Associates, Inc.
57 Morehouse Lane
Red Hook, NY 12571
www.proceedings.com

ISBN: 978-1-61738-392-2

Some format issues inherent in the e-media version may also appear in this print version.

CAMBRIDGE UNIVERSITY PRESS
Cambridge, New York, Melbourne, Madrid, Cape Town,
Singapore, São Paulo, Delhi, Tokyo, Mexico City

Cambridge University Press
32 Avenue of the Americas, New York, NY 10013-2473, USA

www.cambridge.org

Materials Research Society
506 Keystone Drive, Warrendale, PA 15086
http://www.mrs.org

©Materials Research Society 2009

This publication is in copyright. Subject to statutory exception
and to the provisions of relevant collective licensing agreements,
no reproduction of any part may take place without the written
permission of Cambridge University Press.

First published 2009

CODEN: MRSPDH

ISBN: 978-1-61738-392-2

Cambridge University Press has no responsibility for the persistence or
accuracy of URLs for external or third-part Internet Web sites referred to
in this publication and does not guarantee that any content on such Web sites
is, or will remain, accurate or appropriate.

Additional copies of this publication are available from:

Curran Associates, Inc.
57 Morehouse Lane
Red Hook, NY 12571 USA
Phone: 845-758-0400
Fax: 845-758-2634
Email: curran@proceedings.com
Web: www.proceedings.com

TABLE OF CONTENTS

Nb-doped TiO_2 Films for Transparent Conductive Electrodes with Low Resistivity Deposited by DC Megnetron Sputtering Using a TiO_{2-x}-Nb_2O_{5-x} Target 1
Y. Sato, Y. Sanno, N. Oka, T. Kamiyama, Y. Shigesato

NH_3 Doping in MOCVD Growth of ZnO Thin Films 7
T. Zaidi, M. Jamil, A. Melton, N. Li, W. Fenwick, I. Ferguson

Growth of Transparent Conducting Nb-doped Anatase TiO_2 Thin Films on Glass using Seed Layers 15
N. Yamada, T. Hitosugi, S. Nakao, J. Kasai, Y. Hirose, N.L.H. Hoang, T. Shimada, T. Hasegawa

Low-temperature Fabrication of Transparent Conducting Polycrystalline Nb-doped TiO_2 Films by Sputtering 25
N.L.H. Hoang, N. Yamada, T. Hitosugi, J. Kasai, S. Nakao, T. Shimada, T. Hasegawa

Optical and Electrical Properties of Al Doped ZnO Layers Measured by Wide Angle Beam Spectroscopic Ellipsometry 31
C. Major, G. Juhasz, A. Nemeth, Z. Labadi, P. Petrik, Z. Horvath, M. Fried

High Rate Deposition of Al-doped ZnO by Reactive Sputtering: (1) Unipolar Pulsing with Plasma Emission Control 37
K. Hirohata, Y. Nishi, N. Oka, Y. Sato, I. Yamamoto, Y. Shigesato

Favorable Elements for an Indium-based Amorphous Oxide TFT Channel: Study of In-X-O (X=B, Mg, Al, Si, Ti, Zn, Ga, Ge, Mo, Sn) Systems 43
A. Goyal, T. Iwasaki, N. Itagaki, T. Den, H. Kumomi

Impact of Surface Morphology and Polarity on ZnO Optical Emission 49
D. Doutt, Y. Dong, M. Myers, D. Tayim, C. Zgrabik, Z.Q. Fang, D.C. Look, G. Cantwell, J. Zhang, J.J. Song, H.L. Mosbacker, L.J. Brillson

MOCVD Growth of GaN-based Materials on ZnO and Si Substrates 55
W.E. Fenwick, M. Jamil, T. Xu, S. Wang, H. Yu, A. Melton, N. Li, J. Nause, I.T. Ferguson

Growth and Characteristics of ZnO Nanotube Aarrays on Si Substrates by Atomic Layer Deposition in Anodic Aluminum Oxide 63
W. Miao, G. Han, C.V. Thompson, L.K. Tan, C.S. Jin

Scalable Carbon Nanotube Thin Films: Fabrication, Properties and Device Applications 69
L. Hu, Y. Park, D. Hecht, C. Ladous, M. O'Connell, D. Thomas, G. Gruner, G. Irvin, P. Drzaic

Surface Properties of Polycrystalline Transparent Conducting Oxides 75
A. Klein, C. Korber, A. Wachau, R. Schafranek, Y. Gassenbauer, F. Sauberlich, G.V. Rao

Large Electron Mass Anisotropy in Anatase $Ti_{1-x}Nb_xO_2$ Transparent Conductor 86
Y. Hirose, N. Yamada, S. Nakao, T. Hitosugi, T. Shimada, S. Konuma, T. Hasegawa

Characterization of Amorphous Indium-gallium-zinc-oxide (a-IGZO) Films Deposited by DC Magnetron Sputtering with H_2O Introduction 92
T. Aoi, N. Oka, Y. Sato, R. Hayashi, H. Kumomi, Y. Shigesato

Layer-by-layer Self-assembly of Unilamellar Nanosheet Crystallites of Ruthenium Oxides 98
K. Fukuda, H. Kato, W. Sugimoto, Y. Takasu

PMP-MOCVD Grown $Zn_xCd_{1-x}Se$ cladded $Zn_yCd_{1-y}Se$ Quantum Dot Structures Exhibiting Diode Like Characteristics for Electroluminescent Application 104
F. Al-Amoody, A. Rodriguez, E. Suarez, W. Huang, F. Papadimitrakopoulos, F. Jain

Effect of Annealing on Rectifying Contacts on ZnO Thin Films Grown using Pulsed Laser Deposition 108
A. Bhattacharya, R.K. Gupta, P.K. Kahol, K. Ghosh

Luminescence of ZnO Thin Films Grown on Glass by Radio-Frequency Magnetron Sputtering 114
K. Liu, M. Shur, G. Tamulailtis, S. Cho

First-principles Calculation for Effect of Impurities on Electronic States of Amorphous In-Ga-Zn-O 120
H. Omura, T. Iwasaki, H. Kumomi, K. Nomura, T. Kamiya, M. Hirano, H. Hosono

Effect of Thermal Annealing on Deep and Near-band Edge Emission from ZnO Films Grown by Plasma-assisted MBE 126
V. Avrutin, M.A. Reshchikov, N. Izyumskaya, R. Shimada, S.W. Novak, H. Morkoc

Fabrication and Characterization of Indium Tin Oxide Thin Films on Nanoimprinted Glasses 132
Y. Akita, Y. Sugimoto, M. Hosaka, Y. Kato, Y. Ono, O. Sakata, M. Mita, H. Oi, M. Yoshimoto

High Quality ZnO Thin Films for TCOs and Transistors by MOCVD 138
B.I. Willner, S. Sun, G.S. Tompa

Comparison on Optimized Optical Transmission and Electrical Resistivity between Indium Tin Oxide and Gallium Doped Zinc Oxide 146
W. Hsu, F. Meng, C. Lin, K. Liu, T. Cheng, C. Liu, J. Huang, G. Lin

Electrical and Optical Properties of GaN and ZnO Studied by Surface Photovoltage 153
M. Foussekis, A.A. Baski, M.A. Reshchikov

Transparent and Conductive ZnO: Al Powder Prepared by Soft Chemical Route Process and Design of Experiment Technique 159
K. Chiu, Y. Kao, R. Jean

Enhancement-Mode ZnO Thin-Film Transistor Grown by MOCVD 167
J. Jo, J. Yun, H. Kim

Studies on Electrical Transport in p-ZnO/p-Si Heterojunction 172
S. Majumdar, S. Chattopadhyay, P. Banerji

Influence of Single-Wall Carbon Nanotube Length on the Optical and Conductivity Properties of Thin 'Buckypaper' Films 177
D. Simien, J. Fagan, J.F. Douglas, K. Migler, J. Obrzut

Electrical Characteristics and Practical Properties of Amorphous Indium Zinc Oxide Films and Related Materials 187
M. Kasami, K. Yano, F. Utsuno, T. Shibuya, K. Inoue, B. Shinozaki, K. Makise, M. Funaki

Wavelength Dependent Contrast Reversal in Reflectivity of Nickel Alloy Nanofilms 198
M. Syed, J. Wilkerson, A. Barnett, A. Siahmakoun

Study of Current Stability and Fluctuations of Field Emitted Electrons from ZnO Nanostructure 204
K. Uppireddi, B. Yang, P.X. Feng, G. Morell

Transparent Non-volatile Memory Using Pt Nano-particles Embedded in an Amorphous Indium Gallium Zinc Oxide Thin Film Transistor 210
A. Suresh, S. Novak, P. Wellenius, V. Misra, J.F. Muth

Metal Oxide-based (IZO and ZnO) TFTs for Flexible Electronics 216
S.A. Khan, M. Hatalis

Carrier Transport in Homo- and Heteroepitaxial Zinc Oxide Layers 221
K. Ellmer

Optical and Electrical Characteristics of Amorphous InGaZnO After Thermal Annealing 229
S. Taniguchi, N. Yamaguchi, T. Miyajima, M. Ikeda

Author Index

Mater. Res. Soc. Symp. Proc. Vol. 1109 © 2009 Materials Research Society

Nb-doped TiO$_2$ films for transparent conductive electrodes with low resistivity deposited by dc magnetron sputtering using a TiO$_{2-x}$–Nb$_2$O$_{5-x}$ target

Yasushi Sato[1], Yuta Sanno[1], Nobuto Oka[1], Toshihisa Kamiyama[2] and Yuzo Shigesato[1]

[1]Graduate School of Science and Engineering, Aoyama Gakuin University, 5-10-1, Fuchinobe, Sagamihara, Kanagawa 229-8558, Japan

[2]AGC Ceramics Co. Ltd., 5-6-1, Umei, Takasago, Hyogo 676-8655, Japan

ABSTRACT

Nb-doped anatase TiO$_2$ films were deposited on unheated glass by dc magnetron sputtering using a slightly reduced TiO$_{2-x}$–Nb$_2$O$_{5-x}$ target with oxygen flow ratios [O$_2$/(Ar+O$_2$)] in the range from 0.00 to 0.20%. After postannealing in a vacuum (6 × 10^{-4} Pa) at 500 and 600 °C for 1 h, the films were crystallized into the polycrystalline anatase TiO$_2$ structure. The resistivity of the both films decreased to 6.3-6.8 × 10^{-4} Ω·cm with increasing [O$_2$/(Ar+O$_2$)] to 0.10%, where the carrier density and Hall mobility were 1.9-2.0 × 10^{21} cm^{-3} and 4.9-5.0 cm^2·V^{-1}·s^{-1}, respectively. The films exhibited high transparency of over 60-70% in the visible region of light.

INTRODUCTION

Since Furubayashi et al. reported epitaxially grown Nb-doped anatase TiO$_2$ films deposited by pulse laser deposition [1], TiO$_2$-based transparent conductive oxide (TCO) films have been investigated as indium-free alternatives to tin doped indium oxide (ITO) or indium zinc oxide (IZO), due to advantages of resource availability and non-toxicity compared to indium based TCOs. Hitosugi et al. reported that amorphous films were deposited on unheated glass by pulse laser deposition and the films postannealed at 500 °C under pure H$_2$ gas at 1 atm had the lowest resistivity of 4.6 × 10^{-4} Ω·cm with optical transmittance of 60-80% [2].

For the production of large area commercial uniform TCO coatings, sputter deposition is the most promising deposition technique. Gillispie et al. reported an epitaxial Nb-doped anatase TiO$_2$ film with a resistivity of 3.3 × 10^{-4} Ω·cm, which was deposited by rf magnetron sputtering using Nb-doped TiO$_2$ targets [3]. Yamada et al. reported that amorphous films deposited on unheated glass by reactive sputtering and the postannealed at 600 °C under pure H$_2$ gas at 100 kPa exhibited resistivity down to 9.5 × 10^{-4} Ω·cm with optical transmittance of 75% in the visible region [4].

It has also been reported that TiO$_2$ films for optical coatings [5] and photocatalytic films [6] were deposited by dc magnetron sputtering using a slightly reduced TiO$_{2-x}$ target, where the deposition rate was 5-10 times higher than that by reactive sputter deposition using a Ti metal target. Oxidation of the TiO$_{2-x}$ target surface occurs gradually with increasing O$_2$ flow ratio. Therefore, the TiO$_{2-x}$ target enables precise control of the degree of oxidation for TiO$_2$ films deposited by sputtering under O$_2$ flow ratio. We have reported that a Nb-doped TiO$_2$ film with a resistivity of 1.3 × 10^{-3} Ω·cm was successfully deposited by dc magnetron sputtering using a TiO$_{2-x}$ target and Nb$_2$O$_5$ pellets [7].

In this study, Nb-doped TiO$_2$ films were deposited by dc magnetron sputtering using a reduced TiO$_{2-x}$–Nb$_2$O$_{5-x}$ target, and were then postannealed under vacuum at 500-600 °C. The structural, electrical and optical properties of the films were investigated in detail.

EXPERIMENTAL DETAILS

Nb-doped TiO_2 films were deposited on unheated quartz glass substrates by dc magnetron sputtering using a TiO_{2-x}-Nb_2O_{5-x} target (Nb: 9.5 at.%, AGC Ceramics). Using the TiO_{2-x}-Nb_2O_{5-x} target, a stable dc discharge was maintained due to sufficient conductivity, so that charge-up of the target surface was avoided and arcing was suppressed, as for a TiO_{2-x} target [5-7]. In order to determine a suitable deposition condition, oxygen flow ratios $[O_2/(Ar+O_2)]$ were adjusted precisely to 0.0, 0.05, 0.10, 0.15 and 0.20%. After deposition, all the films were postannealed in a vacuum (6×10^{-4} Pa) at 500 and 600 °C for 1 h in the sputtering chamber. The thickness of all the deposited films was kept to approximately 300 nm by control of the deposition time.

X-ray diffraction (XRD) was performed at 40 kV, 20 mA using Cu $K_{\alpha 1}$ radiation (XRD-6000, Shimadzu). The Nb concentration of the films was analyzed using electron probe microanalysis (EPMA; JXA-8100, Jeol), where the compositions represented the cation ratios [Nb/(Ti+Nb)]. The electrical properties of the films were measured by the four-point probe method, and Hall-effect measurements in the van der Pauw geometry (HL-550PC, Bio-Rad). Transmittance and reflectance of the films were measured from 200 to 2500 nm using a spectrophotometer (UV-3100, Shimadzu).

RESULTS and DISCUSSION

Figure 1 shows the deposition rate by dc sputtering, using the TiO_{2-x}-Nb_2O_{5-x} target as a function of the O_2 flow ratio. The variation in deposition rate as a function of O_2 flow ratio did not show a hysteresis curve as in the case of the Ti metal targets [8,9], which implies that oxidization on the TiO_{2-x}-Nb_2O_{5-x} target surface occurs gradually with increasing O_2 flow ratio. This result using the TiO_{2-x}-Nb_2O_{5-x} target is similar to that for the TiO_{2-x} target as previously reported [6,7]. The deposition rate decreases monotonically with increasing O_2 flow ratio from 0.0 to 1.0%, and subsequently remains constant above 5%. The highest deposition rate by dc sputtering of the TiO_{2-x}-Nb_2O_{5-x} target was approximately 20 nm/min when the film was deposited under a 0.10% $O_2/(Ar+O_2)$ flow ratio, which is approximately twice that when using the TiO_{2-x} target [6,7].

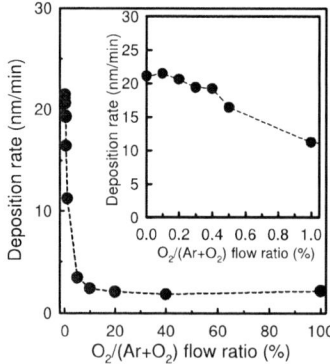

Figure 1. Deposition rates of a Nb-doped TiO_2 film by dc sputtering using a TiO_{2-x}-Nb_2O_{5-x} target, as a function of the $O_2/(Ar+O_2)$ flow ratio.

Nb concentrations of the postannealed films were analyzed by EPMA, as shown in Fig. 2. The Nb concentrations of all the films were found to be almost constant at approximately 8 at.%, which were lower than that in the TiO_{2-x}-Nb_2O_{5-x} target.

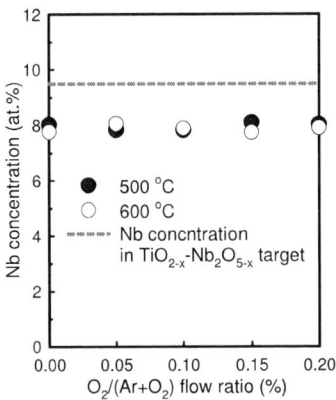

Figure 2. Nb concentrations of Nb-doped TiO_2 films postannealed at 500 and 600 °C as analyzed by EPMA.

Figures 3(a)-(b) show XRD patterns of the postannealed Nb-doped TiO_2 films sputter deposited under various $O_2/(Ar+O_2)$ flow ratios. Although the as-deposited films were entirely amorphous, the postannealed films were polycrystalline anatase TiO_2. No impurities such as rutile TiO_2, Ti-based Magneli phases (Ti_nO_{2n-1}), Nb_2O_5 or NbO_2 were detected in the sample, within the resolution of the XRD measurement. Figure 3(c) shows the d-spacing for the (101) peak of anatase TiO_2, estimated from the XRD peak position as a function of the $O_2/(Ar+O_2)$ flow ratio during sputter deposition. The d-spacing for both the postannealed films decreased with increasing O_2 flow ratio from 0.00 to 0.10%.

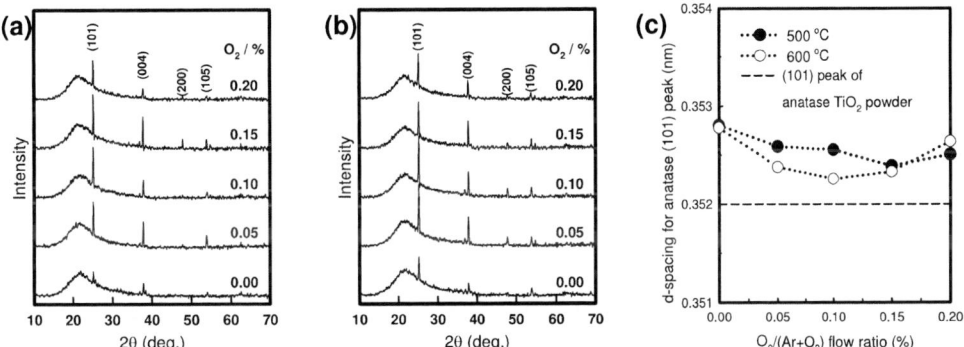

Figure 3. XRD patterns of Nb-doped TiO_2 films prepared by sputter deposition under various $O_2/(Ar+O_2)$ flow ratios and then postannealed at (a) 500 and (b) 600 °C for 1 h. (c) Variation in the d-spacing of the (101) peak for both the postannealed Nb-doped TiO_2 films.

Figure 4 shows the (a) resistivity, (b) carrier density and (c) Hall mobility of the postannealed films, as a function of the $O_2/(Ar+O_2)$ flow ratio during sputter deposition. All of the as-deposited films did not exhibit electrical conductivity. However, after postannealing in a vacuum at 500 and 600 °C, the films for both postannealing conditions became conductive and the resistivity decreased to 6.3-6.8 × 10^{-4} Ω·cm with increasing O_2/(Ar+O_2) flow ratio to 0.10% during sputter deposition. The resistivity then increased to 0.8-1.2 × 10^{-2} Ω·cm with further increase in the O_2/(Ar+O_2) flow ratio to 0.20% during sputter deposition. The variations in resistivity are inversely proportional to the variations in the Hall mobility, that is, the Hall mobility increased with increasing O_2/(Ar+O_2) flow ratio during sputter deposition and reached the highest value of approximately 5 $cm^2·V^{-1}·s^{-1}$ for the 0.10% O_2/(Ar+O_2) flow ratio. In contrast, the mobilities were relatively lower, around 3-4 $cm^2·V^{-1}·s^{-1}$ for 0.00% O_2/(Ar+O_2) flow ratios and approximately 2 $cm^2·V^{-1}·s^{-1}$ for 0.20% O_2/(Ar+O_2) flow ratios. From the XRD results in Figs. 3(a) and 3(b), the intensity of the anatase (101) peak for the films sputtered in 0.00 and 0.20% O_2/(Ar+O_2) flow ratios were lower than those sputtered at 0.10% O_2/(Ar+O_2) flow ratios. Therefore, the decrease in the mobility of the films at 0.00 and 0.20% O_2/(Ar+O_2) flow ratios could be attributed to the degradation in the crystallinity of the film. On the other hand, the carrier density increased slightly to 1.9-2.0 × 10^{21} cm^{-3} with increasing O_2/(Ar+O_2) flow ratio to 0.10%, and a significant decrease in carrier density to 2.3-4.0 × 10^{20} cm^{-3} was observed for increase of the O_2/(Ar+O_2) flow ratio to 0.20%. The efficiency of electrically activated Nb^{5+} at Ti^{4+} sites, i.e. the doping efficiency, for the postannealed films with the lowest resistivities of 6.3-6.8×10^{-4} Ω·cm were estinmated. The doping efficiencies were approximately 80%, which is lower than the one of previously reported epitaxial films [1].

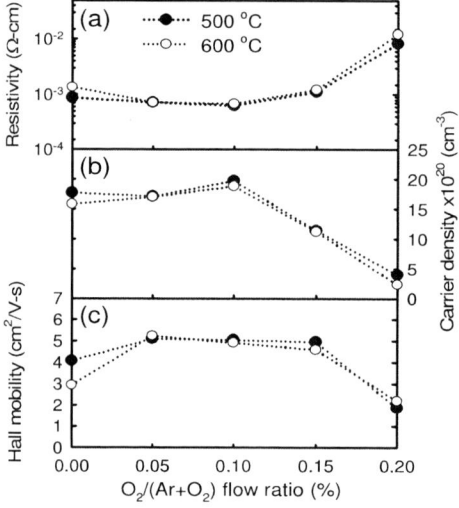

Figure 4. (a) Resistivity, (b) carrier density and (c) Hall mobility of Nb-doped TiO_2 films postannealed at 500 (●) and 600 °C (○), as a function of the O_2/(Ar+O_2) flow ratio during sputter deposition.

Figures 5(a)-(b) show the transmittance and reflectance of films sputtered under various O_2/(Ar+O_2) flow ratios and then postannealed. All the films exhibited 60-70% transmittance in the visible region. The optical band gap (E_g) of the films were estimated from an extrapolation method for highly degenerated oxide semiconductors, based on the assumption of direct allowed transitions to an empty conduction band for anatase TiO_2 [10]. The square roots of the absorption coefficients (α^2) were estimated from the transmittance and reflectance spectra. The E_g values were extrapolated and showed an increase from 3.5 to 3.6 eV with an increase in the carrier density from 2.3-4.0 × 10²⁰ to 1.9-2.0 × 10²¹ cm⁻³. The widening of E_g can be explained by the Burstein-Möss (BM) effect, in which the lowest states in the conduction band are blocked, and transitions can only take place to energies higher than the Fermi energy [11].

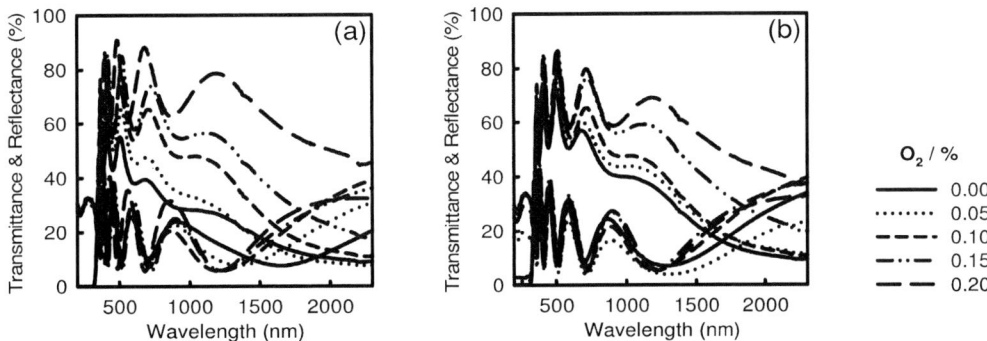

Figure 5. Transmittance and reflectance from 200 to 2500 nm of Nb-doped TiO_2 films sputter deposited under various O_2/(Ar+O_2) flow ratios and postannealed at (a) 500 and (b) 600 °C.

CONCLUSIONS

Nb-doped TiO_2 films were deposited on unheated glass substrates by dc magnetron sputtering using a TiO_{2-x}-Nb_2O_{5-x} target with O_2/(Ar+O_2) flow ratios in the range from 0.00 to 0.20%, and then postannealed in a vacuum at 500 and 600 °C for 1 h. The deposition rate using the TiO_{2-x}-Nb_2O_{5-x} target decreased monotonically with increasing O_2/(Ar+O_2) flow ratio from 0.0 to 1.0%, which implies that oxidation on the TiO_{2-x}-Nb_2O_{5-x} target surface occurs gradually with increasing O_2 flow ratio. EPMA analyses revealed that the concentration of Nb in all the films postannealed remained almost constant around 8 at.%, regardless of the O_2/(Ar+O_2) flow ratio during deposition. XRD measurements showed that the films postannealed at both temperatures formed polycrystalline anatase TiO_2 structure. The lowest resistivities of 6.3 × 10⁻⁴ and 6.8 × 10⁻⁴ Ω·cm were obtained for the films deposited under a 0.10% O_2/(Ar+O_2) flow ratio and postannealed at 500 and 600 °C, respectively. The transmittance of films postannealed was 60-70% in the visible region.

ACKNOWLEDGMENTS

This work was partially supported by a High-Tech Research Center project for private universities with a matching fund subsidy from the Ministry of Education, Culture, Sports, Science and Technology (MEXT) of Japan.

REFERENCES

1. Y. Furubayashi, T. Hitosugi, Y. Yamamoto, K. Inaba, G. Kinoda, Y. Hirose, T. Shimada and T. Hasegawa, *Appl. Phys. Lett.* **86**, 252101 (2005).
2. T. Hitosugi, A. Ueda, S. Nakao, N. Yamada, Y. Furubayashi, Y. Hirose, T. Shimada, and T. Hasegawa, *Appl. Phys. Lett.* **90**, 212106 (2007).
3. M. A. Gillispie, M. F. A. M. van Hest, M. S. Dabney, J. D. Perkins, and D. S, Ginley, *J. Appl. Phys.* **101**, 033125 (2007).

4. N. Yamada, T. Hitosugi, N. L. H. Hoang, Y. Furubayashi, Y. Hirose, S. Konuma, T. Shimada, and T. Hasegawa, *Thin Solid Films* **516**, 5754 (2008).

5. H. Ohsaki, Y. Tachibana, A. Hayashi, A. Mitsui, and Y. Hayashi, *Thin Solid Films* **351**, 57 (1999).

6. Y. Sato, A. Uebayashi, N. Ito, T. Kamiyama, and Y. Shigesato, *J. Vac. Sci. Technol.* **A 26**, 903 (2008).

7. Y. Sato, H. Akizuki, T. Kamiyama, and Y. Shigesato, *Thin Solid Films* **516**, 5758 (2008).

8. M. Yamagishi, S. Kuriki, P. K. Song, and Y. Shigesato, *Thin Solid Films* **442**, 227 (2003).

9. S. Ohno, N. Takasawa, Y. Sato, M. Yoshikawa, K. Suzuki, P. Frach, and Y. Shigesato, *Thin Solid Films* **496**, 126 (2006).

10. D. Kurita, S. Ohta, K. Sugiura, H. Ohta, and K. Koumoto, *J. Appl. Phys.* **100**, 096105 (2006).

11. I. Hamberg and C. G. Granqvist, *J. Appl. Phys.* **60**, 123 (1986).

NH₃ Doping in MOCVD Growth of ZnO Thin Films

Tahir Zaidi[1], Muhammad Jamil[1], Andrew Melton[1], Nola Li[1], William Fenwick[1], and Ian Ferguson[1]

[1]School of Electrical and Computer Engineering, Georgia Institute of Technology, Atlanta, Georgia 30332 USA

ABSTRACT

In this paper effects of NH₃ doping on ZnO thin films grown by metal organic chemical vapor deposition (MOCVD) on c-plane sapphire substrates using diethyl zinc (DEZn) and O_2 precursors and N_2 as the carrier gas have been studied. NH₃ flow rates were varied from 0.1% to 4% in the growth runs. All the runs were done at 500°C at 10 Torr pressure.

The XRD measurements show a single ZnO (002) peak. Raman data for the samples confirms presence of ZnO:N modes at $275 cm^{-1}$, $510 cm^{-1}$ and $575 cm^{-1}$ and $645 cm^{-1}$. The PL results for Zn rich films show weak broad peaks centered at 480nm and 650nm with no ZnO band edge emission, while oxygen rich films show weak ZnO band edge emission and a strong broad orange peak centered at 650nm. Hall effect measurements indicate that all of the as-grown films are highly resistive. Some are weakly p-type with carrier concentration of 4.24×10^{14} cm^{-3} and mobility of 16.55 cm^2/Vs. Annealing in N_2 ambient for 60 minutes at 800°C enhances the PL band edge emission and converts all the films to highly conducting n-type, with carrier concentration on the order of 8×10^{18} cm^{-3}, mobility on the order of 12 cm^2/Vs and resistivity of 0.063 Ω-cm.

INTRODUCTION

Zinc oxide (ZnO) with its wide bandgap (3.37 eV) and high exciton binding energy (~60 meV) is of great interest for use in optoelectronic applications. However, the major limitation it faces is the lack of reproducible p-type doping. While n-type material is easily attainable, p-type material has proved much more difficult to achieve. While several different p-type dopants and growth techniques have been reported, stability and reproducibility are still a problem. [1, 2]. Possible acceptor dopants include group-V (N, P, and As) and group-I elements (Li, Na). Out of these N substituting for O appears to be the most promising choice because of their similar atomic radii and the low activation energy of N compared to other Group V elements [3–7].

N-doped ZnO has been reported with a range of different precursors and growth techniques. MOCVD offers many advantages over other growth methods including simple and accurate doping and thickness control and scalability to production level throughput, however there are very few reports of N-doping of ZnO by MOCVD. One of the major reasons for this is the low solubility of N in ZnO, especially in a CVD process [8].

Optical properties have been studied by room temperature and low temperature photoluminescence (PL). Electrical properties of the doped ZnO thin films were studied using Hall effect measurements. Structural characteristics were determined Raman spectroscopy and X-ray diffraction (XRD). Surface morphology of the films was investigated using field-emission scanning electron microscopy (SEM).

EXPERIMENT

Growth runs for undoped and N-doped ZnO thin films were carried out by metal organic chemical vapor deposition (MOCVD) in a vertical injection chamber using diethyl zinc (DEZn) and O_2 as the zinc and oxygen precursors. NH_3 was used as the dopant source.

Chamber temperature was set at 500°C and pressure at 10 Torr. The NH_3 injection was varied from 0.1% to 4% of the O_2 flow. The VI-II ratio of the precursors (O_2 and Zn) was varied to see its effect on the resulting films. Best results were seen with a VI-II ratio above 500. The NH_3 doping attempts below 500°C and for VI-II ratios below 500 each resulted in increased C in the films, readily forming (NC) complexes which gave a pinkish-brown tint to the films.

The as-grown samples were broken up and subjected to three different annealing conditions: in O_2 ambient at 700°C for 60 minutes; in N_2 ambient at 800°C for 60 minutes; and rapid thermal annealing at 800°C in N_2 ambient. Annealing in NH_3 ambient at 800°C was also attempted but since NH_3 cracks at this temperature it etched the ZnO:N films completely. The samples were characterized for their structural, optical, and electrical and magnetic properties.

DISCUSSION

Structural Characterization

The structural properties were studied by X-ray diffraction (XRD), Raman Spectroscopy and SEM imagery. All three characterization methods showed deteriorating crystal quality with increased N incorporation. All samples grown with NH_3 in excess of 1% were too structurally degraded to be electrically or optically useful.

X-ray diffraction (XRD) ω- 2Θ scans of the undoped and N-doped films show a single (0002) ZnO peak, suggesting a high (002) preferential orientation. For undoped ZnO this peak has a FWHM of 174 arcsec. With increased NH_3 flow the FWHM of the ZnO(002) peak increases and its position shifts (Figure 1a). The shift in peak position was seen to be fairly linear from 0.1% to 4% NH_3 flow (Figure 1b). These results show deteriorating crystal quality and increased strain with increased N incorporation. Figure 2 shows the SEM images of 4% N doped ZnO film. It can be seen that the surface roughness and grain size both increase with increased NH_3 flow during growth.

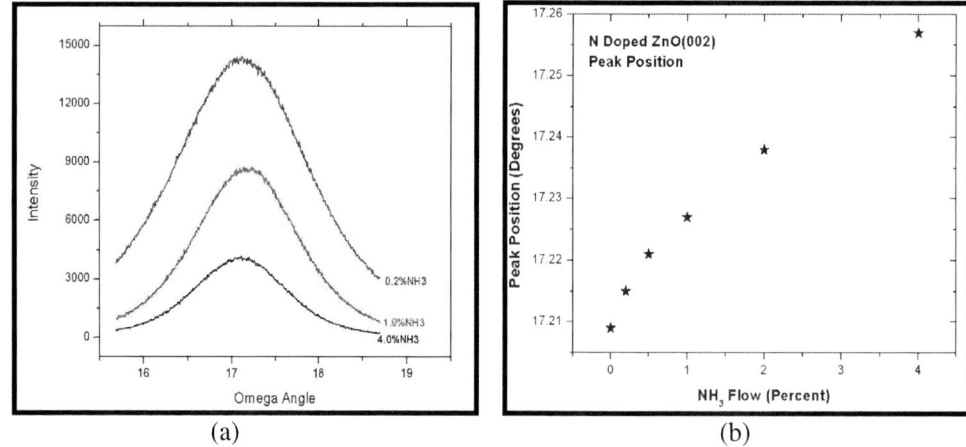

(a)   (b)

Figure 1. (a) XRD of ZnO:N Films showing increasing FWHM with increased NH_3 doping. (b) Peak position of ZnO (002) plotted against NH_3 flow. An almost linear drift in peak position with increased NH_3 flow is clearly seen.

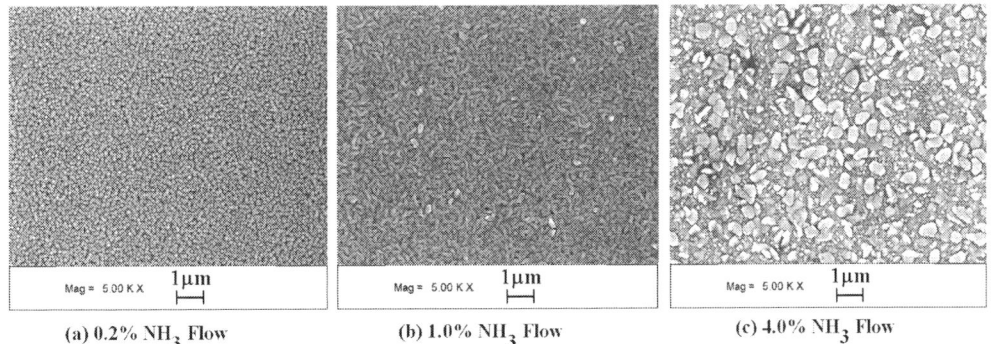

(a) 0.2% NH₃ Flow   (b) 1.0% NH₃ Flow   (c) 4.0% NH₃ Flow

Figure 2. SEM images of ZnO:N films grown with (a) 0.2% NH_3 flow, (b) 1.0% NH_3 flow, and (c) 4.0% NH_3 flow.

Raman spectroscopy (Figure 3) for all doping levels (NH_3 flows from 0.1% to 4%) show ZnO:N modes at 275cm^{-1}, 510cm^{-1}, 575cm^{-1} and 645cm^{-1}. It was seen that increased NH_3 flow resulted in these ZnO:N modes increasing in intensity while the native ZnO modes of 437cm^{-1}, 380cm^{-1}, and 332cm^{-1} were seen to reduce. This also points towards deteriorating crystal quality with N incorporation.

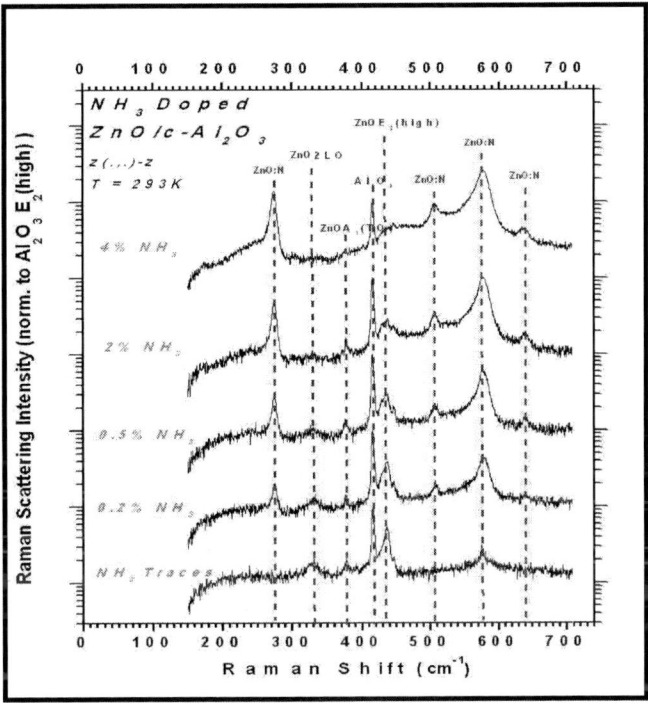

Figure 3. Raman Spectroscopy for ZnO:N films grown with different NH_3 flow rates. We can see a gradual deterioration of crystal quality with increasing NH_3 flow.

Optical Characterization

Figure 4 shows the PL spectra of ZnO:N films obtained for different NH_3 flow rates with VI-II ratio of 500 (Zn rich). It can be seen that ZnO band edge emission is suppressed and there are two broad peaks centered at 480nm and 650nm respectively. These emission peaks have been reported in literature for N-doped ZnO thin films and are attributed to deep acceptor levels [9]. The broad blue-green band emission from the ZnO has been generally explained as the radiative recombination of a photo-generated hole with the electrons at the singly ionized intrinsic oxygen vacancies [10], which form during high temperature growth [11].

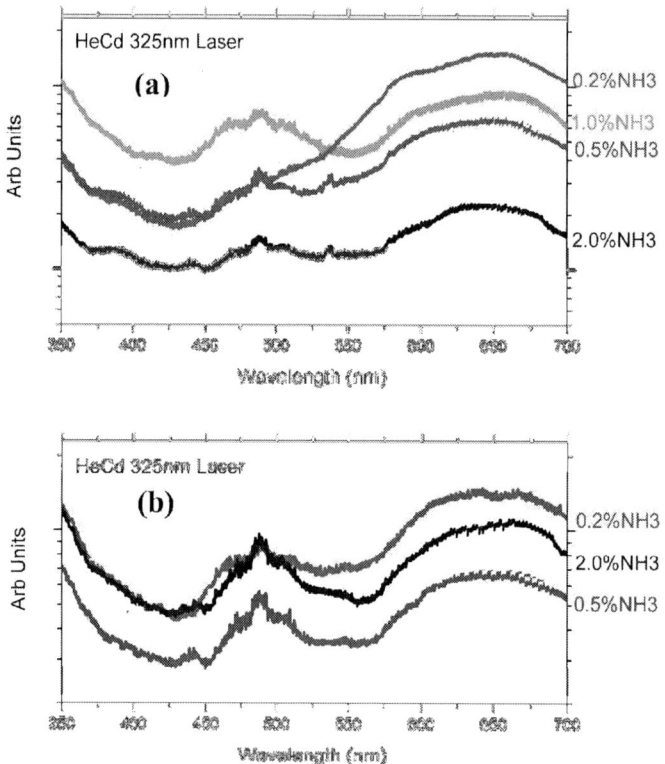

Figure 4. (a) PL spectra of ZnO:N films grown with VI-II ratio of 500 (Zn rich) (b) PL spectra of the same films after annealing in O_2 ambient at 700°C for 60 minutes. No appreciable change in PL was observed with annealing in O_2 ambient.

Figure 5 shows the PL spectra of ZnO:N films obtained for different NH_3 flow rates with VI-II ratio of 1000 (O_2 rich). In this case ZnO band edge emission is visible and a strong broad peak centered at 600nm is also seen. There are different accounts given for this orange-red broad peak in literature [12,13]. One group attributes it to defects, as a similar emission band is seen after N^+ implantation and doping with Li and P [12]. The other proposed explanation is that this peak is the result of excess O_2 in the films and O-H complex formation [13]. The results from this study support the second explanation since this band is seen to be more prominent in O_2 rich films and is also seen to reduce with annealing in N_2 ambient.

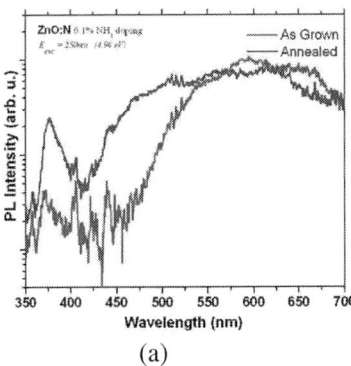 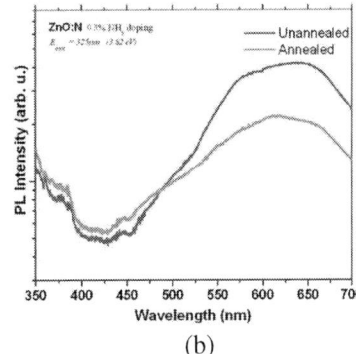

(a)   (b)

Figure 5. PL spectra of ZnO:N films grown with VI-II ratio of 1000 (O_2 rich) both as-grown and annealed in N_2 at 800°C for 60 minutes sample (a) With 0.1% NH_3 doping (b) 0.3% NH_3 doping

Figure 6 shows the low temperature PL spectra of 0.1% and 0.3% NH_3 doped ZnO:N films annealed in N_2 at 800°C for 60 minutes. There are 7 peaks visible in both these spectra which correspond to band edge, N incorporation and defects.

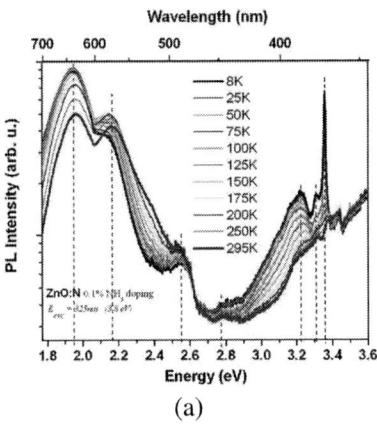 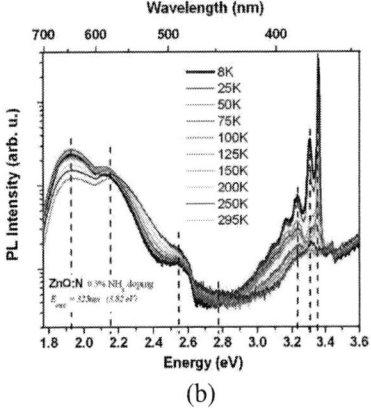

(a)   (b)

Figure 6. Low temperature PL spectra of ZnO:N films grown with VI-II ratio of 1000 (O_2 rich) and annealed in N_2 at 800°C for 60 minutes, sample (a) With 0.1% NH_3 doping (b) 0.3% NH_3 doping

Electrical Characterization

The results of Hall effect measurements are summarized in Table 1. It can be seen that all as grown films are highly resistive. Weak p-type conductivity in as-grown N-doped ZnO thin films with 0.2% NH_3 flow and 1% NH_3 flow was observed. The carrier concentration for these two films is very low with high resistivity and low mobility. Table 2 gives data for samples that underwent N_2 ambient annealing at 800°C for 60 minutes.

Table 1. Hall effect data for as grown and 700°C 60 min O_2 ambient annealed samples.

NH₃ Flow	As Grown				Annealed in O_2 at 700C for 60 min			
	Type	Carrier (cm⁻³)	μ (cm²/Vs)	ρ (Ω cm)	Type	Carrier (cm⁻³)	μ (cm²/Vs)	ρ (Ω cm)
Undoped	n	-1.24 e 17	0.43	115.6				
0.2%	p	9.5 e 12	4.39	1.5 e 5	p	9.7 e 13	34.91	1.84 e 3
0.5%	n	-1.77 e 18	4.74	0.75	n	-1.11 e 18	18.50	0.30
1.0%	p	4.24 e 14	16.55	895				

Table 2. Hall effect data for samples annealed in N2 ambient at 800°C for 60 min.

NH₃ Flow	Type	Carrier (cm⁻³)	μ (cm²/Vs)	ρ (Ω cm)
0.1%	n	-6.43 e 18	19.23	0.049
0.2%	n	-8.39 e 18	11.81	0.063
0.3%	n	-7.87 e 18	15.33	0.051

The mobility of the carriers appears to increase with increasing NH₃ flow rate (Figure 7). Annealing in O_2 ambient at 700°C for 60 minutes causes the mobility to decrease for higher NH₃ doping levels. From this it can be deduced that annealing in O_2 increases the number of holes.

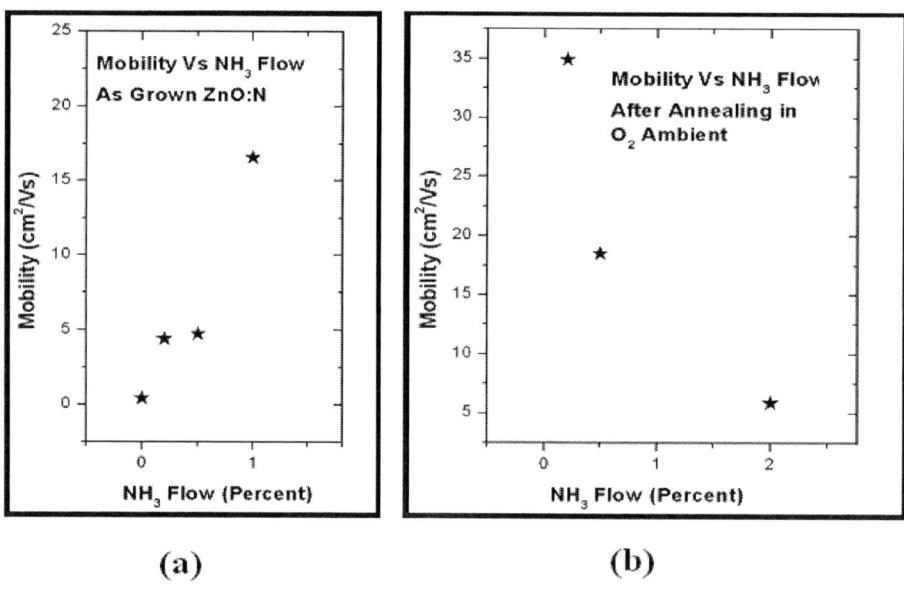

(a)   **(b)**

Figure 7. Plots of mobility vs. NH₃ flow for (a) As-grown ZnO:N films (b) After annealing in O_2 ambient at 700°C for 60 minutes.

Analysis

In MOCVD, low growth temperatures drastically reduce the crystal quality of ZnO films. Furthermore below 500°C excessive C in the films readily formed (NC) complexes with NH_3 giving pinkish-brown tint to the films. For high NH_3 flow rates the end result is an opaque brownish film. While high temperature growth of ZnO thin films by MOCVD improves crystal quality and reduces "hole killer" defects it also results in very low N incorporation. If an attempt is made to increase the N concentration by increased dopant (NH_3) flow, the crystal quality deteriorates rapidly for NH_3 flow beyond 1%. In addition, increased NH_3 at high temperatures gives the same pinkish-brown tint to the films that were observed for the NC complexes formed due to C in the films. High N concentration in ZnO lowers the formation energy and causes undesirable defects, such as $(N_2)_O$ or $(NC)_O$, thereby compensating the N_O acceptor that has been widely reported in literature[8,14-16].

CONCLUSION

ZnO:N thin films were successfully grown on c-plane sapphire using DEZn and O_2 precursors and NH_3 as the doping source. The ideal growth conditions for these films were found to be 500°C and a VI-II ratio above 500. The maximum doping level was determined to be 1% NH_3 beyond which the crystal quality was seen to degrade steadily until the films were nearly amorphous at 4% NH_3 flow. The as-grown films were found to be resistive. Semi-insulating films with resistivity of the order of 2×10^6 Ω-cm were obtained. Passivation due to background carrier concentration, H_2 donors, formation of (O-H) complexes, $(N_2)_o$, and defects prevented the realization of good quality p-type films. The best p-type films obtained had carrier concentration of 4.24×10^{14} cm^{-3} and mobility of 16.55 cm^2/Vs. Annealing in N_2 ambient turned these films into highly conducting n-type with carrier concentration on the order of 8×10^{18} cm^{-3}, mobility on the order of 12 cm^2/Vs and resistivity of 0.063 Ω-cm. Effects of changing VI-II ratio and NH_3 flow rates and growth temperature on the grown films were investigated. XRD measurements show a single ZnO (002) peak; Raman data shows presence of ZnO:N modes at $275cm^{-1}$, $510cm^{-1}$, $575cm^{-1}$ and $645cm^{-1}$; and PL results shows broad peaks at 480nm and 650nm for all the samples.

ACKNOWLEDGEMENTS

This research was supported in part by the US Air Force Office of Scientific Research (AFOSR).

REFERENCES

1. Kaminska E., Przezdziecka E., Piotrowska, A., Kossut, J., Boguslawski, P., Pasternak, I, Jakiela, R., Dynowska, E, *Zinc Oxide and Related Materials Symposium*, Nov. 2006
2. Rogers D J, Teherani F H, Yasan A, Minder K, Kung P and Razeghi M 2006 *Appl. Phys. Lett.* **88** 141918
3. J. Wang, G. Du, B. Zhao, X. Yang, Y. Zhang, Y. Ma, D. Liu, Y. Chang, H. Wang, H. Yang, and S. Yang, J.Cryst. Growth 255, 293 (2003).
4. D.C. Look, D.C. Reynolds, C.W. Litton, R.L. Jones, D.B. Eason, G. Cantwell, Appl. Phys. Lett. 81 (2002) 1830.
5. J.F. Rommeluere, L. Svob, F. Jomard, J. Mimila-Arroyo, A. Lusson, V.Sallet, Y. Marfaing, Appl. Phys. Lett. 83 (2003) 287.
6. W.Z. Xu, Z.Z. Ye, T. Zhou, B.H. Zhao, L.P. Zhu, J.Y. Huang, J. Cryst.Growth 265 (2004) 133
7. J.G. Lu, Z.Z. Ye, F. Zhuge, Y.J. Zeng, B.H. Zhao, L.P. Zhu, Appl. Phys.Lett. 85 (2004) 3134.

8. Y.J. Zeng, Z.Z. Ye. W.Z. Xu, B. Liu, Y. Che, L.P. Zhu, B.H. Zhao, Materials Letters 61 (2007) 41–44

9. Klingshrin C. "ZnO: Material, Physics and Applications", ChemPhysChem 2007, 8, 782 – 803

10. K. Vanhausden, W.L. Warren, C.H. Seager, D.R. Tallant, J.A. Voigt, B.E. Gnade, J. Appl. Phys. 79 (1996) 7983.

11. Z.Q. Chen, S. Yamamoto, M. Maekawa, A. Kawasuso, X.L. Yuan, T. Sekiguchi, J. Appl. Phys. 94 (2003) 4807.

12. C. J. Pan, B. J. Pong, B. W. Chou, G. C. Chi, and C. W. Tu, phys. stat. sol. (c) 3, No. 3, (2006) 611–613.

13. ABDjuri˘ si´ c, Y H Leung, K H Tam,Y F Hsu, L Ding, W K Ge, Y C Zhong, K S Wong, W K Chan, H L Tam, K W Cheah, W M Kwok and D L Phillips, Nanotechnology 18 (2007) 095702,

14. E.-C. Lee, Y.-S. Kim, Y.-G. Jin, K.J. Chang, Phys. Rev., B 64 (2001) 085120.

15. S. Limpijumnong, X.N. Li, S.-H. Wei, S.B. Zhang, Appl. Phys. Lett. 86 (2005) 211910.

16. N.H. Nickel, F. Friedrich, J.F. Rommeluère, P. Galtier, Appl. Phys. Lett. 87 (2005) 211905.

Growth of Transparent Conducting Nb-doped Anatase TiO$_2$ Thin Films on Glass using Seed Layers

N. Yamada[1], T. Hitosugi[1,2], S. Nakao[1,3], J. Kasai[1], Y. Hirose[1,3], N. L. H. Hoang[1,3], T. Shimada[1,3] and T. Hasegawa[1,3]

[1]Kanagawa Academy of Science and Technology (KAST), Kawasaki 213-0012, Japan.
[2]Advanced Institute for Materials Research (AIMR), Tohoku University, Sendai 980-8577, Japan.
[3]Department of Chemistry, University of Tokyo, Tokyo 113-0033, Japan.

ABSTRACT

We report recent progress on sputter-growth of Nb-doped anatase TiO$_2$ (TNO) polycrystalline films, being a promising ITO-alternative transparent conductor. In order to achieve low resistivity (ρ) in TNO, it is necessary to grow anatase phase under reducing atmospheres. However, growth of TNO polycrystalline films on glass under such conditions tends to stabilize rutile phase with higher resistivity. To overcame this difficulty, we have developed a bi-layer technique using a TNO self seed-layer, which prevents the formation of the rutile phase even under reducing growth conditions. As a result, we succeeded in directly fabricating TNO polycrystalline films with ρ of ~1×10^{-3} Ω cm and visible transmittance of 60 ~ 80%, although we still need to further improve these properties towards practical applications. By comparing dc transport properties with optically deduced ones, we discuss material parameters that limit carrier transport in presently obtained TNO polycrystalline films.

INTRODUCTION

Nb-doped anatase TiO$_2$ (TNO) is a new family of transparent conducting oxide (TCO) with excellent conductivity and visible transparency [1-9]. TNO epitaxial films grown by pulsed laser deposition (PLD) technique exhibit resistivity (ρ) of 2 - 3×10^{-4} Ω cm and internal transmittance higher than 90% [1, 2]. These properties are comparable to those of Sn-doped In$_2$O$_3$ (ITO), which is commonly used in various optoelectronic devices, such as flat panel displays, photovoltaics and light emitting diodes. Accordingly, TNO is a promising TCO that has potential to substitute ITO suffering from indium-shortage problem. Recent publications of TiO$_2$-based transparent conducting films are summarized in Table 1 [1 - 12]. As seen from the table, several groups have reported epitaxial growth of TNO films with ρ of the order of 10^{-4} Ω cm by physical vapor deposition techniques, such as PLD and magnetron sputtering (MSP).

In order to establish TNO as a practical TCO material, it is highly desirable to develop a productive fabrication process of polycrystalline films on glass substrates. Recently, we have shown that conductive TNO films can be obtained under strongly reducing conditions [6 - 8]. That is, introduction of oxygen deficiencies during film deposition may be a key to achieving highly conductive TNO. There are two different routes to fabricating TNO polycrystalline films on glass. One is to crystallize amorphous phase by annealing under reducing atmosphere (thermal annealing method) [6 - 9]. The other is direct synthesis of polycrystalline films from vapor phase onto heated (T_s= 200 - 400 ºC) glass. In the former route, it is easier to obtain strongly reduced anatase phase on glass, resulting in low ρ values of 6 - 8×10^{-4} Ω cm [6 - 9]. Meanwhile, directly synthesized polycrystalline films show two or more orders of magnitude higher ρ [10 - 12] than those of annealed ones, as seen from Table 1, because direct growth of reduced anatase phase on glass is difficult, as discussed later. Thus, the development in the direct growth of highly conducting polycrystalline films on glass is a challenging issue.

Here, we report on our attempts to fabricate conductive TNO films directly on glass by using seed-layers, which stabilized anatase phase. We also discuss carrier transport properties of directly synthesized TNO polycrystalline films.

Table I. Recent publications of TNO transparent conductor grown by various methods.

	Author	ρ @ RT (Ωcm)	Fabrication	Year	Ref
Epitaxial films	Furubayashi et al.	2.3×10^{-4}	PLD	2005	1
	Hitosugi et al	2.5×10^{-4}	PLD	2005	2
	Kurita et al.	3×10^{-4}	PLD	2006	3
	Gillispie et al.	3.3×10^{-4}	MSP	2007	4
	Zhang et al.	4.0×10^{-4}	PLD	2007	5
Polycrystalline	Hitosugi et al.	4.6×10^{-4}	**PLD + annealing**	2007	6
	Yamada et al.	9.5×10^{-4}	**MSP + annealing**	2007	7
	Sato et al.	1.3×10^{-3}	**MSP + annealing**	2008	8
	Hoang et al.	6.4×10^{-4}	**MSP + annealing**	2008	9
	Guo et al.	1.9×10^{3}	**MSP (as-grown)**	2006	10
	Hitosugi et al.	1.5×10^{-1}	**PLD (as-grown)**	2007	11
	Gillispie et al.	1.8×10^{-2}	**MSP (as-grown)**	2007	12

EXPERIMENT

Epitaxial and polycrystalline TNO films were grown by rf magnetron sputtering technique. Single crystalline LaAlO$_3$ (LAO) (100) substrates were used for epitaxial growth, while polycrystalline films were deposited on alkali-free glass (Corning Eagle2000) substrate. The films were grown at various substrate temperatures ranging from 250 to 450 °C. To improve systematicity, polycrystalline and epitaxial films were simultaneously grown in the same deposition run. A disk of oxygen-deficient Ti$_{0.94}$Nb$_{0.06}$O$_{1.95}$ was used as a target. Sputtering was conducted in a mixture of Ar and O$_2$ at various O$_2$/(Ar+O$_2$) ratios [$\equiv f(O_2)$] under a total pressure of 1.0 Pa. The RF power applied to the target was kept constant at 100 W during sputtering.

The Nb content of Ti$_{1-x}$Nb$_x$O$_2$ films was determined to be $x = 0.063 \pm 0.002$ by using Rutherford backscattering spectrometry. Structural properties were characterized by X-ray diffraction (XRD) and cross-sectional transmission electron microscopy (TEM) measurements. Field-emission scanning electron microscope (SEM) was used to observe surface morphology. Carrier transport properties, including resistivity (ρ), carrier density (n_e) and Hall mobility (μ_H), were determined with standard Hall bar geometry. Optical transmittance and reflectance spectra in a photon energy (hv) range of 0.5 - 4.5 eV were measured using a UV-VIS-IR spectrophotometer.

RESULTS AND DISCUSSION

Growth parameter based phase diagrams

Figures 1(a) and 1(b) show growth parameter based phase diagrams of TNO films on LAO and glass substrates, respectively. On LAO, epitaxial films of anatase phase were obtained in a region of $T_s > 250°C$ and $f(O_2) > 0.2\%$. Mixtures of anatase and rutile phase grow at $f(O_2) < 0.2\%$, regardless of T_s. The mixed phase exhibit ρ values higher than 10^{-3} Ω cm, because the rutile phase has one order of magnitude higher resistivity [14]. Conductive anatase phase with $\rho < 5×10^{-4}$ Ω cm appeared in the hatched region of $T_s > 400°C$ and $0.25\% < f(O_2) < 0.5\%$, where high carrier density $n_e > 1.4×10^{21}$ cm^{-3} and Hall mobility $\mu_H > 10$ cm^{2}V^{-1}s^{-1} were achieved (Fig. 3). Film growth under more oxidizing condition, $f(O_2) > 0.5\%$, results in decrease in both n_e and μ_H, and, thus, increase in ρ, which is presumably due to the formation of acceptor-like defects, as reported in literature [5, 15]. As shown in Fig. 1(b), on the other hand, the anatase phase TNO on glass was obtained in a narrower region, which does not include the hatched region, where highly conductive anatase TNO epitaxial films were grown on LAO. The anatase phase on glass substrate obtained at relatively low T_s and high $f(O_2)$ showed poor conductivity, possibly due to the formation of acceptor-like defects, as mentioned above.

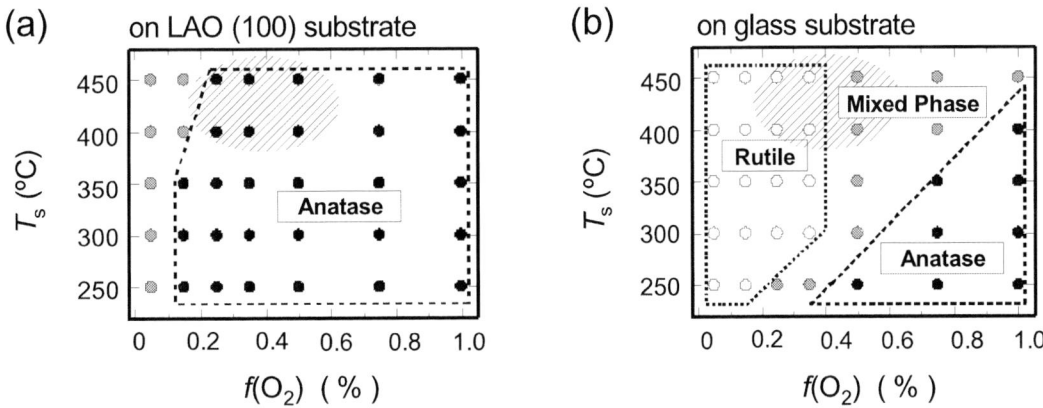

Figure 1. Growth parameter based phase-diagram of TNO films on (a) LaAlO$_3$ (100) and (b) glass substrates.

Fig. 1(b) gives a clear answer to the question why highly conductive TNO polycrystalline films have not been directly deposited on glass. From Figs. 1(a) and (b), it is evident that anatase phase is stabilized on LAO through epitaxial interaction between substrate and film even under reducing growth conditions. As a consequence, modification of the glass surface is needed to grow anatase phase on glass.

Seed-layer technique

To modify the glass surface for stabilizing anatase phase, we formed a 10-nm-thick anatase phase TNO polycrystalline layer, as an anatase template (hereafter denoted as a seed-layer), at $T_s = 250°C$ and $f(O_2) = 1.0\%$. Then, we grew the second TNO layer (denoted as a top-layer) with a thickness of about 170 nm on the template at $T_s = 400°C$ at $0\% < f(O_2) < 0.5\%$ [Fig. 2(b)]. As seen from Fig. 2(a), single phase anatase TNO polycrystalline layers with good crystallinity were successfully grown on the template even under a highly reducing atmosphere of $f(O_2) = 0.05\%$. It is noted that rutile phase grows on bare glass and on LAO [Fig. 1(a)] under such a reducing condition. This implies that the template behaves as a seed-layer and promotes the preferential growth of anatase phase. The present seed-layer method enables us to grow anatase phase in a wider range of $f(O_2) \geq 0.05\%$.

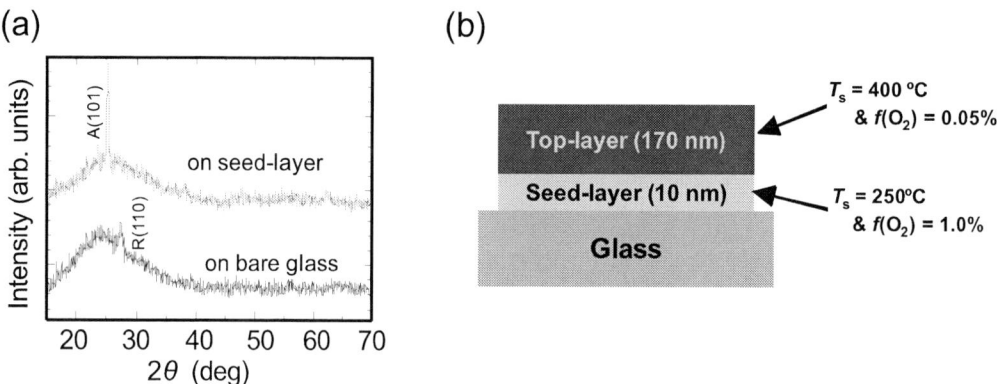

Figure 2. (a) XRD patterns of TNO polycrystalline films on seed-layer and bare glass. (b) Bi-layer structure proposed in this study.

Figures 3(a) and 3(b) are typical bright- and dark-field cross sectional TEM images of an anatase phase polycrystalline film on seed-layer, taken at the same location of the film. As seen from Fig. 3, each grain grows along the perpendicular direction, *i.e.*, from substrate/film interface to film surface. This suggests that top-layers grow in an epitaxial manner on the seed-layer, resulting in stabilization of anatase phase through epitaxial interaction.

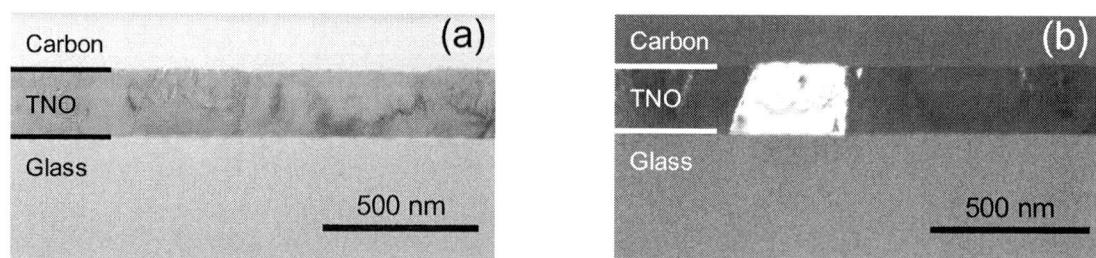

Figure 3. Typical cross-sectional TEM (a) bright- and (b) dark-field images.

Electronic transport properties (ρ, n_e, and μ_H at room temperature) of TNO polycrystalline films fabricated by using the seed-layer method are plotted in Figs. 4(a) - (c), as functions of $f(O_2)$ during growth of top-layer (open marks). For comparison, transport properties of TNO on LAO are also shown (closed marks). As seen from Fig. 2(a), ρ of the order of 10^{-3} Ω cm is obtained at $f(O_2) < 0.15\%$ for polycrystalline films on seeded glass. The minimum ρ value is as low as 1.1×10^{-3} Ω cm at $f(O_2) = 0.05\%$, where $n_e = 1.6 \times 10^{21}$ cm^{-3} and $\mu_H = 3.6$ cm^2 V^{-1} s^{-1} are attained. The minimum ρ is one or more orders of magnitude smaller than the reported values obtained for TNO polycrystalline films directly deposited on glass [10 - 12]. Nevertheless, the minimum ρ value does not reach to the order of 10^{-4} Ω cm, although n_e is sufficiently high (> 1.5×10^{21} cm^{-3}). This implies that electron mobility limits the resistivity of TNO films obtained by the seed-layer technique. Indeed, the presently observed μ_H value is significantly lower than those of epitaxial films (13 cm^2 V^{-1} s^{-1}) and polycrystalline films synthesized by the thermal annealing method (~8 cm^2 V^{-1} s^{-1}) [9]. Further improvement of μ_H is necessary to achieve low ρ of the order of 10^{-4} Ω cm. Moreover, it should be noted here that μ_H of polycrystalline films depends heavily on $f(O_2)$ during top-layer growth. That is, μ_H of polycrystalline films steeply decreases with increasing $f(O_2)$, while μ_H of epitaxial films shows a week $f(O_2)$ dependence. The μ_H values of the polycrystalline films at $f(O_2) = 0.25 - 0.35\%$ are lower than 1 cm^2 V^{-1} s^{-1}, while epitaxial films show higher μ_H values $11 - 13$ cm^2 V^{-1} s^{-1} in the same $f(O_2)$ region.

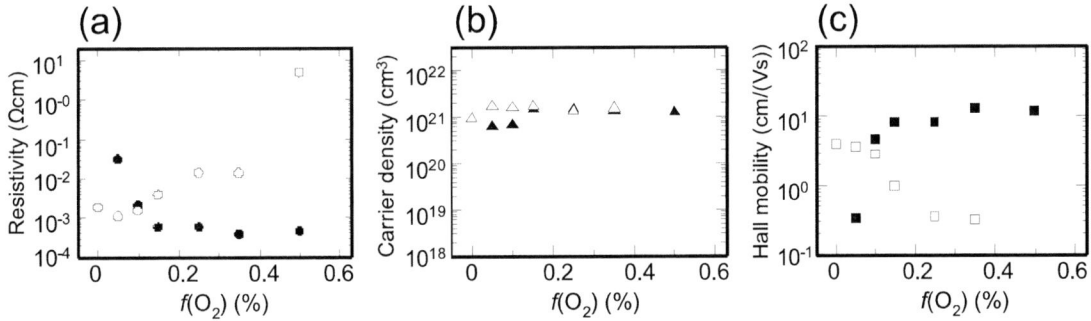

Figure 4. Room temperature (a) resistivity, (b) carrier density and (c) Hall mobility of TNO epitaxial (closed marks) and polycrystalline films on seed-layer (open marks) as functions of $f(O_2)$.

In general, carrier transport in polycrystalline films suffers from grain boundary scattering. Accordingly, it is likely the small μ_H and its strong $f(O_2)$ dependence seen in our polycrystalline films are attributed to grain boundary effects [16]. In the next section, we discuss dc transport properties in relation with grain boundary effects.

Grain boundary scattering in TNO polycrystalline films

Transport properties evaluated by optical measurements, based on the Drude analysis, are considered to reflect intra-grain properties. By comparing dc and optical transport conductivity, therefore, we can separate the intra- and inter-grain components. According to earlier studies [7, 16, 19], dielectric response, $\varepsilon(\omega) = \varepsilon_1(\omega) + i\sigma_1(\omega)/(\varepsilon_0\omega)$, of highly conductive TNO in the near infrared region can be well described by the Drude model,

$$\varepsilon_1(\omega) = \varepsilon(\infty) - \frac{\omega_p^2}{\omega^2 + \gamma^2} \qquad (1)$$

$$\sigma_1(\omega) = \frac{\varepsilon_0 \omega_p^2 \gamma}{\omega^2 + \gamma^2}, \qquad (2)$$

where ω is the frequency of the incident light, $\varepsilon_1(\omega)$ and $\sigma_1(\omega)$ the real part of the complex dielectric function and conductivity, respectively, $\varepsilon(\infty)$ the high-frequency dielectric constant, ε_0 the static dielectric constant of vacuum, ω_p the plasma frequency, γ the scattering rate of carrier. We conducted Drude-fits to transmittance $T(hv)$ and reflectance $R(hv)$ spectra of polycrystalline TNO films, in order to extract intra-grain transport properties, where fitting parameters, ω_p and γ were adjusted so as to reproduce experimental $T(hv)$ and $R(hv)$. For simplicity, $\varepsilon(\infty)$ is set to a literature value of 5.9 [20] for all samples. As shown in Fig. 5, good fits are obtained over $hv = 0.5 - 2.5$ eV.

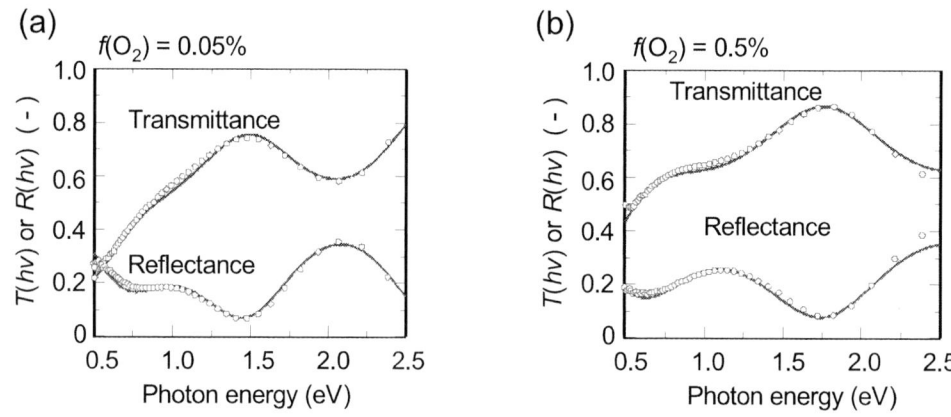

Figure 5. Optical transmittance and reflectance of TNO polycrystalline films grown at (a) $f(O_2)$ = 0.05% and (b) $f(O_2)$ = 0.35% on seed-layer. Open marks represent experimental values and continuous lines are result of the Drude fits.

Figure 6(a) shows $f(O_2)$ dependence of ω_p. As seen from the figure, ω_p gradually decreases with increasing $f(O_2)$. This tendency is consistent with the decrease in n_e with increasing $f(O_2)$ [Fig. 4(b)], since ω_p is expressed as $\omega_p^2 = e^2 n_e / (\varepsilon_0 m^*)$. The effective mass values m^*, estimated from ω_p and n_e, are $0.8 - 0.9 m_0$ (m_0 is the free electron mass), independent of $f(O_2)$, as shown in Fig. 6(b). Effective mass of tetragonal anatase TNO is known to be anisotropic [19, 21], and the m^* values obtained here correspond to averages over principal directions. For $Ti_{0.94}Nb_{0.06}O_2$, $m^*_{[100]} = 0.6 m_0$ (perpendicular to c-axis) and $m^*_{[100]} = 3.7 m_0$ (parallel to c-axis) have been reported [16, 19]. Assuming that effective mass of polycrystalline films (m^*_{poly}) is expressed as the sum of $1/m^*_{[100]}$ and $1/m^*_{[001]}$ with 2:1 ratio, *i.e.*, $3/m^*_{poly} = 2/m^*_{[100]} + 1/m^*_{[001]}$, we obtain $m^*_{poly} \sim 0.8 m_0$. As seen from Fig. 6(b), the m^*_{poly} values observed here are consistent with the estimation.

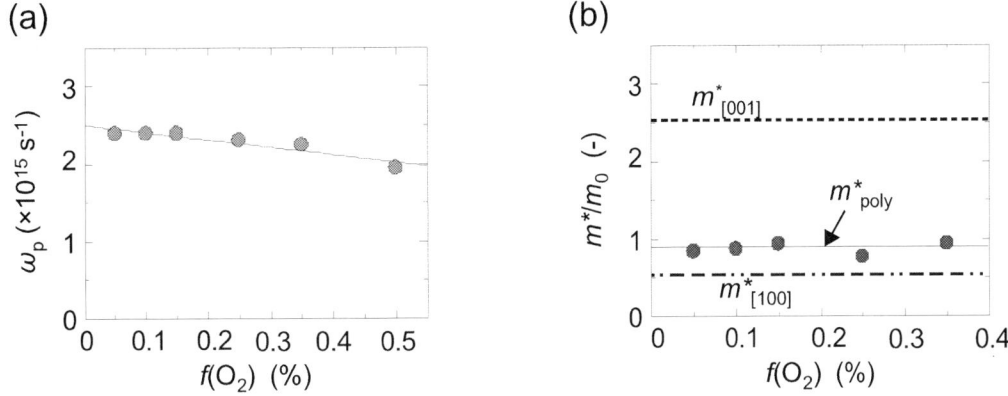

Figure 6. (a) Plasma frequency ω_p and (b) effective mass m^*/m_0 of TNO polycrystalline films as functions of $f(O_2)$.

In Fig. 7, optically deduced intra-grain mobility $\mu_{opt} = e / (\gamma \cdot m^*)$ and μ_H of polycrystalline films are compared as functions of $f(O_2)$. Notably, the two mobility values shows significantly different $f(O_2)$ dependences: μ_{opt} is nearly constant, ~4 $cm^2V^{-1}s^{-1}$, independent of $f(O_2)$, while μ_H steeply decreases with $f(O_2)$. Moreover, μ_H is always smaller than μ_{opt}. For $f(O_2) \geq 0.25\%$ films, μ_H is approximately one order of magnitude smaller than the intra-grain mobility μ_{opt}. The smaller μ_H than μ_{opt} is thought to be a consequence of grain boundary scattering [17, 18]. Hence, the steep decrease of μ_H with $f(O_2)$ implies that grain boundary scattering becomes more significant with increasing $f(O_2)$. We speculate that at higher $f(O_2)$ an amount of oxygen atoms adsorbed at grain boundaries might increase, resulting in the evolution of depletion layers at grain boundaries, *i.e.*, the enhancement of barrier height and widening of depletion layers, which hinders carrier transport across the grain boundaries. In fact, based on *in-situ* conductivity measurements and surface photovoltaic spectroscopy for sputter-deposited anatase TiO_2 polycrystalline films, Rothschild *et al.* have reported the growth of grain boundary barriers upon air-annealing in an oxidizing atmosphere [22].

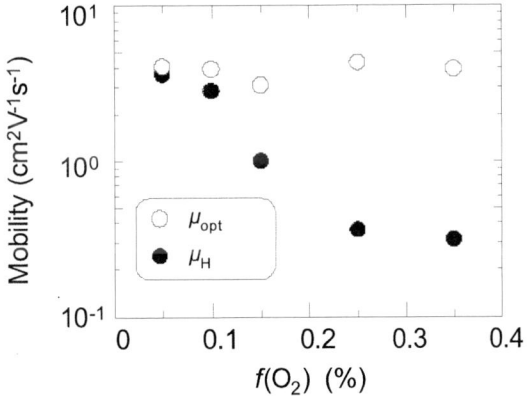

Figure 7. Comparison of Hall mobility μ_H and optically derived in-grain mobility μ_{opt} as functions of $f(O_2)$.

Figure 8(a) is a magnified view of the TEM image shown in Fig. 3(a). The polycrystalline film consists of approximately 200-nm-width grains, and each grain shows a characteristic subgrain structure of 10 - 20 nm. The structure did not depend on $f(O_2)$ during growth of the top-layer. Such a subgrain structure is more clearly visible from plan view SEM micrograph, as shown in Fig. 8(b), which is quite similar to the grain-subgrain structure observed for ITO on glass [23]. The subgrain structure aggregate near the film surface; *i.e.*, the top-layer which functions as the conductive layer has inhomogeneous subgrain structure. The size of subgrains is smaller than or comparable to those of TiO_2 nanocrystalline films, in which grain boundary scattering dominates carrier transport [22]. Thus, carrier scattering at subgrain boundaries might be significant in our polycrystalline films grown under the optimal condition for epitaxial films.

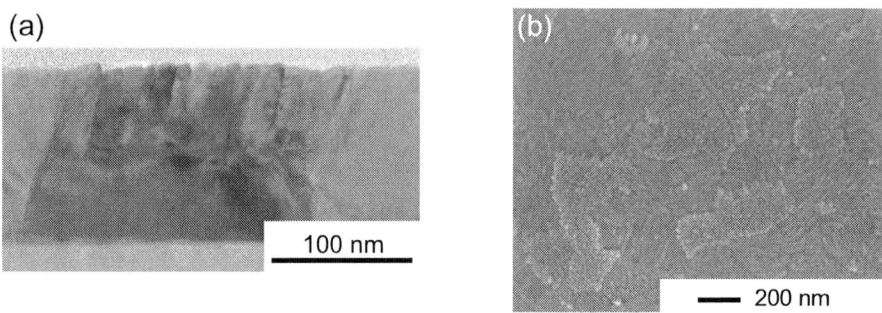

Figure 8. (a) Magnified view of cross-sectional TEM image shown in Fig. 3, and (b) typical surface morphology of TNO polycrystalline films observed by using SEM.

In order to reduce ρ further, therefore, efforts to suppress grain boundary scattering is required. In addition, intra-grain mobility ($\sim$4 cm^2 V^{-1} s^{-1}) deduced from the Drude analysis is still lower than those of epitaxial films (13 cm^2 V^{-1} s^{-1}). Hence, the enhancement of intra-grain mobility is also needed to improve the performance of TNO polycrystalline films synthesized by the method proposed in this study.

CONCLUSIONS

In conclusion, we developed a novel method using seed-layer for the direct growth of transparent conducting TNO polycrystalline films on glass, where a thin anatase TNO layer formed at T_s = 250 °C and $f(O_2)$ = 1.0% is utilized as the seed-layer. The TNO polycrystalline film on the seed-layer under an optimal condition exhibited ρ = 1.1×10^{-3} Ω cm and sufficient transparency in the visible region. The proposed seed-layer method may provide a cost-effective route for the growth of transparent conducting TNO polycrystalline films on glass substrates. The TNO polycrystalline films obtained in this study have fairly low μ_H (at most 3.6 cm^2 V^{-1} s^{-1}) as compared to epitaxial films (13 cm^2 V^{-1} s^{-1}) and polycrystalline films synthesized by the thermal annealing method ($\sim$8 cm^2 V^{-1} s^{-1}). From the Drude analysis, it was found that grain boundary scattering was an important factor that governs the electric transport in the presently obtained films. Furthermore, the intra-grain mobility (4 cm^2 V^{-1} s^{-1}) was still lower than that of epitaxial films. Therefore, to attain low ρ of the order of 10^{-4} Ω cm towards practical use of TNO, the enhancements of intra-grain mobility as well as the suppression of grain boundary scattering, through improvement of intra-grain structure and crystallinity, is further needed.

ACKNOWLEDGMENTS

The authors are grateful to S. Konuma of Kanagawa Academy of Science and Technology (KAST) for TEM observations. We also thank H. Odaka and I. Hayashi of Asahi Glass for supplying the sputtering target and SEM observation. This work was supported by MEXT Elements Science and Technology Project, Grant-in-Aid for Young Scientist (B), Global COE Program (Chemistry Innovation through Cooperation of Science and Engineering), Asahi Glass Research Collaboration Program and NEDO.

REFERENCES

1. Y. Furubayashi, T. Hitosugi, Y. Yamamoto, K. Inaba, G. Kinoda, Y. Hirose, T. Shimada, and T. Hasegawa, *Appl. Phys. Lett.* **86**, 252101 (2005).
2. T. Hitosugi, Y. Furubayashi, A. Ueda, K. Itabashi, K. Inaba, Y. Hirose, G. Kinoda, Y. Yamamoto, T. Shimada, and T. Hasegawa, *Jpn. J. Appl. Phys.* **44**, L1063 (2005).
3. D. Kurita, S. Ohta, K. Sugiura, H. Ohta and K. Koumoto, *J. Appl. Phys.* **100**, 096105 (2006).
4. M. A. Gillispie, F. A. M. van Hest, M. S. Dabney, J. D. Perkins and D. S. Ginley, *J. Appl. Phys.* **101**, 033125 (2007).
5. S. X. Zhang, S. Dhar, W. Yu, H. Xu, S. B. Ogale and T. Venkatesan, *Appl. Phys. Lett.* **91**, 112113 (2007).
6. T. Hitosugi, A. Ueda, S. Nakao, N. Yamada, Y. Furubayashi, Y. Hirose, T. Shimada and T. Hasegawa, *Appl. Phys. Lett.* **90**, 212106 (2007).
7. N. Yamada, T. Hitosugi, N. L. H. Hoang, Y. Furubayashi, Y. Hirose, T. Shimada and T. Hasegawa, *Jpn. J. Appl. Phys.* **46**, 5275 (2007).
8. Y. Sato, H. Akizuki, T. Kamiyama, and Y. Shigesato, *Thin Solid Films* **516**, 5758 (2008).
9. N. L. H. Hoang, N. Yamada, T. Hitosugi, J. Kasai, S. Nakao, T. Shimada, and T. Hasegawa, *Appl. Phys. Express* **1**, 115001 (2008).
10. S. Y. Guo, W. N. Shafarman and A. E. Delahoy, *J. Vac. Sci. Technol A* **24**, 1524 (2006).
11. T. Hitosugi, A. Ueda, Y. Furubayashi, Y. Hirose, S. Konuma, T. Shimada and T. Hasegawa, *Jpn. J. Appl. Phys.* **46**, L86 (2007).
12. M. A. Gillispie, M. F. A. M. van Hest, M. D. Dabney, J. D. Perkins and D. S. Ginley, *J. Mater. Res.* **22**, 2832 (2007).
13. Y. Furubayashi, T. Hitosugi and T. Hasegawa, *Appl. Phys. Lett.* **88**, 226103 (2006).
14. S. Na-Phattalung, M. F. Smith, K. Kim, M. -H. Du, S. -H. Wei, S. B. Zhang, and S. Limpijumnong, *Phys. Rev. B* **73**, 125205 (2006).
15. T. Hitosugi, N. Yamada, S. Nakao, K. Hatabayashi, T. Shimada and T. Hasegawa, *J. Vac. Sci. Technol. A.* **26**, 1027 (2008).
16. Y. Furubayashi, N. Yamada, Y. Hirose, Y. Yamamoto, M. Otani, T. Hitosugi, T. Shimada and T. Hasegawa, *J. Appl. Phys.* **101**, 093705 (2007).
17. R. Clanget, *Appl. Phys.* 2, 247 (1973).
18. A. S. Gilmore, A. Al-Kaoud, V. Kaydanov, and T. R. Ohno, *Mater. Res. Soc. Symp. Proc.* **666**, F3.10.1 (2001).

19. Y. Hirose, N. Yamada, S. Nakao, T. Hitosugi, T. Shimada, and T. Hasegawa, *Phys. Rev. B*, submitted.
20. R. J. Gonzalez, R. Zallen, and H. Berger, *Phys. Rev. B* **55**, 7014 (1997).
21. T. Hitosugi, H. Kamisaka, K. Yamashita, H. Nogawa, Y. Furubayashi, S. Nakao, N. Yamada, A. Chikamatsu, H. Kumigashira, M. Oshima, Y. Hirose, T. Shimada, and T. Hasegawa, *Appl. Phys. Express* **1**, 111203 (2008).
22. A. Rothschild, Y. Komem, A. Levakov, N. Ashkenasy, and Y. Shapira, *Appl. Phys. Lett.* **82**, 574 (2003).
23. Y. Shigesato and D. C. Paine, *Thin Solid Films* **238**, 44 (1994).

Mater. Res. Soc. Symp. Proc. Vol. 1109 © 2009 Materials Research Society

Low-temperature Fabrication of Transparent Conducting Polycrystalline Nb-doped TiO$_2$ Films by Sputtering

N. L. H. Hoang[1], N. Yamada[2], T. Hitosugi[2,3], J. Kasai[2], S. Nakao[2], T. Shimada[1,2], T.Hasegawa[1,2]
[1]Department of Chemistry, University of Tokyo, Tokyo 113-0033, Japan
[2]Kanagawa Academy of Science and Technology (KAST), Kanagawa 213-0012, Japan
[3]Advanced Institute for Materials Research (AIMR), Tohoku University, Sendai 980-8577, Japan

ABSTRACT

This paper presents a low-temperature (~300°C) process for preparing transparent conducting anatase Nb-doped TiO$_2$ (TNO) films on glass by sputtering. An amorphous film composed of an oxygen-rich bottom layer and oxygen-deficient top layer was deposited at room temperature. This film was then annealed in a reducing atmosphere in order to crystallize anatase. The oxygen-rich bottom layer behaved as a seed layer during crystallization of the top layer, resulting in lower crystallization temperature and significant improvement in crystallinity. The TNO polycrystalline films obtained by post-deposition annealing at 400°C exhibited resistivity of 6.4×10^{-4} Ω cm and absorption of less than 10% in the visible region. The low-temperature process developed here was applied to fabrication of TNO films on plastics and glass with low glass-transition temperature.

INTRODUCTION

Transparent conducting oxides (TCOs) are among the key materials supporting optoelectronics technology, and sputter-deposited Sn-doped In$_2$O$_3$ (ITO) has been widely used as a practical TCO material because of its excellent resistivity ρ (~2×10^{-4} Ωcm) and transparency in the visible region. However, rapid growth of new optoelectronic devices, including blue light-emitting diode, vertical cavity surface emitting laser (VCSEL) and solar cell, requires the development of new TCOs with advanced properties that conventional TCOs do not possess [1], such as high work function [2] and durability against atomic hydrogen [3].

Recently, Nb-doped anatase TiO$_2$ (Ti$_{1-x}$Nb$_x$O$_2$; TNO) in both epitaxial and polycrystalline thin film form was found to exhibit low ρ of the order of 10^{-4} Ωcm and high transmittance of 60 ~ 90 % in the visible region [4-6]. This leads to expectation that TNO has sufficient potential as a next-generation transparent conductive oxide (TCO). Sputtering, which is suitable for low-cost and uniform coating on large-area substrates, has been established as a standard technique for preparing TCO films. Consequently, a sputter-based procedure for transparent conducting TNO films on glass is desirable. There is also a strong demand for suppression of processing temperature, because it allows us to deposit TNO films on plastics and thus substantially expands the application fields of TNO. In our previous studies, however, polycrystalline TNO films with low ρ were obtained only by annealing amorphous films at temperatures exceeding 500 °C.

In this paper, a novel low temperature (~300°C) process for preparing highly conductive TNO films is introduced. TNO films were fabricated on conventional substrates, such as glass with low glass transition temperature and plastics.

EXPERIMENT DETAILS

Polycrystalline TNO films were crystallized from amorphous ones which were sputtered-deposited on unheated non-alkaline glass substrates (Corning #1737). The substrate temperature during deposition was confirmed to be in a range of 70 - 80°C. A ϕ2-in disk with $Ti_{0.963}Nb_{0.037}O_{2-\delta}$ composition was used as a target after heat-treatment in an argon atmosphere with the purpose of introducing oxygen vacancies. The base pressure prior to each deposition run was ~5 × 10^{-5} Pa. Sputter deposition was performed in a gas mixture of argon and oxygen at various flow ratios $f(O_2)$ = [O_2/ (Ar + O_2)] under a total pressure of 1.0 Pa. The RF power applied to the target was kept constant at 120 W during sputtering. Before each deposition, the target surface was sputter-cleaned by pure argon for 10 min, and then pre-sputtered for 5 min under the film deposition conditions. The as-deposited amorphous films were crystallized by annealing in a rapid thermal annealing furnace, where the temperature was raised at a heating rate of 100°C/min. The Nb content of the $Ti_{1-x}Nb_xO_2$ films was determined to be x = 0.040 ± 0.001 by Rutherford backscattering spectrometry (RBS). Carrier transport properties were measured using standard Hall bar geometry while structural properties were characterized by X-ray diffraction.

RESULTS AND DISCUSSION

In order to determine the crystallization temperature (T_{crys}) as a function of $f(O_2)$, the ρ values of amorphous films grown at different $f(O_2)$ were measured in a heating-cooling cycle under pure hydrogen atmosphere (1×10^5 Pa), as shown in Fig. 1(a). The temperature (T) was first raised to 500°C at a rate of 3 °C/min, held for 60 min, and then cooled to room temperature at 3 °C/min. As seen in Fig. 1(b), each ρ-T curve exhibits an abrupt drop in ρ around 300 – 340 °C, corresponding to a phase transformation from amorphous to polycrystalline anatase [5]. The solid line with white dots in Fig. 1(b) indicates that T_{crys} tends to be suppressed with increasing $f(O_2)$. For example, $f(O_2)$ = 5% film shows T_{crys} = 300 °C, which is 40 °C lower than that of $f(O_2)$ = 0.05% film. Continuing to raise $f(O_2)$ up to 100% suppresses T_{cyrs} down to 250 °C. On the contrary, the ρ value increases by one order of magnitude with increasing $f(O_2)$ from 0 to 5%. Therefore, it is difficult to decrease T_{crys} while keeping low ρ.

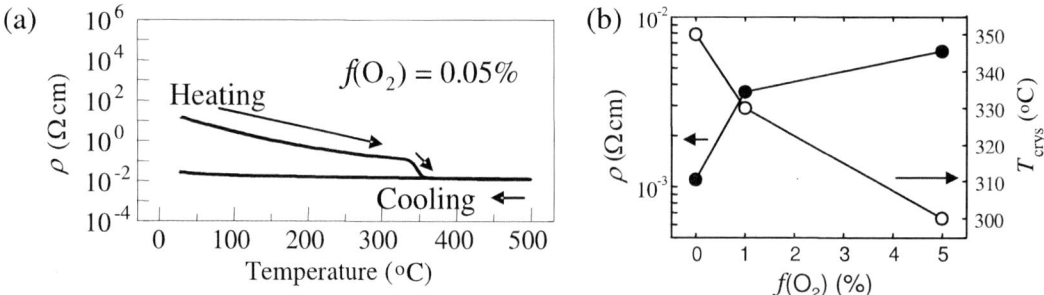

Figure 1. (a) Resistivity of as-deposited films of $f(O_2)$ = 0.05% measured *in situ* during the heating and subsequent cooling treatments, and (b) T_{crys} and resistivity ρ as functions of $f(O_2)$.

To overcome this difficulty, we propose to use an amorphous film with a double-layered structure illustrated in Fig. 2(a). The oxygen-rich bottom layer with low T_{crys} (300 °C) was expected to act as a nucleation center, from which crystallization of the top oxygen-deficient layer (T_{crys} = 350 °C)) was initiated. The double-layered amorphous film was found to undergo crystallization at around 300 °C, which was identical to T_{crys} of the bottom layer, as shown in Fig. 2(b). This proves that crystallization process propagates from the bottom layer to the top one at 300 °C. The polycrystalline film obtained by annealing at 300 °C exhibited ρ of 9×10^{-4} Ω cm, which is equivalent to that of the film crystallized at 400 °C ($7 \times 10^{-4}\Omega$ cm). Hence, it is evident that the proposed double-layer method is quite effective for fabricating TNO polycrystalline films at low temperature without decreasing electrical conductivity. We applied a similar process to a film with $Ti_{0.94}Nb_{0.06}O_2$ composition and attained a lower ρ of 6.4×10^{-4} Ω cm by annealing at 400°C in vacuum. These films showed high optical transmittance (60-80%) and low absorption (less than 10%) in the visible region.

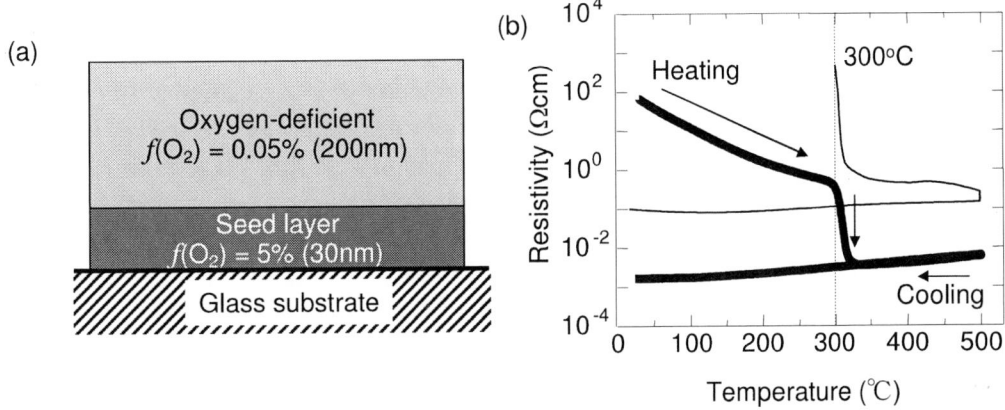

Figure 2. (a) Structure of double-layered film. (b) Resistivity of double-layer films (bold line) and single-layer film deposited at $f(O_2) = 5\%$ (thin line) measured *in situ* during heating and subsequent cooling treatments.

From the results in Fig. 1(b) and Fig. 2(b), it is expected that T_{crys} of double-layered film could be further reduced by using a higher oxygen-rich bottom layer. We fabricated TNO films in a wide $f(O_2)$ range from 0% to 100% for the seed layer. Fig. 3 shows that T_{crys} of the double-layered film decreases with increasing $f(O_2)$ of the bottom layer. We succeeded in suppressing the processing temperature to 250°C by using a bottom layer prepared at $f(O_2) = 100\%$, although the observed ρ value was somewhat higher: $\rho = 2.9 \times 10^{-3}$ Ωcm.

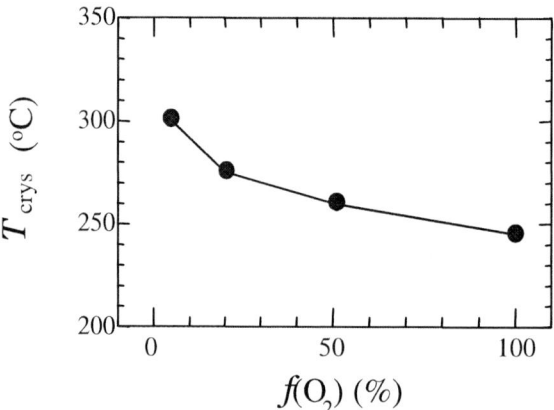

Figure 3. T_{crys} of double-layered films as a function of $f(O_2)$ of the seed layer.

TNO on low glass-transition temperature substrates

Using the low-temperature process described above, highly-conductive TNO films were deposited on polyimide and soda-lime glass with a low glass-transition temperature of 550 °C. $\rho = 1.9 \times 10^{-3}$ Ωcm on polyimide and $\rho = 7.0 \times 10^{-4}$ Ωcm on soda-lime glass were achieved when annealing at 300 and 400°C in hydrogen atmosphere. The ρ value on soda-lime glass is comparable to that on non-alkaline glass. There is a possibility that the seed layer may act as a barrier which blocks sodium or impurity diffusion from glass and plastic substrates. This barrier effect could provide explanations on the excellent conductivity obtained even in TNO films on soda-lime glass.

Polarized optical microscope (POM) analysis

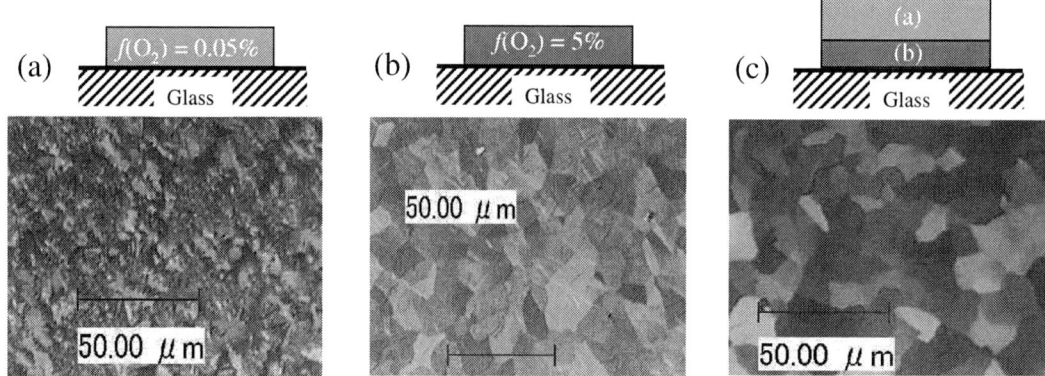

Figure 4. POM images of TNO films crystallized from (a) $f(O_2)$=0.05% , (b) $f(O_2)$=5 %, and (c) double-layered amorphous films.

Because TiO_2 polycrystalline films have optical anisotropy, the crystalline grains can be visualized by using POM. Figs. 4(a) and (b) show POM images of the TNO films crystallized from oxygen-deficient [$f(O_2)$=0.05%] and oxygen-rich [$f(O_2)$=5 %] single-layered amorphous layers, respectively. Fig. 4(c) is a POM image observed for the double-layer film. By comparing Figs. 4(a) with 4(b), it is evident that the grain size tends to enlarge with increasing $f(O_2)$. The grain size of the double-layered film [Fig. 4 (c)] is almost identical to that of $f(O_2)$ = 5% film [Fig. 4(b)]. Larger grain size, i.e. less number of grain boundaries, might contribute to lower resistivity in the double-layered film, because grain boundary scattering dominates the resistivity value of polycrystalline TNO films.

Carrier transport properties at low temperatures

Fig. 5 plots resistivity ρ, carrier concentration n_e, and Hall mobility μ_H of the double-layered TNO film as functions of temperature (T). The figure also includes the corresponding data reported for PLD-deposited polycrystalline TNO on glass [5], and sputter-deposited epitaxial TNO on LaAlO$_3$ (LAO) (100) [7] for comparison. The $\rho - T$ curve of the present TNO film shows metallic ($d\rho/dT > 0$) behavior with residual resistivity of 3.2×10^{-4} Ω cm at 10 K. As seen from Fig. 5(b), n_e is almost independent of temperature, so that the present film can be regarded as a degenerate semiconductor. The μ_H (300 K) value obtained here, 8.0 cm^2 V^{-1} s^{-1}, is compatible to that of PLD-deposited film and is close to that of sputter-deposited epitaxial one, 13 cm^2 V^{-1} s^{-1}. This implies that the present polycrystalline TNO films, prepared based on the double-layered technique, possess excellent transparent conducting properties.

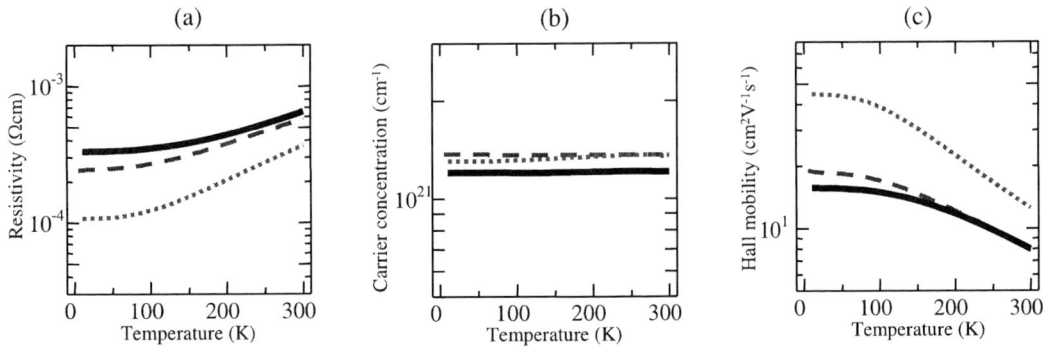

Figure 5. (a) Resistivity, (b) Carrier concentration, (c) Hall mobility of the double-layered TNO film (bold line) plotted as functions of temperature. The figures also include the corresponding data reported for sputter-deposited epitaxial film on LAO (dotted line), and PLD-deposited TNO polycrystalline film on glass (dashed line).

SUMMARY

In summary, the research group has developed a sputter-based low-temperature fabrication process for transparent conducting anatase Nb-doped TiO$_2$ (TNO) polycrystalline films on glass. It was found that the crystallization temperature of amorphous TNO decreased with increasing oxygen gas flow ratio $f(O_2)$, while the ρ value had a positive linear relation with $f(O_2)$. On the basis of these facts, the use of an oxygen-rich layer as a nucleation center for the oxygen-deficient conductive layer is recommended. The double-layered amorphous film, composed of $f(O_2)$ = 5% bottom and $f(O_2)$ = 0.05% top layers, was confirmed to undergo crystallization at 300 °C. The polycrystalline film showed resistivity of $\rho = 7 \times 10^{-4}$ Ω cm, which is lower than those of single-layered films, $\rho \sim 1 \times 10^{-3}$ Ω cm. XRD and POM measurements revealed that the use of oxygen-rich bottom layer substantially improved crystallinity and grain size, resulting in suppression of ρ. By using the present low-temperature procedure, TNO films was successfully fabricated on polyimide and soda-lime glass at processing temperatures of 300 and 400 °C, respectively.

ACKNOWLEDGMENTS

This work was supported by New Energy and Industrial Technology Development Organization (NEDO), Ministry of Education, Culture, Sports, Science and Technology (MEXT) Elements Science and Technology Project, and Grant-in-Aid for Young Scientist (B).

REFERENCES

1. R. G. Gordon, MRS Bull. **25**, 52 (2000).
2. J.H. Lim, D.K. Hwang, H.S. Kim, J.Y. Oh, J.H. Yang, R. Navamathavan, and S.J. Park, Appl. Phys. Lett. **85,** 6191 (2004).
3. M. Kambe, K. Sato, D. Kobayashi, Y. Kurokawa, S. Miyajima, M. Fukawa, N. Taneda, A. Yamada, and M. Konagai, Jpn. J. Appl. Phys. **45**, L291 (2006).
4. Y. Furubayashi, T. Hitosugi, Y. Yamamoto, K. Inaba, G. Kinoda, Y. Hirose, T. Shimada and T. Hasegawa, Appl. Phys. Lett. **86**, 252101 (2005).
5. T. Hitosugi, A. Ueda, S. Nakao, N. Yamada, Y. Furubayashi, Y. Hirose, T. Shimada and T. Hasegawa, Appl. Phys. Lett. **90**, 212106 (2007).
6. N. Yamada, T. Hitosugi, N. L. H. Hoang, Y. Furubayashi, Y. Hirose, T. Shimada, and T. Hasegawa, Jpn. J. Appl. Phys **46**, 5275 (2007).
7. N. Yamada, T. Hitosugi, T. Hasegawa, J. Vac. Soc. Jpn. **51**, No. 9, 607 (2008) (in Japanese).

Optical and Electrical Properties of Al Doped ZnO Layers Measured by Wide Angle Beam Spectroscopic Ellipsometry

C. Major[1], G. Juhász[1], A. Nemeth[1], Z. Labadi[1], P. Petrik[1], Z. Horváth[2], M. Fried[1]

[1] Research Institute for Technical Physics and Materials Science, H-1525 Budapest, POB 49, Hungary

[2] Research Institute for Solid State Physics and Optics, H-1525 Budapest, POB 49, Hungary

ABSTRACT

Al-doped ZnO (ZAO) thin film was deposited by reactive magnetron sputtering to be used as transparent conductive oxide. The sample was deposited at given value of plasma power, Ar/O_2 ratio, and target voltage on a 95mm diameter Si substrate. The specific resistance values were measured along the center line of the sample at 9 points prior to spectroscopic ellipsometry measurements. Wide angle beam (WAB) spectroscopic ellipsometry (SE) measurement was carried out in the range of 320-570nm to obtain the spectra of the mentioned nine points. WAB ellipsometers uses non-collimated illumination and gives multiple-angle-of-incidence and multiwavelength information. Our aim was to make our WAB ellipsometer suitable for spectral measurements and to obtain the spectra of many points of an entire sample simultaneously. The results show that WAB spectroscopic ellipsometry (WAB-SE) mixed with an appropriate model dielectric function (MDF) has the capability to make "in line" control in solar cell fabrication.

INTRODUCTION

Zinc oxide is an optical material of substancial interest of technological research due to its highly versatile properties to be exploited in optoelectronics [1], gas sensing [2], surface acoustic wave applications [3], UV light emitters [4] and varistors [5]. ZnO is a wide band gap semiconductor with a gap energy of about 3.4 eV at room temperature and regularly used as transparent conductive layer, thus a promising material for solar cells.

Spectroscopic ellipsometry provides a non-destructive technique for "in line" or "real time" control for solar cell production [6-8]. Usually, spectrosopic ellipsometry measurements are carried out with a millimetre sized beam however if the lateral distribution of optical, electrical properties and thickness of a sample surface is of interest, measurements can be carried out point by point by mechanical scanning. This is usually not a practical solution to obtain the spectra of many points because it takes much time.

For this purpose, WAB ellipsometry instruments have been developed by our group. The WAB imaging techniques based on detector matrix can speed up the measurements. In these cases the surface is illuminated by a non-collimated light beam and the reflected light is detected mostly by a CCD camera [9-12]. In this way, many points of the surface are measured simultaneously and the measured values are evaluated as an image to determine the lateral distribution. Disadvantage of the method is that measurements can be carried out at limited number of wavelengths, successively.

One of our aims was to develop a new instrument (based on WAB ellipsometry) which has the capability to obtain 50 spectral points (over the range of 320-570nm) with 10 mm lateral and 6nm spectral resolution of many sample points simultaneously. Second goal was to apply an appropriate MDF which has the capability to make "in situ" control of optical and electrical properties of ZAO films in conveyor-type solar cell production.

EXPERIMENT

To reach our purpose, a redesign of the optics of our WAB ellipsometer [12] was necessary. Figure 1 shows the side view of the new optical arrangement.

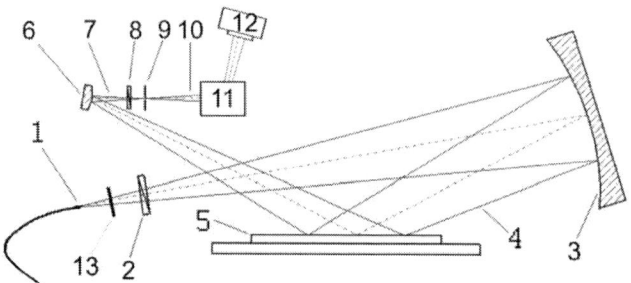

Figure 1, Optical layout of our WAB spectroscopic ellipsometer.

In this promising opportunity a rectangular (narrow) aperture (13) is placed in the light path close to the fiber end that provides a thin, long illuminated line on the sample (5) surface from end to end. After reflection the beam propagates up to a cylindric mirror (6) which is placed in the light path for correction of the aberration (mostly astigmatism) caused by the tilted spherical mirror. The corrected beam goes through the analyzer (8) and a pinhole (9), and reaches the corrector-disperser optics (11). It contains a cylindric, spherical lens pair and a concave optical grating. The dispersed beam is detected by a CCD camera (12). In the given configuration the lateral resolution better than 10 mm (over a 20 cm long line) and 50 spectral points of each lateral point is detectable simultaneously over the used spectral range. The necessary angle-of-incidence and the optical element-effect calibration is made via well-known and optimized structures such as silicon/silicon-dioxide samples.

ZAO layer was deposited by DC reactive magnetron sputtering. The target was a high purity Zn 99.95%, alloyed with 2% Al. ZnO deposition was made onto a 95mm diameter silicon substrate in Ar/O_2 atmosphere. In order to stabilize the glow discharge, pulse signal was added to the DC voltage. The target pulse parameters were changed in order to reach stable plasma discharge with low target voltage. The deposition was carried out without bias voltage and substrate preheating.

Since the refractive index and extinction coefficient in the interband transition region of ZnO strongly depend on the electronic energy band structure, the appropriate model dielectric function (MDF) suggested by Adachi [13] seem to be a good choice to model the dielectric function of the layer. Using the Kramers-Kronig transformation and assuming the conduction and valence bands are parabolic we obtain the expression [13]:

$$\varepsilon(E) = \sum_{\alpha=A,B,C} A_{0\alpha} \cdot E_{0\alpha}^{-1.5} \cdot f(\chi_{0\alpha}) \qquad\qquad (e.g.,1)$$

with

$$A_{0\alpha} = \frac{4}{3}\left(\frac{3}{2}\mu_{0\alpha}\right)^{1.5} P_{0\alpha}^2 \qquad\qquad (e.g.,2)$$

$$f(\chi_{0\alpha}) = \chi_{0\alpha}^{-2}\left[2 - (1+\chi_{0\alpha})^{1/2} - (1-\chi_{0\alpha})^{1/2}\right] \qquad\qquad (e.g.,3)$$

where

$$\chi_{0\alpha} = (E + i\Gamma)/E_{0\alpha} \qquad\qquad (e.g.,4)$$

In eqs. 1-4 $P_{0\alpha}^2$ is the squared momentum-matrix element, μ_0 is the combined density of states mass and Γ is the broadening parameter of the $E_{0\alpha}$ gap energy.

It is well known the optical spectra dramatically changes due to the excitonic interaction in the neighborhood of the lowest-direct band edge of semiconductors [13]. The discrete series of the exciton states at E_0 gap can be written with the Lorentzian line shape:

$$\varepsilon(E) = \sum_{\alpha=A,B,C}\sum_{n=1}^{\infty} \frac{A_{0\alpha}^{n\alpha}}{E_{x0}^{n\alpha} - E - i\Gamma} \qquad\qquad (e.g.,5)$$

where $A_{0\alpha}^{n\alpha}$ is the discrete-exciton strength parameter and $E_{x0}^{n\alpha}$ is the discrete-exciton energy which can be obtained from

$$E_{x0}^{n\alpha} = E_{0\alpha} - \frac{G_{0\alpha}^{3D}}{n^2} \qquad\qquad (e.g.,6)$$

$G_{0\alpha}^{3D}$ is the 3D-exciton Rydberg energy.

The continuum exciton contribution to $\varepsilon(E)$ can be written as [13],

$$\varepsilon(E) = \sum_{\alpha=A,B,C} \frac{A_{0\chi}^{C\alpha} E_{0\alpha}^{C1}}{4G_{0\alpha}^{3D}(E-i\Gamma)^2} \ln\frac{(E_{0\alpha})^2}{(E_{0\alpha})^2 - (E+i\Gamma)^2} \qquad\qquad (e.g.,7)$$

where $A_{0\chi}^{C\alpha}$ is the continuum-exciton strength parameter and $E_{0\alpha}^{C1}$ is the ground-state exciton energy. In case of ZnO polycrystals - like in our case - these equations are getting more simple - namely A, B and C indices mean the same because of anizotropy.

The MDF parameters were fitted with a MATLAB script, because of its high performance built-in functions. As the excitonic interaction dramatically changes the spectra in the neighborhood of the gap energy, we made the assumption that the first oscillator (see eq. 1.) can model the full spectra except around the band edge. The initial parameters of this oscillator were obtained from the literature [13], and a random search was done in a narrow range of the parameter space in the wavelength region excluded the band edge. Afterwards the Levenberg-Marquardt algorithm was used, that provided a numerical solution to the problem of minimizing a nonlinear function over a space of parameters of the function. Next step, discrete and continuum exciton oscillators were added to obtain better fit in the whole spectrum, including the gap region. Four phase optical model: ambient/surface roughness (Bruggeman effective medium approximation)/ZnO layer/substrate was applied to evaluate the measurements in this analysis. The angle of incidence was changing from 69.13° to 70,1° with a step about 0.1° point by point, depending on the lateral position on the sample.

DISCUSSION

Figure 2 shows a photo of the sample where the measured points are indicated by numbers from 1 to 9. WAB-SE measurements of ψ and Δ at the measured points are shown in Figure 3 and Figure 4 respectively. These points have different physical properties such as transparency, conductance and thickness.

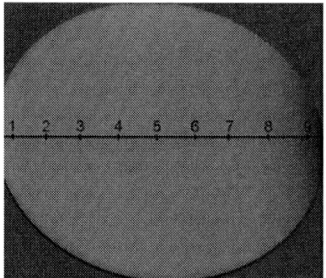

Figure 2, Measured points on the sample.

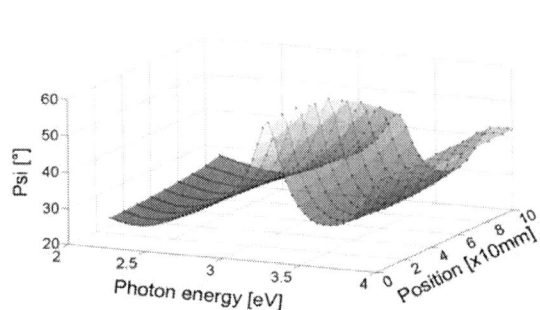

Figure 3, Spectra of measured Psi.   Figure 4, Spectra of measured Delta.

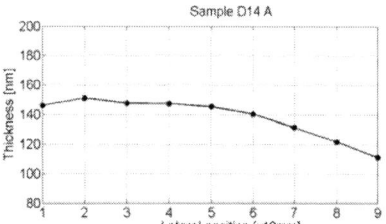

Figure 5, Thickness distribution.   Figure 6, Specific resistance distributon.

Figure 5 and Figure 6 show the thickness and specific resistance distribution respectively. It is well visible in Figure 2 that the right side of the sample acts in dark grey which is deep

34

brown in fact. In other word sample has no vertical symmetry in resistance and color. Change in color is probably caused by not only the decreasing thickness, rather the changing optical properties. Evaluation of the measurements show that the parameters of the used MDF react to the changing physical properties. Table I contains the relevant MDF parameters and physical properties. The Mean Square Error (MSE) was better than the expected in every case.

Table I, Relevant, fitted MDF parameters and physical properties.

Point Nr.	Resistance [Ohm *cm]	Thickness [nm]	E Gap [eV]	Disc. exciton amplitude [eV ^1.5]	Cont. exciton amplitude [eV^1.5]
1	1.02 e-3	146±0.3	3.648±0.0013	0.0226±0.0003	0.272±0.0062
2	1.10 e-3	152±0.3	3.641±0.0012	0.0170±0.0002	0.282±0.0060
3	1.16 e-3	147±0.3	3.610±0.0080	0.0163±0.0002	0.290±0.0043
4	1.26 e-3	147±0.2	3.640±0.0017	0.0185±0.0005	0.295±0.0071
5	1.64 e-3	144±0.4	3.606±0.0023	0.0156±0.0004	0.304±0.0058
6	2.00 e-3	140±0.5	3.607±0.0031	0.0140±0.0002	0.315±0.0041
7	2.42 e-3	133±0.3	3.606±0.0016	0.0140±0.0003	0.336±0.0066
8	3.23 e-3	121±0.3	3.601±0.0021	0.0085±0.0003	0.344±0.0058
9	5.36 e-3	113±0.2	3.598±0.0014	0.0023±0.0004	0.362±0.0057

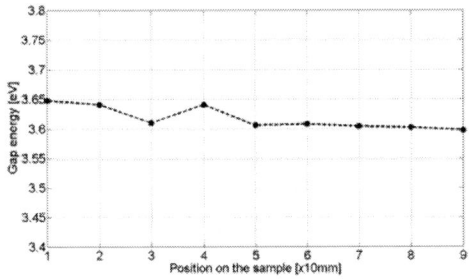

Figure 7, Gap energy distribution

Figure 8, Disc. exciton amplitude distribution

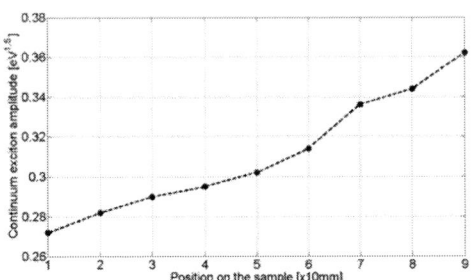

Figure 9, Cont. exciton distribution

It is clear that a small but significant change in resistance is in correlation with the small change of the gap energy because of the Burnstein-Moss effect [14]. The gap energy changes from 3.5 to 3.95 eV during a 3-order of magnitude specific resistance change. Furthermore the strength parameter of excitons are also in accordance with the changing resistance [15]. Important to note that decreasing discrete excitons and increasing continuum excitons are probably caused by structural differences. They are not in connection directly with conductance. We would like to note that it was not a goal of this paper to explore the reasons but additional investigations are on way to clear out these questions.

CONCLUSIONS

We succesfully designed a WAB spectrosopic ellipsometer which has the capability to measure 50 spectral points of many sample points in only one measuring cycle. The present realization of the device allows us to measure maximum 20 cm long samples. The lateral resolution better than 10mm in the present configuration. Advantageous feature that the optical layout is arbitrarily scalable. After ellipsometric measurements evaluation we detected characteristic differences in the used MDF. Thus our WAB-SE alloyed with an appropriate MDF could be an effective monitoring tool in the ZAO growth process.

REFERENCES

1. C. Agashe, O. Kluth, G. Schöpe, H. Siekmann, J. Hüpkes, B. Rech, Thin Solid Films, 442, 167-172, 2003
2. G. S. Devi, V. B. Subrahmanyam, S. C. Gadkari, S. K. Gupta, Analytica Chimica Acta 568, 1-2, 41-46, 2006
3. A. Talbi, F. Sarry, M. Elhakiki, L. Le Brizoual, O. Elmazria, P. Nicolay, P. Alnot, Sensors and Actuators A: Physical, 128, 78-83, 2006
4. S. Y. Myong, K. S. Lim, Organic Electronics, 8, 1, 51-56, 2007
5. M. Wang, K. Hu, B. Zhao N. Zhang, Ceramics International, 33, 2, 151-154, 2007
6. J. Koh, H. Fujiwara, Y. Lu, C.R. Wronski, R.W. Collins, Thin Solid Films 313-314, 469-473, 1998
7. H. Fujiwara, J. Koh, Y. Lee, C. R. Wronski, R. W. Collins, J. Appl. Phys. 84, 4, 1998
8. M. Wakagi, H. Fujiwara, R.W. Collins, Thin Solid Films 313-314, 464-468, 1998
9. Patent pending, P104255-1389, 2008
10. M. Fried, Z. Horváth, G. Juhász, T. Mosoni and P. Petrik, "Divergent Illumination Optical Testing Device"; Proc. Of 13rd Advanced Equipment Control / Advanced Process Control Conference, October 6–11, Banff, Canada, poster, 2001
11. G. Juhasz, Z. Horvath, C. Major, M. Fried, PSS (c), 5, 5, 1081-1084, 2008
12. C. Major, G. Juhasz, Z. Horvath, O. Polgar, M. Fried, PSS (c), 5, 5, 1077-1080, 2008
13. H. Yoshikawa, S. Adachi, Jpn. J. Appl. Phys. 36, 6237-6243, 1997
14. M. Suchea, S. Christoulakis, N. Katsarakis, T. Kitsopoulos, G. Kiriakidis Thin Solid Films, 515, 6562-6566, 2007
15. C. Major, A. Nemeth, M. Fried, Z. Labadi, I. Barsony, submitted to Thin Solid Films

High Rate Deposition of Al-doped ZnO by Reactive Sputtering: (1) Unipolar Pulsing with Plasma Emission Control

Kento Hirohata[1], Yasutaka Nishi[1], Nobuto Oka[1], Yasushi Sato[1], Isao Yamamoto[2] and Yuzo Shigesato[1]

[1]Graduate School of Science and Engineering, Aoyama Gakuin University, 5-10-1 Fuchinobe, Sagamihara, Kanagawa 229-8558, Japan
[2]Nissan Motor Co., 560-2 Okatsukoku, Atsugi, Kanagawa 243-0192, Japan

ABSTRACT

Al-doped ZnO (AZO) films were deposited on unheated substrates and substrates heated at 200°C by reactive sputtering with a Zn–Al alloy target using mid-frequency pulsing and a plasma control unit with a feedback system that used the optical emission intensity of the atomic O* line at 777 nm; this feedback system enabled precise control of the deposition rate in the transition region. The deposition rates were about 10-20 times higher than those used for depositing films by conventional sputtering using oxide targets. The lowest resistivity of 3.8×10^{-4} Ωcm was obtained for AZO films deposited on the substrate heated at 200°C with a sputter power of 4 kW. In this case, a deposition rate of 385 nm/min was achieved.

INTRODUCTION

Recently, there has been a strong demand from industry for sputter deposition of transparent conductive oxide (TCO) films at much higher deposition rates than conventional processes. A reactive sputtering process using metal or alloy targets is a promising technique for achieving high deposition rates of TCO films at low cost because they permit a higher sputter power to be applied to relatively cheap metallic targets. However, the reactive sputtering process is highly nonlinear and the deposition rate exhibits hysteresis with respect to the reactive gas flow [1-3]. This behavior is due to changes in the oxidation state of the target surface; these changes significantly affect the deposition rate with increasing O_2 flow. Therefore, precise control of the oxidation of the target surface during reactive sputtering is essential for achieving high-rate deposition of TCO films with good reproducibility.

In this study, we investigated high-rate deposition of Al-doped ZnO (AZO) films. These films have attracted much attention as a promising TCO alternative to Sn-doped In_2O_3 and amorphous In_2O_3-ZnO. Deposition was performed using a reactive sputtering system with unipolar pulsing and a plasma control unit (PCU). The effect of various deposition parameters on the structure and optoelectrical properties of the AZO films was investigated in detail.

EXPERIMENTAL

A schematic illustration of the reactive sputtering system used in this study is shown in Figure 1. This system consists of a magnetron cathode (RM400, FEP), hidden anodes, a dc power source (Pinnacle, Advanced Energy), which was controlled by a mid-frequency pulse unit (UBS-C2,

FEP), and a PCU in which the plasma emission intensity at the target was monitored as a feedback parameter for controlling the reactive gas flow. A planar Zn-Al alloy target (Al: 1.5 wt%, 130 mm × 400 mm) was connected to the pulse unit, which was operated in unipolar pulse mode. The duty cycle (i.e., the ratio of the on period to the cycle period) was adjusted to reduce arcing on the alloy target caused by high power densities. The hidden anode was designed to ensure that the discharge was stable without causing surface poisoning during deposition. Ar (purity: 99.999%) and O_2 (purity: 99.999%) were introduced as sputtering and reactive gases, respectively. The O_2 gas flow was precisely controlled by piezoelectric values, which were under the control of the PCU. An optical emission detector, which consisted of a photomultiplier and an optical filter, was used to convert the plasma emission of the atomic O* line at 777 nm from an optical intensity into a photovoltage. The PCU adjusted the piezoelectric value to precisely and rapidly control the O_2 flow until the photovoltage reached a value equal to the set points for the optical emission intensity (OEI). A high deposition rate could be achieved by selecting set points in the transition region between the metallic and oxide states on the target surface [4,5]. The chamber was evacuated to a pressure of less than 8.0×10^{-4} Pa. The total gas pressure was 0.5 Pa. Two different sputtering powers were used (2 and 4 kW). The set points of the OEI of the O* line at 777 nm were selected to give a photovoltage of 0.7-5.0 V.

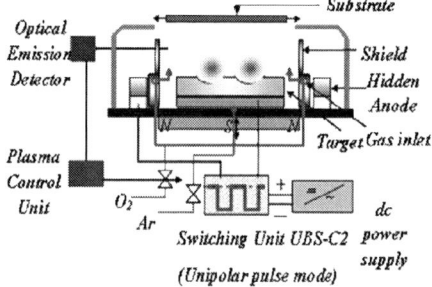

Figure 1. Schematic diagram of reactive magnetron sputtering with a unipolar unit and a plasma control unit.

Film thickness was measured using a surface profiler (Dektak[3], Sloan Tech.). The Zn/Al weight ratio of the films was analyzed by electron probe microanalysis (EPMA; JXA-8100, Jeol); the compositions were expressed in terms of the weight ratio, Al/(Zn+Al). X-ray diffraction (XRD) was conducted at 40 kV and 20 mA using Cu Kα_1 radiation (XRD-6000, Shimadzu). Resistivity, Hall mobility and carrier density were measured by a four-point probe method and a Hall-effect measurement in the van der Pauw geometry (HL-5500PC, Accent). The transmittance and reflectance of the films were measured using an optical spectrophotometer (UV-3150, Shimadzu) in the spectral range of 200-2500 nm.

RESULTS AND DISCUSSION

Figure 2 shows the relationships between the OEI of the O* line at 777 nm and various O_2 flow ratios with and without PCU control for applied sputtering powers of 2 and 4 kW. When the O_2 flow ratio was controlled without using the PCU, the variation in the OEI was nonlinear; the OEI increased (or decreased) dramatically with increasing (or decreasing) O_2 flow ratio in the transition region. This indicates that the deposition rate could not be controlled in the transition region. By contrast, when the O_2 flow ratio was controlled using the PCU, there was a one-to-one correspondence between the OEI and the O_2 flow ratio, enabling the deposition rate to be precisely controlled in the transition region.

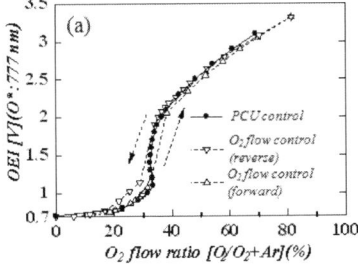 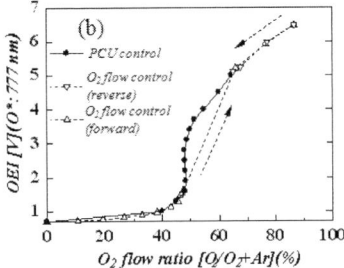

Figure 2. Relationships between OEI of O* line at 777 nm and various O_2 flow ratios with and without PCU control for applied sputtering powers of (a) 2 kW and (b) 4 kW.

Figure 3 shows the variation in the deposition rates of the films deposited using PCU as a function of the OEI. The deposition rates obtained using a unipolar pulsing system exhibit similar trends to those obtained using a bipolar pulsing system as previously reported by us [6-9]. The deposition rates of films deposited on unheated glass decrease with increasing OEI, indicating as the OEI is increased the surface of the Zn-Al alloy target changes from being metallic to oxidized. The deposition rate obtained at a sputtering power of 4 kW is about 2-3 times higher than that obtained at 2 kW. The deposition rates of films deposited on glass heated at 200°C exhibit different trends from those of films deposited on unheated glass. Specifically, the deposition rates are relatively low at low OEIs for the films deposited on heated glass. The highest deposition rates of AZO films deposited on unheated glass and glass heated to 200°C were about 10-20 times higher than those of films deposited by conventional dc magnetron sputtering using an AZO ceramic target. Table I summarizes the deposition rates of AZO films deposited by various sputter deposition techniques in this and previous studies.

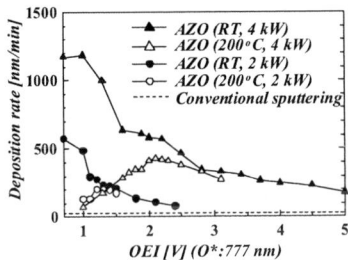

Figure 3. Deposition rates of films deposited at 2 and 4 kW using PCU control as a function of OEI.

Figure 4. Zn/Al weight ratios of the films deposited on unheated glass at 2 and 4 kW at various OEIs.

Zn/Al weight ratios of AZO films deposited under various OEIs were measured by EPMA. As Figure 4 shows, the Zn/Al weight ratios of the films deposited on unheated glass at sputtering powers of 2 and 4 kW are almost constant at around 1.1 wt.%, irrespective of the OEI used. By contrast, the Zn/Al weight ratios of the films deposited on glass heated at 200°C at sputtering powers of 2 kW and 4 kW decrease from 2.8-3.6 to 1.1 wt.% with increasing OEI. This result can be explained by the higher vapor pressure of Zn metal than ZnO at such a high temperature, which gives rise to partial re-evaporation of the Zn metal atoms. This explains why the deposition rates are lower at lower OEIs, as shown in Figure 3.

Table I. Deposition rates of AZO films deposited by various sputter deposition techniques in this and previous studies

Film	Substrate temperature(°C)	Deposition rate (nm/min)	Resistivity (Ωcm)	Power density (W/cm²)	Target-substrate distance(mm)	Total gas pressure (Pa)
AZO, MFRMS[6] Bipolar system	200	528	3.0×10^{-4}	9.3	90	0.35
AZO, DCRMS[7] Bipolar system	300	66	4.0×10^{-4}	1.0	90	0.15
AZO, MFRMS[8] Bipolar system	300	290	3.9×10^{-4}	6.2	115	0.50
AZO, MFRMS[9] Bipolar system	300	242	4.4×10^{-4}	6.2	115	0.50
AZO, MFRMS * Unipolar system	unheated	235	2.0×10^{-3}	4.4	115	0.50
AZO, MFRMS * Unipolar system	200	206	5.2×10^{-4}	4.4	115	0.50
AZO, MFRMS * Unipolar system	unheated	580	1.4×10^{-3}	8.9	115	0.50
AZO, MFRMS * Unipolar system	200	385	3.8×10^{-4}	8.9	115	0.50
AZO, MFRMS[10] Unipolar system	unheated	230	1.3×10^{-3}	4.4	115	0.50
AZO, MFRMS[10] Unipolar system	200	160	3.9×10^{-4}	4.4	115	0.50
AZO, MFRMS[10] Unipolar system	unheated	525	1.1×10^{-3}	8.9	115	0.50
AZO, MFRMS[10] Unipolar system	200	390	3.8×10^{-4}	8.9	115	0.50
AZO, GFS[11]	unheated	270	8.2×10^{-4}	5.9	105	50
AZO, GFS[11]	200	200	5.2×10^{-4}	5.9	105	50
AZO, DC+RFMS[12]	150	62	5.0×10^{-4}	(DC 80 W +RF 150 W)	50	0.40
AZO, DCMS	300	30	6.5×10^{-4}	1.1	60	0.50

Figures 5(a)-(d) show XRD patterns of the AZO films deposited at various OEIs. The XRD patterns of AZO films deposited at higher OEIs clearly show a peak due to ZnO with (001) orientation, whereas those of the films deposited on the unheated substrate using OEIs lower than 2.0 V with 4 kW exhibit low intensity ZnO and Zn peaks. On the other hand, XRD patterns of the AZO films deposited at 200°C have a ZnO (002) peak over the whole range of OEIs used in this experiment. These data indicate that the AZO films on both unheated glass and glass heated at 200°C glass possess a polycrystalline ZnO structure with the preferred orientation of the c-axis being perpendicular to the film surface.

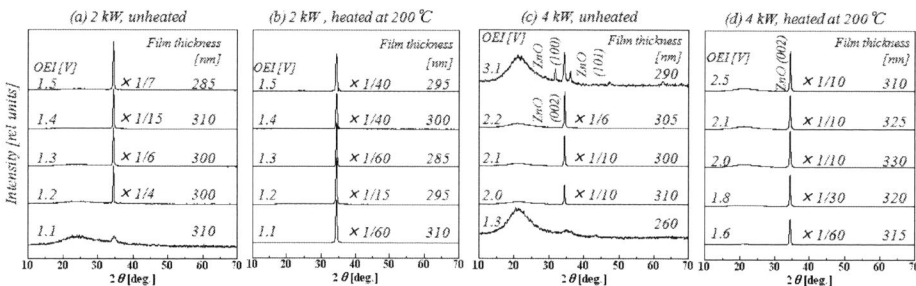

Figure 5. XRD patterns of the AZO films deposited with sputtering powers of 2 and 4 kW at various OEIs.

Figure 6(a) shows the variations in the electrical properties of the AZO films deposited on the unheated glass and the glass heated at 200°C at 2 kW as a function of the OEI. The lowest resistivity of the films deposited on glass heated at 200°C at an OEI of 1.1 V was 5.0×10^{-4} Ωcm; the deposition rate under these conditions was 130 nm/min. On the other hand, Figure 6(b) shows the variations in the electrical properties of the AZO films deposited on unheated glass and glass heated at 200°C at 4 kW as a function of the OEI. The lowest resistivity for the films deposited on glass heated at 200°C at an OEI of 1.9 V was 3.8×10^{-4} Ωcm; the deposition rate under these conditions was 385 nm/min. The resistivities of the films deposited under both sets of conditions increase with increasing OEI because of the reduction in the carrier density, which is thought to be caused by extinction of oxygen vacancies.

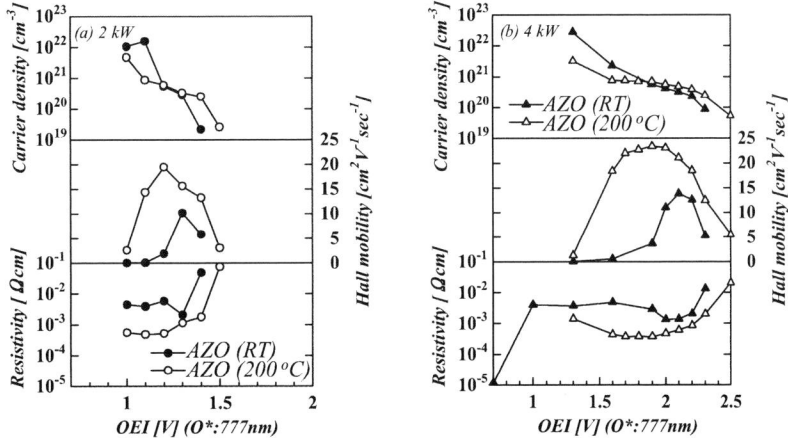

Figure 6. Variations in the electrical properties of the AZO films deposited on unheated glass and glass heated at 200°C as a function of OEIs at sputtering powers of (a) 2 kW and (b) 4 kW.

Figure 7 shows the transmittance and reflectance of the AZO films deposited on unheated glass and on heated substrates at sputtering powers of (a) 2 kW, unheated, (b) 2 kW, heated, (c) 4 kW, unheated and (d) 4 kW heated. AZO films deposited on unheated glass at a sputtering power of 2 kW with OEIs less than 1.3 V and at a sputtering power of 4 kW with OEIs of less than 2.0 V were not transparent. On the other hand, AZO films deposited on glass heated at 200 °C at a sputtering power of 2 kW with OEIs less than 1.1 V and at a sputtering power of 4 kW with OEIs of less than 1.6 V were not transparent, implying that these films contained some low-valence oxides. By contrast, as shown in Figure 7, the AZO films deposited at high OEIs exhibited high transparencies of 80-85% in the visible region.

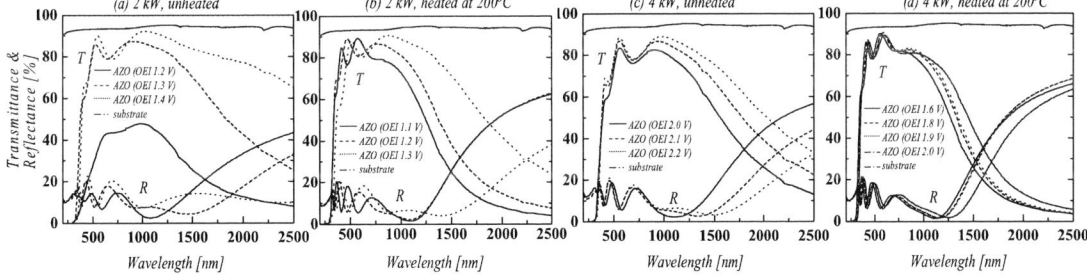

Figure 7. Transmittances and reflectances of AZO films deposited on unheated and heated substrates at a sputtering power of 2 kW and 4 kW. (a) 2 kW, unheated (OEIs of 1.2-1.4V), (b) 2 kW, heated (OEIs of 1.1-1.3 V), (c) 4 kW, unheated (OEIs of 2.0-2.2V) and (d) 4 kW, heated (OEIs of 1.6, 1.8-2.0 V).

CONCLUSION

AZO films were deposited on glass by reactive sputtering using a Zn–Al alloy target with mid-frequency pulsing and a PCU with a feedback system that used the OEI of the atomic O* line at 777 nm. In the case of the films deposited on a substrate heated at $200^{\circ}C$ at a sputtering power of 2 kW, the lowest resistivity was 5.0×10^{-4} Ωcm at a deposition rate of 130 nm/min. On the other hand, films deposited on a substrate heated at $200^{\circ}C$ at a sputtering power of 4 kW, the lowest resistivity was 3.8×10^{-4} Ωcm at a deposition rate of 385 nm/min. Both of these AZO films had optical transmittances of over 80% in the visible region. The obtained data demonstrate that these films should be acceptable for some optoelectrical applications. Therefore, these results indicate that unipolar pulsing using a PCU is a promising technique for producing AZO films at high deposition rates.

ACKNOWLEDGE

The authors would like to thank Drs. Y. Iwabuchi and M. Yoshikawa of Bridgestone Corporation for helpful discussions. This work was partially supported by a High-Tech Research Center project for private universities with matching fund subsidy from the Ministry of Education, Culture, Sports, Science and Technology (MEXT) of the Japanese Government.

REFERENCES

[1] S. Berg, T. Larsson, C. Nender and H.O. Blom, J. Appl. Phys. 63, 887 (1988).

[2] S. Schiller, U. Heisig, C. Korndorfer, G. Beister, J. Reschke, K. Steinfelder and J. Strumpfel, Surf. Coat. Technol. 33, 405 (1987).

[3] S. Berg, T. Larsson and H.O. Blom, J. Vac. Sci. Technol. A4, 594 (1986).

[4] S. Ohno, D. Sato, K. Kon, P.K. Song, M. Yoshikawa, K. Suzuki, P. Frach and Y. Shigesato, Thin Solid Films 445, 207 (2003).

[5] S. Ohno, Y. Kawaguchi, A. Miyamura, Y. Sato, P.K. Song, M. Yoshikawa, P. Frach and Y. Shigesato, Sci. & Technol. 7, 56 (2006).

[6] B. Szyszka, Thin Solid Films 351, 164 (1999).

[7] S. Jäger, B. Szyszka, J. Szczyrbowski and G. Bräuer, Surf. Coat. Technol. 98, 1304 (1998).

[8] M. Kon, P.K. Song, A. Mizukami, K. Suzuki, P. Frach and Y. Shigesato, Jpn. J. Appl. Phys. 41, 814 (2002).

[9] M. Kon, S. Ohno, P. K. Song, P. Frach and Y. Shigesato, Jpn. J. Appl. Phys. 42, 263 (2003).

[10] Y. Nishi, K. Hirohata, N. Oka, Y. Sato, I. Yamamoto and Y. Shigesato, in Proceedings 2008 MRS Fall Meeting in Boston.

[11] H. Takeda, Y. Sato, Y. Iwabuchi, M. Yoshikawa and Y. Shigesato, Thin Sold Films (in press).

[12] T. Minami, Y. Ohtani, T. Miyata and T. Kuboi, J. Vac. Sci. Technol. A25, 1172 (2007).

[13] P.K. Song, Y. Shigesato, M. Kamei and I. Yasui, Jpn. J. Appl. Phys. 38, 2921 (1999).

[14] Y. Ohhata, F. Shinoki and S. Yoshida, Thin Solid Films 59, 255 (1979).

Mater. Res. Soc. Symp. Proc. Vol. 1109 © 2009 Materials Research Society

Favorable Elements for an Indium-based Amorphous Oxide TFT Channel: Study of In-X-O (X=B, Mg, Al, Si, Ti, Zn, Ga, Ge, Mo, Sn) Systems

Amita Goyal[1], Tatsuya Iwasaki[1], Naho Itagaki[1], Tohru Den[1], and Hideya Kumomi[1]

[1]Canon Inc., 3-30-2, Shimomaruko, Ohta-ku, Tokyo 146-8501, Japan

ABSTRACT

Various kinds of thin film transistors (TFTs) with In-X-O (X= B, Mg, Al, Si, Ti, Zn, Ga, Ge, Mo and Sn) channel layer were studied using combinatorial technique. The channel layers were deposited by sputtering on unheated substrates and post annealing of TFTs was carried out at 300°C in air for 1 hour. Most of the In-X-O TFTs showed the switching behavior except for TFTs with In-Mo-O and In-Sn-O channel layer. High TFT performance is obtained in the case of In-Ge-O ($\mu_{FE} \sim 6 cm^2/V$-s, on/off $\sim 10^{10}$), In-Zn-O ($\mu_{FE} \sim 26 cm^2/V$-s, on/off $\sim 10^{10}$), In-Si-O ($\mu_{FE} \sim 3 cm^2/V$-s, on/off $\sim 10^9$). The electron mobility of In-X-O shows inverse correlation with the electron effective mass, m_e, of X-O, except when X is a transition metal element.

INTRODUCTION

Recently, transparent amorphous oxide semiconductor (TAOS) thin film transistors have become popular candidates for applications in large area electronics such as flat panel displays. These materials foster the advantages of low temperature fabrication, high mobility, and uniformity, and thus can realize the goal of light weight high performance display systems. The above benefits have generated much interest in study of TAOS materials in greater detail both by industry and academia[1,2].

Many new materials such as $InGaZnO_4$ [1,2,3], In-Zn-O[4,5], Ga-Zn-Sn-O[6], Zn-Sn-O[7] have been proposed as TAOS for TFT channel layer. $InGaZnO_4$ [2, 3] has been widely studied material owing to its room temperature fabrication and high mobility (6-9 cm^2/Vs). The electrical conduction in these amorphous oxides is very efficient since the conduction band is comprised of heavy metal cations such as In^{3+}. The large radii and overlap between the adjacent spherical ns orbitals creates conduction paths in amorphous network, which are insensitive to the bonding angles [8]. Thus a large mobility can be achieved. It is very important to understand the dependence of the TFT performance on the channel material. By far only a limited range of binary multi-cation oxides have been studied in such a detail. In this paper we report new In-X-O TAOS materials like In-X-O (X= B, Mg, Al, Si, Ti, Zn, Ga, Ge, Mo, and Sn) as channel layer materials for TFT application. Further their compositional dependence is studied.

EXPERIMENTAL DETAILS

Thin film deposition- At first the In-X-O film was sputter deposited using the technique of combinatorial rf (radio frequency) sputtering [2]. For this purpose, two ceramic targets of In_2O_3 and one ceramic target of single-cation oxide (X-O) were placed on three independent cathodes faced slantingly to the substrate on an anode. The targets were co-sputtered in Ar and O_2 mixing gases without intentional heating of the substrate to obtain a compositionally graded film on a thermally oxidized silicon wafer (3 inch in diameter). The thickness uniformity is critical as it effects the TFT properties, and was controlled by supplying independent rf power to each cathode. Therefore, the thin film that has uniform thickness (thickness variation <10 %) and a wide compositional distribution across the 3-inch substrate can be obtained in a single process run. The electrical properties of the thin film were evaluated by resistivity measurements using four probe technique and hall measurements in van der Pauw mode. The structural properties were analyzed by obtaining a x-ray diffraction (XRD) pattern. For compositional analysis an XRF (x-ray fluorescence) measurements were made.

Thin film transistor- A bottom-gate top-contact type TFT structure was fabricated as shown in the schematic drawing of Figure 1. The TFT consisted of Si gate electrode (conductivity $\sim 1 \times 10^2$ S cm^{-1}) with a thermally-oxidized SiO_2 as gate insulator, a room-temperature sputter deposited In-X-O channel layer (thickness ~25nm), and top-contact source and drain electrodes of Au/Ti bilayer deposited using Electron beam deposition. Photolithography techniques were used for patterning various component structures of the TFT. The channel width and length were 150 μm and 10 μm, respectively. The maximum process temperature was 120 °C. Post annealing was done at 300°C in air for 1hr. Combinatorial technique was employed to study the channel compositional dependence of the In-X-O TFTs in the margin of 0.05< X/(In+X)≤ 0.6 [2].

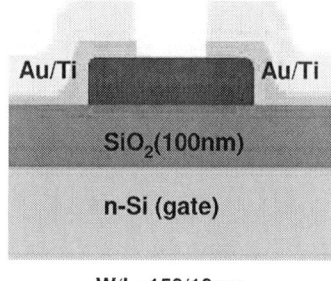

W/L=150/10um

Figure 1 – Schematic drawing of the In-X-O channel layer TFT.

Output and transfer characteristics of the TFT were measured using the Keithley 4200-SCS semiconductor characterization system. The threshold voltage, V_{th} (V), is determined from the slope and horizontal axis intercept of a linear fitting of $\sqrt{I_{DS}} - V_{GS}$ plot, where I_{DS} and V_{GS} are source-to-drain current and gate-to-source voltage, respectively. The field-effect saturation mobility, μ_{sat}, (cm^2/V·s) was calculated with the determined V_{th} using the following equation in the saturation region

$$I_{DS} = \left(\frac{\mu_{sat} W \varepsilon_o \varepsilon_r}{2Ld} \right) \left(V_{GS} - V_{th} \right)^2 \qquad (1)$$

where ε_0 is the dielectric constant of vacuum, d is the thickness of gate insulator, ε_r is relative dielectric constant of the gate insulator, W is the channel width, L is the channel length. For thermally-oxidized SiO_2, ε_r=3.9.

RESULTS and DISCUSSION

1) <u>Physical properties of channel layer materials</u>

Structural properties- Amorphous states are formed by sputtering for all of In-X-O systems, by selecting their composition. Figure 2 illustrates an approximate value for the upper limit expressed as In/(In+X) $\geq$ 0.4 for which an amorphous In-X-O can be obtained in as deposited state. It can be seen that In-Ge-O, In-B-O, In-Al-O, and In-Mg-O have a broader composition margin within which an amorphous oxide can be obtained. Such an compositional margin shrinks in descending order as given by In-Ga-O, In-Ti-O, In-Zn-O/In-Mo-O and In-Sn-O. The grain boundaries in a polycrystalline/crystalline phase can adversely affect the TFT performance. Thus an amorphous film is desired for good and uniform TFT performance as compared to polycrystalline films.

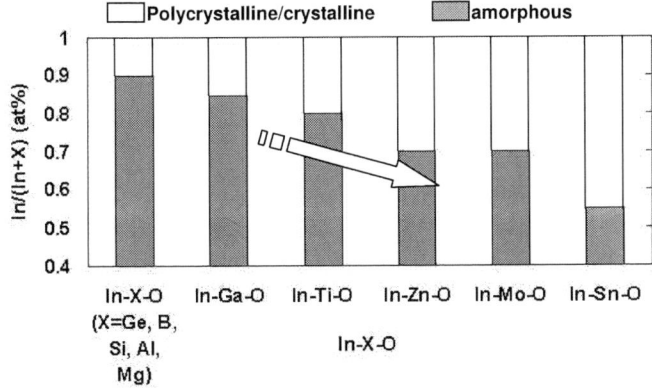

Figure 2- Compositional margin for In/(In+X)>0.4 for an amorphous/crystalline phase.

Resistivity- Figure 3 illustrates the resistivity, ρ ($\Omega\cdot cm$), of In-Ge-O as a function of Ge content. It can be seen that ρ increases with the increase in Ge content up to Ge ~ 50 at%, after which it becomes >10^7. The compositional margin expressed as 0<Ge/(In+Ge)$\leq$ 0.6 can be divided into three regions. There is a compositional margin (approx 0.1$\leq$Ge/(In+Ge) $\leq$0.5) for which a semiconducting behavior can be obtained. Below and above this margin, the In-Ge-O is conductive and insulating, respectively. Similar studies of resistivity vs. X/(X+In) were made for 0<X/(In+X)$\leq$ 0.6, where it was observed that in general the resistivity increased with X content except for case of X=Zn, and Sn.

Hall mobility, μ_{Hall}, (cm^2/V·s) measurements were made on various In-X-O in van der Pauw configuration. The composition was set to In/(In+X)~0.9 where X= Ge, Mg, Al, Si, Ti, Zn, Ga, and Ge. Medvedeva [9] reported that the effective mass of the muti-component oxides could be an "effective" average over the effective masses of the corresponding single-cation oxides.

According to this report, since the intrinsic properties are determined by the local symmetry, i.e. the nearest neighbors, the effective mass averaging should also apply to amorphous conductive oxides. Considering the approach , the electron effective mass, m_e, of amorphous X-O materials in decreasing order is given by $m_{e(Zn-O)} > m_{e(Ga-O)} \geq m_{e(Ge-O)} > m_{e(Mg-O)} > m_{e(Al-O)} > m_{e(Si-O)}$. The hall mobility in the present study, μ_{Hall}, of In-X-O (X= Zn, Ge, Ga, Mg, Al, Si) was plotted as a function of m_e of X-O. It can be seen from Fig. 4 that the μ_{Hall} of these binary-cation oxides has an inverse correlation with m_e of X-O.

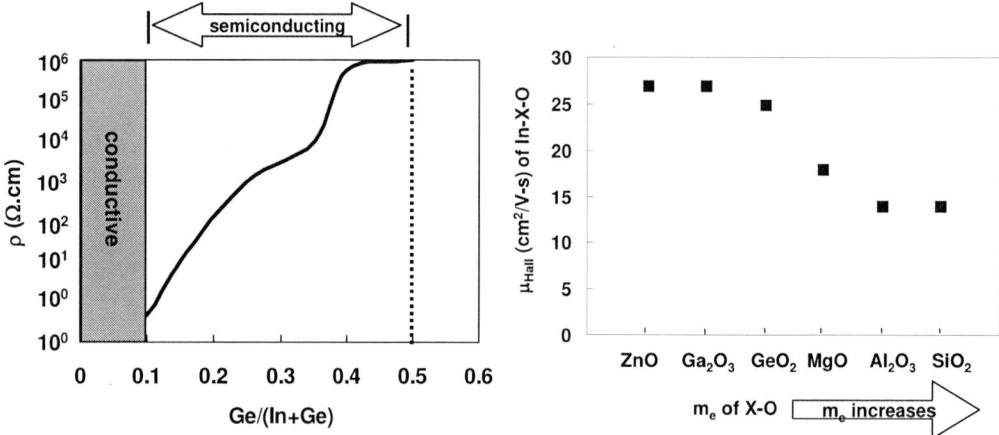

Figure 3 – Resistivity, $\rho(\Omega \cdot cm)$, as a function of Ge content.

Figure 4 – In-X-O hall mobility, μ_{Hall}, as a function of effective mass, m_e, of X-O

The trend in these experimental results is consistent with the theoretical calculations of Medvedeva [9]. Thus it is expected that on averaging the effective masses of In-O and X-O to obtain the averaged m_e of amorphous In-X-O a trend of inverse correlation between μ_{Hall} and m_e of In-X-O will be obtained. In the case of In-Ti-O and In-Mo-O, the effective mass and also the conduction mechanism could be affected by the angular dependence between the adjacent d orbitals. The hall mobility obtained for amorphous In-Ti-O was small relative to other In-X-O (μ_{Hall}~6cm^2/V·s) under this study.

2) TFT characteristics

TFT characteristics for each In-X-O system and their compositional dependence were studied using the combinatorial technique. For most channel materials (except In-Mo-O and In-Sn-O), TFTs switch successfully (ON/OFF current ratio> 10^7). Table 1 lists the on/off current ratio and field-effect saturation mobility, μ_{sat}, of as deposited In-Zn-O channel layer TFT and various In-X-O (X= B, Mg, Al, Si, Ti, Ga, Ge, Mo and Sn) channel layers TFTs after post annealing at 300°C in air. It can be seen that In-Zn-O, In-Ge-O, In-Si-O show good TFT performance with a high mobility and high on/off ratio. Figure 5b illustrates the transfer characteristics of In-Zn-O (In:Zn=4:6), In-Ge-O (In:Ge>0.7), and In-Si-O (In:Si>0.7) TFTs listed in table 1. TFTs of In-Al-O, In-Mg-O showed relatively poor performance with either lower mobility or lower on/off ratio. TFTs with channel layers of In-Sn-O did not show switching behavior as the off-current could not be suppressed by the gate-field effectively. When X is a transition metal element, the TFTs either show no switching operation as in the case of In-

Mo-O or poor performance as in the case of In-Ti-O. Thus the *d* orbital of these elements may not be suitable for In-X-O TAOS TFTs. In the TFTs listed in Table 1, the compositional ratios are In:Zn=4:6, 0.6 < In/(In+X) <0.9 (X= Ge, Ga, Mg, Al, Ti, B, Si).

Table 1 – TFT characteristics of In-X-O channel layer TFTs.

In-X-O	μ_{sat} (cm^2/V-s)	ON/OFF current ratio	Group of X
In-Zn-O	26.5	10^{10}	1B
In-Ge-O	6	10^{10}	4A
In-Si-O	3	10^9	4A
In-Ga-O	6	10^9	3A
In-B-O	4	10^{10}	3A
In-Mg-O	2	10^8	2A
In-Al-O	2.7	10^9	3A
In-Ti-O	<1	10^7	4B
In-Mo-O	-	-	6B
In-Sn-O	-	-	4A

The highest performance was achieved for In-Zn-O TFT (μ~26cm^2/V·s, S=0.2V/dec) [4]. The compositional study showed the best performance at Zn:In=60:40. We found another candidate of In-Ge-O for which high TFT performance could be obtained in both as deposited and post annealed conditions. This is clearly illustrated in the transfer characteristics of as-fabricated (S~0.9 V/dec., μ_{sat}= 3.4cm^2/V-s, on/off >10^9) and post-annealed In-Ge-O TFT (S~0.36 V/dec, μ_{sat}= 3.4cm^2/V-s, on/off ~ 10^{10}) for In:Ge~65:35 [Fig. 5a].

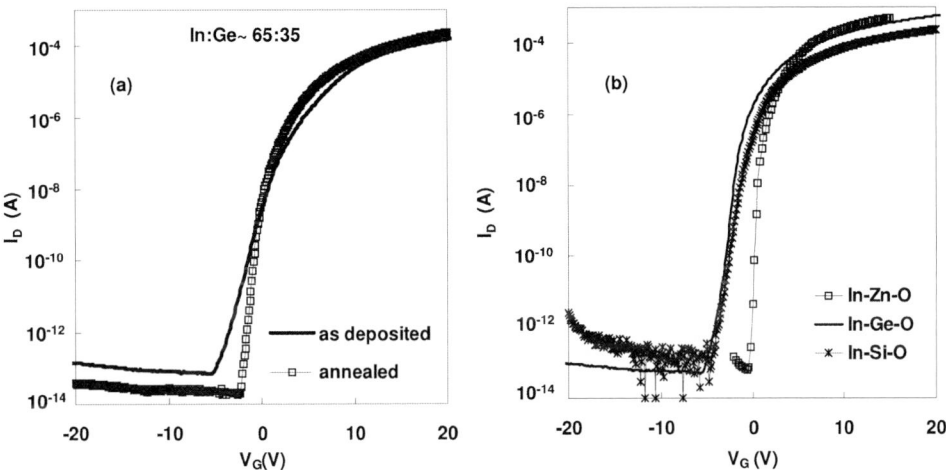

Figure 5– Transfer characteristics of (a) In-Ge-O TFT as-fabricated and after post-annealing, (b) In-Zn-O, In-Ge-O, and In-Si-O channel layer TFTs.

The field-effect saturation mobility, μ_{sat}, is related to hall mobility, μ_{Hall}, and shows similar behavior of inverse correlation with m_e of X-O. Thus a higher μ_{sat} could be obtained for a smaller m_e (data not shown here) as expected from results shown in Fig. 4.

It is also found that subthreshold swing is influenced by atomic radii of the X in In-X-O. This suggests that atomic radii of the X could be related to the interfacial trap states in In-X-O TFTs, though this aspect needs more detailed study.

It seems that B, Mg, Al, Ge, Ga, Al, and Si belong to the same class as Ga in In-based TAOS materials, thus are effective to control the carrier concentration and attain lower off currents in the TFTs; whereas Zn and Sn behave differently. Mo and Ti (Transition metal elements) do not seem to be suitable for TAOS based TFTs.

CONCLUSIONS

Various In-X-O channel layer materials were studied. It is found that most of In-X-O can be used for TFT channels, though the TFT characteristics depend on "X" in In-X-O. The electron hall mobility of In-X-O channel materials is inversely related to the electron effective mass of X-O, estimated by the method of "the effective mass averaging". This indicates that effective mass of simple oxide (X-O) is useful to design the elemental combination in TAOS materials. The subthreshold swing is likely to be influenced by the ionic radii of X in In-X-O. It is concluded that a good TFT performance can be obtained in case of In-Zn-O, In-Ge-O, In-Si-O system. In particular, In-Ge-O seems to be a promising candidate for TFT channel layer material with high TFT performance ($\mu_{FE} \sim 6 cm^2/V\text{-}s$, on/off $\sim 10^{10}$) and wide compositional margin. The wide availability of group-four elements of periodic table like Si and Ge under this study and their high effectiveness for TFT applications marks them as promising new candidates for In-based TAOS materials for TFT channel.

REFERENCES

1. K. Nomura, H. Ohta, A. Takagi, T. Kamiya, M. Hirano, and H. Hosono, *Nature_London,* **432**, 488 (2004).
2. T. Iwasaki, N. Itagaki, T. Den, H. Kumomi, **Appl. *Phys. Letters*** 90, 242114 (2007).
3. H. Yabuta, M. Sano, K.Abe, T. Aiba, T. Den, H. Kumomi, K. Nomura, T. Kamiya, H. Hosono, **Appl. *Phys. Letters*** 89, 112123(2006).
4. N. Itagaki, T. Iwasaki, H. Kumomi, T. Den, K. Nomura, T. Kamiya, H. Hosono, *Phys. Stat.* **Sol., 205**, 1915 (2008).
5. P. Barquinha, A. Pimental, A. Marques, L. Pereira, R. Martins, E Fortunato, *Journal of Non-Crys. solids*, **352**, 1749 (2006).
6. E. Fortunato, L. Pereira, P. Barquinha, A. Rego, G. Gonçalves, A.Vilà, J.Morante, R. Martins., **Appl. *Phys. Letters,*** **92** , 222103 (2008).
7. D. Hong, H. Q. Chiang, J. F. Wager, *Journal of Vac. Soc and Tech.* **24**, L23 (2006).
8. H. Hosono, *Journal of Non Crys. Solids*, **352**,851(2006).
9. J. E. Medvedeva *EPL*, **78**, 57004 (2007).

Mater. Res. Soc. Symp. Proc. Vol. 1109 © 2009 Materials Research Society

Impact of Surface Morphology and Polarity on ZnO Optical Emission

D. Doutt,[1] Y. Dong,[2] M. Myers,[1] D. Tayim,[3] C. Zgrabik,[1] Z.Q.Fang,[4] D.C. Look,[4] G. Cantwell,[5] J. Zhang,[5] J.J. Song,[5] H.L. Mosbacker,[1] L.J. Brillson,[1,5]

[1]Department of Physics, The Ohio State University, Columbus, OH 43210

[2]Dept. of Electrical & Computer Engineering, Ohio State University, Columbus, OH 43210

[3]Columbus School for Girls, Bexley, OH 43209

[4]Z.Q.Fang and D.C. Look, Wright State University, Dayton, OH 45433

[5]ZN Technology, Inc., La Brea, CA 92821

ABSTRACT

We have used nanoscale depth-resolved cathodoluminescence spectroscopy (DRCLS), atomic force microscopy (AFM), and Kelvin probe force microscopy (KPFM) to measure the spatial distribution of native point defects within the outer few hundred nanometers of the surface for leading ZnO growth and surface finishing techniques. These studies reveal dramatic differences between polar surfaces and show how surface morphology and polarity impact the efficiency of near band edge (NBE) optical emission. ZnO crystals display wide variations in defect concentration and their variation with depth. ZnO crystals grown by melt, hydrothermal, and vapor-phase transport methods show mutually independent native point defect optical emissions at 2.1, 2.5 and 3.0 eV that vary by orders-of-magnitude at and below their free surfaces. Within the first few tens of nm, the relative efficiency of NBE emission increases dramatically for surface roughness below threshold values of unit cell dimensions and is related to electronically-active defect states at surface asperities. UHV-clean Zn-polar surfaces exhibit significantly higher NBE emission and lower defects than O-polar surfaces of the same crystals along with 0.2 vs. 0.4 nm minimum rms roughness, respectively. In turn, these morphology and defect differences produce orders-of-magnitude variation in near band edge optical emission efficiency.

INTRODUCTION

Advances in ZnO growth and processing are enabling optoelectronic applications, yet control of ZnO's optical emission properties at the nanoscale remain to be explored. ZnO thin films and nanostructures are of particular interest for device applications, emphasizing the importance of controlling charge transport, recombination, and light emission in thicknesses of only hundreds of nm or less. In order to understand the parameters that dominate these physical processes, we have used nanoscale depth-resolved cathodoluminescence spectroscopy (DRCLS), atomic force microscopy (AFM), and Kelvin probe force microscopy (KPFM) to measure the spatial distribution of native point defects within the outer few hundred nanometers of the surface for leading ZnO growth and surface finishing techniques. This approach allows us to

49

correlate surface roughness and crystal face polarity with the densities of near band edge and deep level defect emissions. Earlier studies have revealed orders of magnitude variations in native point defect densities extending hundreds of nm into the bulk of ZnO crystals grown by different techniques or even for the same technique and growth source [1]. Melt- and vapor-phase-grown ZnO exhibit uniformly distributed defect emission from the surface to the bulk, whereas hydrothermal ZnO from different sources displays up to hundred-fold increases in defect density from <5 to ~1000 nm below the surface. Here we report on the major variations in defect densities within and across the near-surface region and their correlation with surface morphology and polarity.

EXPERIMENTAL DETAILS

We used an ultrahigh vacuum (UHV) JEOL 7800F field emission scanning electron microscope coupled with a 10 K specimen holder, a parabolic mirror to collect luminescence, a monochromator and a UV-vis photomultiplier to perform DRCLS measurements. For a review of the DRCLS technique, see, for example, Reference [2]. Here the incident electron beam excites depth- and laterally-resolved free electron-hole pairs that recombine radiatively to produce optical emissions that are characteristic of near band edge (NBE) and band-to-defect electronic transitions. We used a Park Systems XE-70 AFM/SPM resting on a Herzan TS-150 active vibration isolation table inside an acoustic enclosure for AFM and KPFM mapping. In KPFM mode, an external Stanford SR830 lock-in amplifier and an applied DC bias feedback loop provide an efficient on-pass scan that measures both the topography and surface potential simultaneously. See, e.g., Reference [3].

RESULTS

DRCLS spectra of two ZnO surfaces grown hydrothermally illustrate the dramatic differences that can occur between ZnO crystals. Figures 1(a) and (b) show NBE emissions with phonon replicas along with deep level defect emissions centered at 2.1 and 2.4-2.5 eV. In Fig. 1(a), the 2 keV (25 nm peak excitation depth) spectra exhibit defect emissions that are higher than those at 5 keV (105 nm depth). Note the significant variations in defect intensities at different excitation depths. Corresponding AFM and KPFM maps show relatively high surface roughness associated with asperities (c) such as those circled along with large (~200 meV) potential variations (e) associated with those asperties.

In contrast, the 2 keV spectra in Fig. 1(b) show much lower 2 keV near-surface defect emissions on an absolute scale as well as relative to the NBE peak intensity. Further below this surface, defect emissions increase by over an order of magnitude [4] to the same levels as in Fig.1(a). Subsurface defect emissions, although high, are quite uniform for this crystal. This contrast between surface and bulk is due to an electrochemical etch polish that decreases surface roughness 3-fold. Consistent with the low surface defect emissions, AFM maps in (d) show a relatively smooth, low asperity surface and relatively small (<50 meV) potential variations. The lack of correspondence between low potential (dark) areas and surface morphology indicates the presence of defects below rather that at the ZnO surface. The variations in potential in Fig. 1(e) can in fact be modeled in terms of charged sites at the asperity edges to provide measures of near-surface defect and free carrier concentration [4].

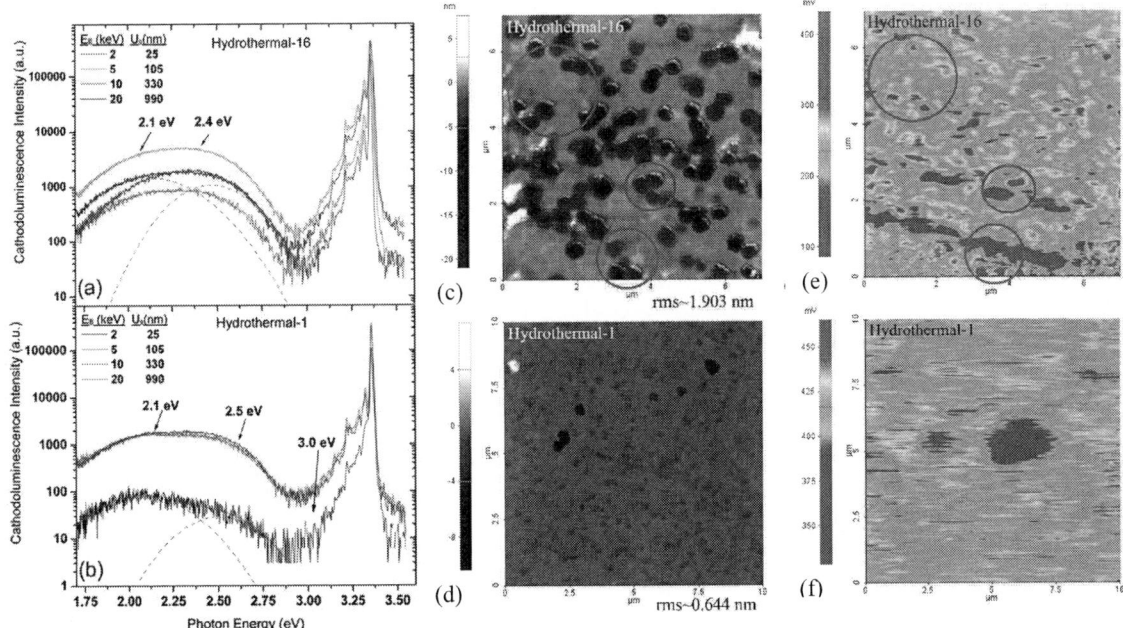

Figure 1. DRCLS of hydrothermally-grown ZnO showing (a) high and (b) low near-surface defect emissions. Corresponding AFM maps show (c) and (d) low surface roughness. Corresponding KPFM maps show large potential variations near morphological asperities, e.g., circled regions, in (e) and lower variations below the surface in (f).[4]

The surface roughness and associated electrically-active defects have a pronounced effect on the efficiency of NBE emission. Figure 2 illustrates the ratio of 2 keV NBE (3.3 eV) to defect (2.1-2.5 eV) emissions in the top 25 nm of a wide variety of ZnO crystals grown by vapor phase transport, melt, and hydrothermal techniques. In general, crystals that exhibit high I(3.3 eV)/I(2.1-2.5 eV) ratios have the highest absolute near-surface emission intensities. There is a

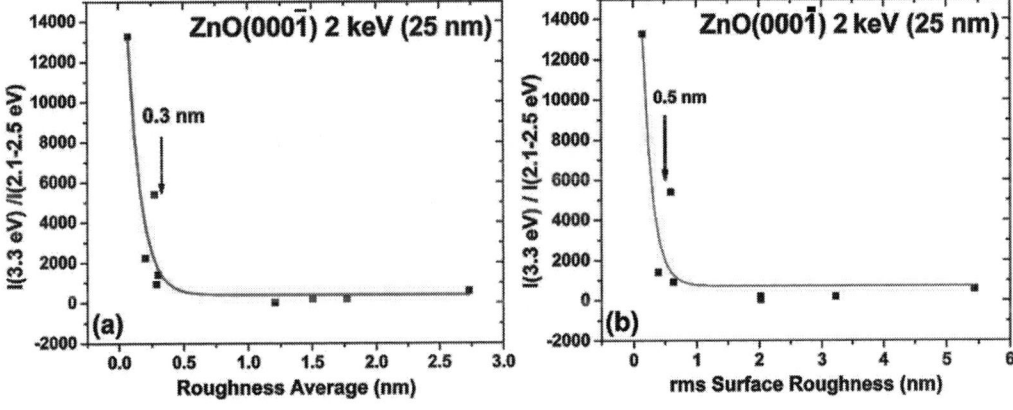

Figure 2. Threshold dependence of I(3.3 eV)/I(2.1-2.5 eV) versus (a) roughness average and (b) rms surface roughness at ~ 0.5 nm.[3]

clear correlation between this ratio and both (a) roughness average and (b) rms surface roughness. Both exhibit a dramatic rise by over an order of magnitude for roughness below 0.5 nm, i.e., single unit cell dimensions. This demonstrates a correlation between electrically-active native point defects and above-monolayer surface roughness features.

There are also remarkable differences in defect concentrations and morphological features between the two polar surfaces of ZnO. Figure 3 illustrates DRCL spectra from (0001)Zn and (000$\bar{1}$)O surfaces. Although NBE and defect features appear similar in Fig. 3(a), I(3.3 eV) is 4-5 times higher for Zn vs. O face,[5] both before and after ROP treatment to clean the surface and remove near-surface native point defects [6] and I(3.3 eV)/I(2.5 eV) is > 2 times lower for the Zn vs. O face. This shows that the Zn face has fewer 2.5 eV defects, attributed to O vacancies.[7]

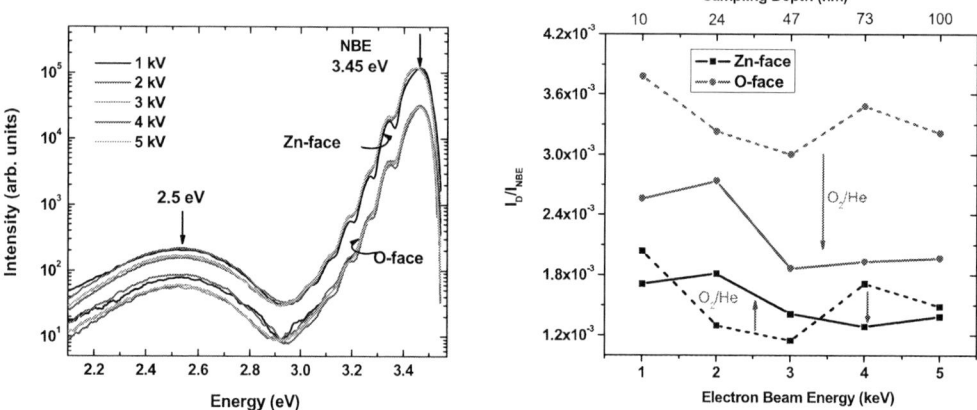

Figure 3. (a) DRCL spectra of Zn- and O-polar faces vs. depth and (b) corresponding I(3.3 eV)/I(2.5 eV) at different sample depths before (dashed) and after (solid) O$_2$/He ROP treatment.

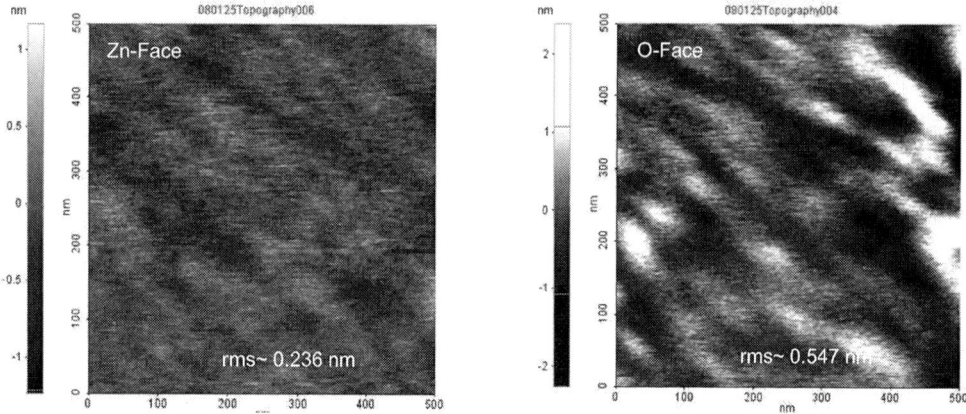

Figure 4. AFM topography maps of Zn (top) and O (bottom) surfaces before (left) and after (right) ROP treatment. The O-face surfaces are rougher both before and after ROP treatment.

Consistent with the results shown in Fig. 1, AFM maps show that the O face is consistently rougher than the Zn face. Figure 4 shows polar surfaces of ZnO grown by vapor phase transport. Rms roughness is ~0.236 nm before and after ROP treatment for the Zn face, whereas it is more than double for the O face, increasing further with ROP treatment. Other surface regions exhibit similar roughness values. The increase for ZnO($000\bar{1}$)O surfaces is consistent with the higher reactivity of its sp^3 bond termination.[8]

Pronounced differences in defects are also evident with extended ROP treatment to reduce native point defects further. A 4-hour ROP treatment affects the electronic states of the Zn- and O-polar surfaces differently. Figure 5(a) shows that long-time (>4 hr) O$_2$ plasma cleaning generates increased emission at 2.1 eV corresponding to Zn vacancies (V_{Zn})[9], but only at the Zn face. Extended ROP cleaning of the O face introduces no additional 2.1 eV V_{Zn} feature. This contrast suggests that the O plasma extracts Zn from the Zn face preferentially.

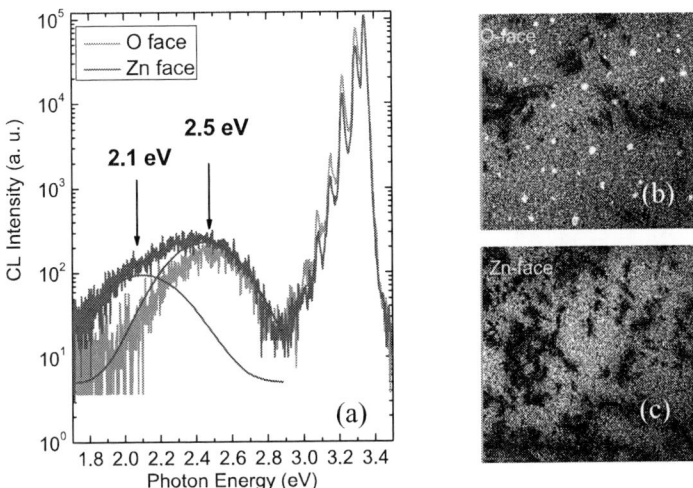

Figure 5. (a) 2 keV CL of ZnO Zn and O faces after 4 hours ROP treatment. CL maps (40 μm scale bar) of NBE intensity on (b) O face and (c) Zn face.

CL mapping reveals another striking difference between the Zn- vs. O-polar surfaces. On both surfaces, NBE emission intensity varies across the surface, consistent with the results of Figs. 2 and 4. The Fig. 5(b) O-face map displays additional, < 5 μm, bright spots spaced approximately 20-40 μm apart. Figure 5(c) shows that the NBE emission extending across the surface for the Zn face. However, the small spot emission is quenched. These localized areas of high NBE emission may be associated with dislocations, albeit at much lower density than typically reported for ZnO. These localized features are currently under investigation.

DISCUSSION

Our DRCLS measurements have identified how surface morphology and polarity impacts the efficiency of band gap optical emission. AFM scans of surface morphology reveal large variations in surface roughness, asperities, and extended features related to the growth method, subsequent polishing and etching. UHV-clean Zn-polar surfaces exhibit significantly higher NBE emission and lower defects than O-polar surfaces of the same crystals along with 0.2 vs. 0.4 nm rms roughness, respectively. KPFM maps of surface and subsurface electric potential acquired simultaneously exhibit systematic correlations that depend on the distribution of native point defects measured by DRCLS. When defect emissions are low, surface roughness is low and morphology matches its respective KPFM potential map. When DRCLS emissions vary with depth, surface morphology and potential maps can be strikingly different. Indeed, electric potential can vary by hundreds of meV across micron square areas, emphasizing the impact of sub-surface defects on electronic properties. Detailed depth correlations show that chemomechanical polishing can reduce outer layer roughness and native defects while sub-surface defect and potential features remain. Furthermore, pronounced surface asperities with large KPFM changes can be associated with native point defects. The relative strength of near band edge to deep level defect emissions exhibits a dramatic threshold dependence on surface roughness: optical emission efficiency increases over ten-fold as roughness decreases to unit cell dimensions, emphasizing the role of surface polishing and etching for high efficiency emitters.

CONCLUSIONS

These results stress the importance of monolayer-scale surface roughness to minimize surface recombination of free carriers by defects. AFM surface roughness values can be a predictive tool for assessing surface optoelectronic quality, e.g., NBE emission intensity, especially for nanoscale optical devices. Zn-polar ZnO systematically shows stronger near band edge emission and lower oxygen vacancy defect emission, consistent with its smoother surface compared with the O face. These results highlight the value of near-surface, depth-resolved probes in assessing crystal quality. Furthermore, surface and near-surface states due to morphology and polarity provide a guide to understanding and controlling and optical emission.

ACKNOWLEDGMENTS

The authors gratefully acknowledge support by the National Science Foundation, Grant No. DMR-0803276 (Verne Hess).

REFERENCES

[1] L.J. Brillson, H.L. Mosbacker, D. Doutt, M. Kramer, Z.L. Fang, D.C. Look, G. Cantwell, J. Zhang, and J.J. Song, Superlattices and Microstructures, in press (2008) and references therein.
[2] L.J. Brillson, J. Vac. Sci. Technol. **B 19**, 1762 (2001).
[3] D.R. Doutt, C. Zgrabik, H.L. Mosbacker, and L.J. Brillson, J. Vac. Sci. Technol. **26**, 1477 (2008).
[4] D. Doutt, H.L.Mosbacker, G. Cantwell, J.Zhang, J.J. Song, and L.J. Brillson, Appl. Phys. Lett., in press (2008).
[5] Y. Dong, Z-Q. Fang, D.C. Look, G. Cantwell, J. Zhang, J.J. Song, and L.J. Brillson, Appl. Phys. Lett. **93**, 172111 (2008).
[6] H.L. Mosbacker, Y.M. Strzhemechny, B.D. White, P.E. Smith, D.C. Look, D.C. Reynolds, C.W. Litton, and L.J. Brillson, Appl. Phys. Lett. **87**, 012102 (2005).
[7] L.J. Brillson, H.L. Mosbacker, M.J.Hetzer, Y.Strzhemechny, G.H.Jessen, D.C.Look, G.Cantwell, J. Zhang, and J.J.Song, Appl. Phys. Lett. **90**, 102116 (2007).
[8] H.C. Gatos, P.L. Moody, and M. C. Lavine, J. Appl. Phys. **31**, 212 (1960).
[9] H.L. Mosbacker, C. Zgrabik, A. Swain, S. El Hage, M. Hetzer, D.C.Look, G.Cantwell, J. Zhang, J.J.Song, and L.J. Brillson, Appl. Phys. Lett. **91**, 072102 (2007).

Mater. Res. Soc. Symp. Proc. Vol. 1109 © 2009 Materials Research Society   1109-B08-10

MOCVD Growth of GaN-based Materials on ZnO and Si Substrates

William E. Fenwick[1], Muhammad Jamil[1], Tianming Xu[1], Shen-Jie Wang[1], Hongbo Yu[1], Andrew Melton[1], Nola Li[1], Jeff Nause[2], and Ian T. Ferguson[1]

[1]School of Electrical and Computer Engineering, Georgia Institute of Technology (GIT), 777 Atlantic Dr., Atlanta, GA 30332-0250, United States

[2]Cermet, Inc. 1019 Collier Rd., Atlanta, GA 30318

ABSTRACT

The structural, optical, and electrical properties of MOCVD-grown GaN thin films on ZnO and Si substrates have been investigated in this work. ZnO is an interesting substrate material because of its physical properties, which are similar to those of GaN. Si is also of great interest as a substrate for MOCVD growth of GaN because of its low cost and its dominance in the microelectronics industry. The major issues with GaN growth on ZnO are the stability of the ZnO substrate in a hydrogen atmosphere and the diffusion of zinc and oxygen into the GaN layer, while the major issue with growth on Si is its lattice and thermal mismatch with GaN, leading to tensile stress and cracking in the GaN layers. Atomic layer deposition (ALD)–grown Al_2O_3 has been investigated as a transition layer for GaN growth on ZnO and Si to address these issues and allow for growth of higher quality material. The Al_2O_3 layer is shown to improve quality in the GaN layers compared to those on bare substrates.

INTRODUCTION

The development of high-brightness light emitting diodes (LEDs) in the last decade has made LEDs a promising technology in the lighting industry. These devices are most often based on thin films of gallium nitride and its alloys with indium ($In_xGa_{1-x}N$) and aluminum ($Al_xGa_{1-x}N$) deposited on substrates of either sapphire (Al_2O_3) or SiC. Other substrates are desirable, however, either because of their physical properties (ZnO), or because of their low cost and application in other areas (Si). The physical properties of ZnO are very similar to those of GaN, making the integration of these two materials promising for new device technologies. Si, while presenting serious challenges in terms of physical properties, is the most inexpensive substrate material. Si is also particularly interesting as a substrate because of its dominance in the microelectronics industry.

ZnO has been the focus of intense research recently because of its physical properties such as bandgap ($E_g = 3.37$ eV) and exciton binding energy (~60 meV), which offer promise for more efficient emitters at room temperature. The similarity of ZnO and GaN has also led to investigations into the integration of these two materials.[1-4] The near lattice match of GaN and ZnO should allow for the growth of low defect density GaN, in turn leading to more efficient GaN-based devices. The use of ZnO as a substrate will also allow the removal of the substrate by wet etching, allowing for new approaches to both light extraction and thermal management in LEDs. The fundamental problems of interdiffusion at the interface and ZnO instability in H_2 remain to be solved, however, if ZnO is to become a significant substrate material in the GaN-based LED market.

A significant amount of work has also been done toward MOCVD growth of GaN on Si in the last decade, and, if successful, this work will have two major benefits. First, Si substrates are cheaper than other commercially used substrates, such as sapphire and SiC. This reduction in substrate cost will lead to a reduction in overall cost of production of GaN-based LEDs. Second, MOCVD growth of high quality GaN on Si will allow for monolithic integration of GaN-based optoelectronics with Si-based microelectronics, opening the door for a wide range of novel applications in communications, computation, and related areas.[5, 6] There are major technical challenges that must be overcome, however, if the integration of these technologies is to have an impact on the microelectronics and optoelectronics industries in the future.

EXPERIMENT

All GaN thin films in this study were grown using a modified commercial MOCVD system with a vertical injection, rotating disk configuration. The Al_2O_3 layers in this study were deposited using a custom-built ALD system. Structural properties of the thin films were investigated using a Philips X'Pert Pro MRD four circle diffractometer and a Renishaw micro-Raman system with a 488 nm Ar-ion laser. A PSIA atomic force microscope (AFM) was used to study surface morphology. Photoluminescence spectra were taken with two different experimental setups. A 325 nm Melles-Griot HeCd laser with an Acton Spectra Pro 2300i monochromator and PIXIS 100 CCD camera was used for temperature-dependent measurements. A 248 nm Ne-Cu laser with a CVI monochromator and a Hamamatsu photomultiplier tube (PMT) was used for room temperature measurements. Electrical properties were studied using a HEM2000 Hall Effect Measurement system from EGK Co., Ltd.

The ALD-grown Al_2O_3 films were first deposited using trimethyl aluminum (TMAl) and H_2O as Al and O precursors, respectively. The structural properties and surface morphology of these layers were investigated. Al_2O_3 films ranging in thickness from 2 nm to 200 nm were investigated in this study. GaN thin films were also grown by MOCVD on bare ZnO (0001) and Si (111) substrates to provide a baseline process from which to work when growing on Al_2O_3 substrates. ZnO substrates were cleaned with acetone for 3 minutes and methanol for 3 minutes, then rinsed in deionized water and blown dry with N_2. All Si substrates used in this work were cleaned before deposition by dipping in HF for 1 minute to remove the native oxide layer and then rinsing in deionized water for 1 minute. These layers were grown on both simple AlN buffer layers and buffer structures containing LT-AlN interlayers. The growth process was then transferred to Al_2O_3/ZnO and Al_2O_3/Si substrates. The properties of these films were studied to determine the effect of the Al_2O_3 layer on the subsequent GaN layer, and to develop an MOCVD growth process for GaN on sacrificial substrates that can consistently yield device-quality material.

DISCUSSION

GaN layers were grown on both ZnO and Si substrates in this work. The use of an ALD-grown Al_2O_3 interlayer was also investigated for MOCVD growth of GaN on Si. High quality GaN layers were obtained on both ZnO and Si, and the ALD-grown Al_2O_3 interlayer was shown to increase crystal quality for GaN layers on Si.

MOCVD Growth of GaN on ZnO

An MOCVD process for growth of high-quality GaN on ZnO substrates was developed. The first step in this development process was to investigate the stability of the ZnO substrates in a hydrogen atmosphere. ZnO substrates were exposed to H_2 at typical GaN growth temperatures, and the etch rate was measured. Figure 1 shows the temperature dependence of the ZnO etch

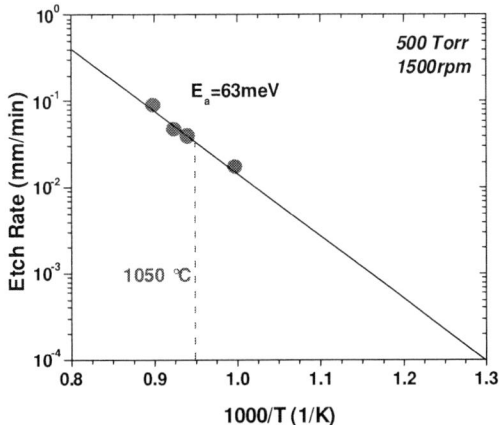

Figure 1. Etch rate of ZnO substrate in H_2.

rate in H_2. The dependence of the etch rate in H_2 on reciprocal temperature shows an Arrhenius dependence with a calculated activation energy of 63 meV. An extrapolation to lower growth temperatures (~600 °C) suggests that the ZnO substrate is still unstable at relatively low growth temperatures, necessitating the development of a near H_2-free approach to GaN growth on ZnO.

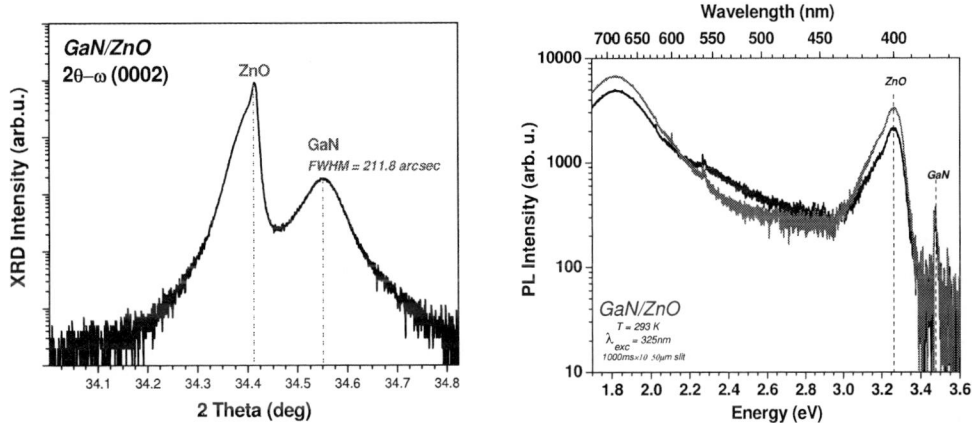

Figure 2. (a) XRD 2q-w scan of a GaN epilayer on ZnO (0001) substrate. (b) RT-PL results showing strong ZnO bandedge and weak GaN bandedge luminescence.

The first sample was grown with a low temperature GaN layer of ~20 nm, and then increased in temperature up to ~750 °C for the main GaN layer. This low growth temperature was used for the main GaN layer because of the instability of ZnO in a hydrogen atmosphere at typical GaN growth temperatures, as mentioned previously. Figure 2(a) shows XRD data from a 2Θ−ω scan, where the linewidth of the GaN peak is 211.8 arcsec. RT-PL measurements show a strong ZnO bandedge, as expected, and a relatively weak GaN bandedge, Figure 2(b). The GaN bandedge luminescence is reduced most likely due to diffusion of Zn and O defects. Reabsorption by the ZnO substrate may also contribute to the degradation of the GaN luminescence. The origin of the defect peak near 700 nm also remains unclear, though it may be due to the formation of a second phase at the GaN/ZnO interface or to the interdiffusion of impurities at the interface. This interdiffusion creates a major problem for nitride growth on ZnO, as shown in Figure 3. SIMS was done on an $In_xGa_{1-x}N$ layer on ZnO in order to study impurity concentrations. Figure 3 shows SIMS impurity concentrations for an $In_xGa_{1-x}N$ layer on ZnO with a 30 nm LT-GaN buffer layer. $Al_xGa_{1-x}N$/GaN superlattices were studied in order to prevent diffusion of Zn and O into the GaN layer as observed via SIMS.

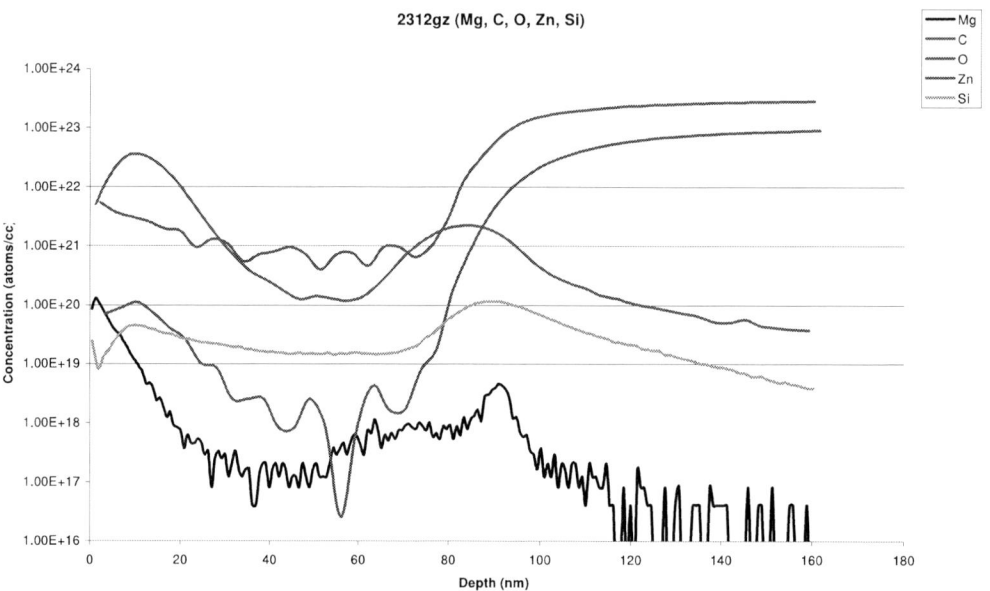

Figure 3. SIMS results showing diffusion of Zn and O from the substrate into the GaN epilayer.

MOCVD Growth of GaN on Si

GaN layers were first grown using the simple AlN buffer layer grown at ~1060 °C as described above and a 40 nm-thick low-temperature GaN (LT-GaN) interlayer inserted on the AlN layer in order to relieve strain and increase crystal quality in the GaN layer, similar to GaN growth on sapphire. However, the use of the LT-GaN interlayer led to poor crystal quality as observed by XRD, with a (0002) rocking curve linewidth of about 1200 arcseconds. The linewidth of the near bandedge emission, however, was 37 meV, which is near the typical linewidth of ~30 meV for GaN on sapphire.

RT-PL showed an improvement using a thinner (25nm) LT-GaN layer on the AlN buffer layer, with a decrease in linewidth to 30 meV, though XRD showed no improvement in structural quality compared to the sample with a 45nm LT-GaN layer. Also, both samples using LT-GaN interlayers also exhibited cracking, suggesting that the use of the LT-GaN interlayer does not sufficiently alleviate tensile strain in the material.

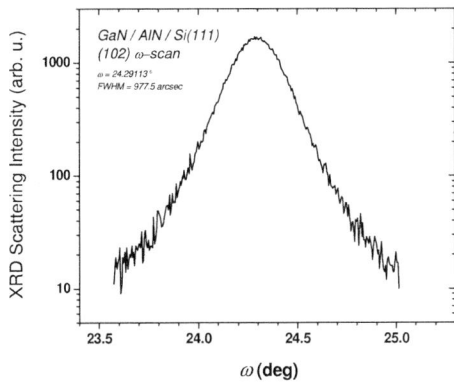 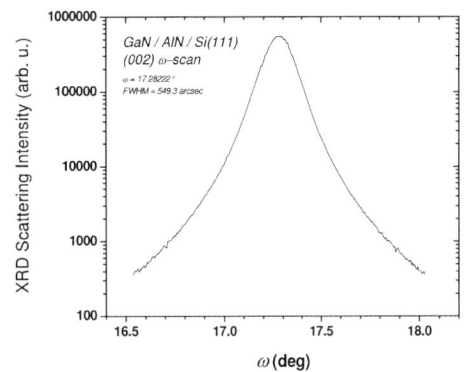

Figure 4. (a) XRD rocking curve scan of the GaN (0002) peak with a linewidth of 549.3 arcsec. (b) XRD rocking curve scan of the GaN (10-12) peak with a linewidth of 977.5 arc sec

The next step was to remove the LT-GaN interlayer completely and grow a high temperature GaN layer directly on the HT-AlN buffer layer. Removal of the LT-GaN layer resulted in a marked increase in crystal quality. An XRD ω-scan of the (002) reflection is shown in Figure 4(a) for a 1.5 μm-thick GaN layer grown at high temperature directly on a HT-AlN buffer layer. Figure 4(b) shows a rocking curve scan of the GaN (102) reflection with a linewidth of 679arcsec, suggesting high crystalline quality GaN layers.

RT-PL spectra of the GaN layers grown with a simple AlN buffer layer show a redshift in the bandedge luminescence, which is most likely due to tensile strain in the layers. Figure 5 shows a typical RT-PL spectrum taken from a GaN epilayer on Si. The linewidth of the bandedge emission is 54.8meV. The ratio of bandedge luminescence to yellow luminescence is also relatively low (BL/YL = 5.396) compared to GaN on sapphire, which may be an indication of a higher defect density in the layers on Si.

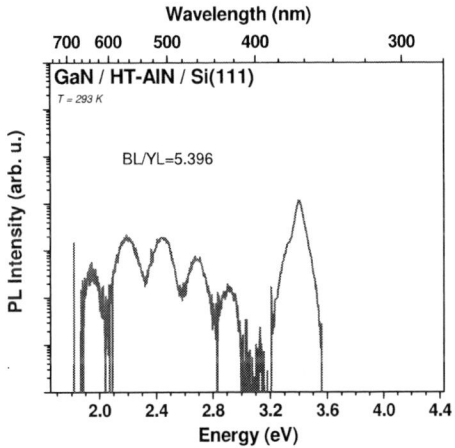

Figure 5. RT-PL spectra (taken with 248 nm Ne-Cu laser) of GaN layer on bare Si with a HT-AlN buffer layer.

GaN layers were grown on Al_2O_3/Si substrates after the development of a GaN growth process on bare Si. The first step taken in this work was the investigation fundamental growth parameters and structures to lay a foundation for future work. Both LT-GaN and HT-AlN were investigated as buffer layers on the Al_2O_3/Si substrate. GaN growth on sapphire typically proceeds with the growth of a thin (20-30 nm) LT-GaN buffer layer before the high quality GaN layer, while growth on Si usually proceeds with a HT-AlN buffer layer ranging from 50 nm to 200 nm in thickness.[7, 8] Both approaches were investigated here, with a HT-GaN layer grown under the same conditions on all buffer layers.

The LT-GaN buffer layers were deposited at 745 °C with a nominal thickness of 25-30 nm, while the HT-AlN buffer layers were deposited at a growth temperature of 1060 °C and thickness of 100 nm. Both annealed and unannealed Al_2O_3/Si substrates were used in this study, and the samples grown with the HT-AlN layer consistently yielded higher quality material, as observed by XRD and RT-PL.

The layer on HT-AlN is much smoother than the layer on LT-GaN. The thickness of the LT-GaN layer is similar to that of LT-GaN buffer layers used on sapphire. However, the polycrystalline nature of the Al_2O_3 film on Si leads to a much rougher surface on the GaN buffer layer. Figure 6 shows SEM images of both a LT-GaN buffer layer and a HT-AlN buffer layer on 10 nm Al_2O_3 annealed at 1100 °C for 90 seconds. The HT-AlN buffer layer is much smoother than the LT-GaN layer. This rough surface also affects the growth rate of the GaN layers grown on LT-GaN buffers.

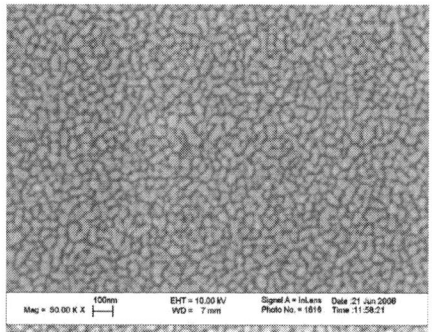 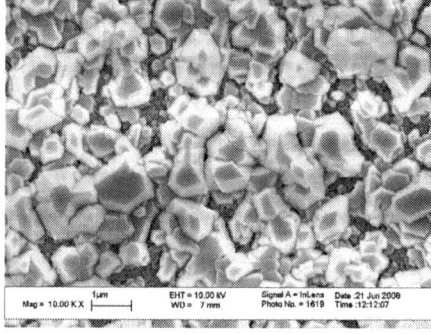

Figure 6. SEM images of a HT-AlN buffer layer on Al_2O_3/Si (left), and a LT-GaN buffer layer on Al_2O_3/Si (right).

HT-AlN buffer layers were investigated further because of the smoother surface and higher growth rate of the subsequent GaN layer. The effect of AlN thickness on structural quality of the GaN layer was studied. The structural quality of the topmost GaN layer, as observed by XRD, increased with increasing AlN thickness. However, above a thickness of 100 nm, there was only a small decrease in (0002) FWHM. 100 nm AlN buffer layers were used in subsequent studies for this reason.

High quality GaN layers were subsequently grown on these Al_2O_3/Si substrates using a HT-AlN buffer layer to study the effects of the oxide interlayer on GaN crystal quality. Table 1 summarizes the properties of the GaN layers grown on both bare Si and Al_2O_3/Si in this work. Material quality improves on the Al_2O_3/Si substrates, with structural and optical properties approaching those of GaN on sapphire. These results shows great promise toward GaN-based devices on Si.

Table 1. Strucural, optical, and surface properties of GaN layers grown on both bare Si and Al_2O_3/Si substrates.

	XRD		PL		AFM
	(002) FWHM (arcsec)	(102) FHWM (arcsec)	FWHM (meV)	BL/YL	RMS roughness (Å)
HT-AlN buffer	549.3	977.5	49.3	5.396	5.67
LT-AlN interlayers	436.8	1041.9	46.9	5.521	3.99
5nm Al_2O_3/Si	378.6	849.5	46.5	7.395	3.93
10nm Al_2O_3/Si	433.9	1344.6	47.7	4.497	5.65
20nm Al_2O_3/Si	416.6	740.1	43.4	28.223	3.70

CONCLUSIONS

High quality GaN layers were grown on ZnO and Si substrates by MOCVD, and their optical and structural properties were investigated. An ALD-grown Al_2O_3 interlayer was studied to improve crystal quality of GaN thin films on Si. GaN layers grown with this oxide interlayer showed improved structural and optical quality compared to GaN layers grown on bare Si. This shows promise toward GaN-based devices on both ZnO and Si substrates.

ACKNOWLEDGMENTS

This work was funded by the U.S. Department of Energy and by DARPA.

REFERENCES

[1] D. C. Kim, W. S. Han, B. H. Kong, H. K. Cho, and C. H. Hong, Physica B **401-402**, 386 (2007).

[2] D. P. Norton, Y. W. Heo, M. P. Ivill, K. Ip, S. J. Pearton, M. F. Chisolm, and T. Steiner, Materials Today **June 2004**, 34 (2004).

[3] A. Ougazzaden, D. J. Rogers, F. H. Teherani, T. Moudakir, S. Gautier, T. Aggerstam, S. O. Saad, J. Martin, Z. Djebbour, O. Durand, G. Garry, A. Lusson, D. McGrouther, and J. N. Chapman, Journal of Crystal Growth **310**, 944 (2008).

[4] U. Ozgur, Y. I. Alivov, C. Liu, A. Teke, M. A. Reshchikov, S. Dogan, V. Avrutin, S.-J. Cho, and H. Morkoc, Journal of Applied Physics **98**, 041301 (2005).

[5] C.-H. Hsieh, M.-Y. Ke, G.-A. Shih, T.-Y. Chiu, and J.-J. Huang, IEEE Photonics Technology Letters **19**, 662 (2007).

[6] S. V. Sorokin, I. V. Sedova, A. A. Toropov, G. P. Yablonskii, E. V. Lutsenko, A. G. Voinilovich, A. V. Danilchyk, Y. Dikme, H. Kalisch, B. Schineller, M. Heuken, and S. V. Ivanov, IEEE Electronics Letters **43**, 162 (2007).

[7] P. R. Hageman, S. Haffouz, V. Kirilyuk, A. Grzegorczyk, and P. K. Larsen, physica status solidi (a) **188**, 4 (2001).

[8] D. K. Kim, Solid-State Electronics **51**, 4 (2007).

Mater. Res. Soc. Symp. Proc. Vol. 1109 © 2009 Materials Research Society 1109-B09-09

Growth and Characteristics of ZnO Nanotube Aarrays on Si Substrates by Atomic Layer Deposition in Anodic Aluminum Oxide

Wang Miao[1], Gao Han[2], Carl V. Thompson[1,3], Lee Kheng Tan[2], and Chua Soo Jin[1,2,*]

[1] *Singapore-MIT Alliance, E4-04-10, 4 Engineering Drive 3, Singapore 117576*
[2] *Institute of Materials Research and Engineering (IMRE), 3 Research Link, Singapore 117602*
[3] *Department of Materials Science and Engineering, MIT, Cambridge, MA 02139, USA*
* Corresponding author: elecsj@nus.edu.sg

ABSTRACT

Atomic layer deposition (ALD) is a gas-phase thin film deposition method characterized by sequential surface-saturating reactions combined with alternately dosing the precursors on substrates. In this study, ZnO nanotube arrays were successfully deposited in anodic aluminum oxide (AAO) templates on Si substrates using diethylzinc (DEZn) and H_2O as precursors and nitrogen gas as a carrier gas. Each cycle is separated by nitrogen purges. The substrate temperature was kept at 170°C. The morphology, optical properties and crystallinity of the nanotube arrays were analyzed using scanning electron microscope, photoluminescence, transmission electron microscope and X-ray diffraction. By varying the number of ALD cycles, a shift of UV peak position can be detected from the ZnO film on Si substrate.

INTRODUCTION

ZnO is an attractive compound semiconductor with a direct energy band gap of 3.37 eV and exciton bonding energy of 60 meV. It is one of a few materials which can be utilized to make optical devices emitting blue-ultraviolet light at room temperature. The other outstanding applications of ZnO include gas sensors[1], solar cells[2], UV detectors[3] and thin film transistors[4] etc. The gas sensing, photon-to-electron conversion efficiency and photonic performance would be further enhanced by reducing the dimensions of ZnO structures because of surface area increase and quantum confinement effect. 1D ZnO nanotubes are therefore very promising nanostructures.

In order to grow ZnO thin films and nanostructures, various deposition techniques have been used. Atomic Layer Deposition (ALD) is a very promising technique especially for uniform films with high aspect ratios. ALD is an analog of CVD technique based on sequential self-limiting gas-solid reactions, which relies on consecutive surface reactions and utilizes critical purge steps to prevent interactions between reactive precursors. This technique is applicable for manufacturing conformal inorganic material layers with thickness down to nanometer range. Especially, ALD can produce high quality films at lower temperature than other classical methods including CVD, which makes it very attractive.

Anodized aluminum oxide (AAO) template is a very useful and effective approach to realize controlled growth of nanostructure, because its structural parameters such as pore size and inter-hole distance can be controlled by tuning its anodization conditions.

In this study, ordered ZnO nanotube arrays with uniform size controlled by AAO template were deposited by ALD on Si substrate using diethyl zinc (DEZn) as a metal precursor and DI water as a reactant. The structural and optical properties were characterized.

EXPERIMENT

AAO template was prepared by a two-step anodization as previously reported[5]. In brief, an 1 μm thick Al film was deposited onto a Si (100) substrate by electron beam evaporation. The first step anodization was carried out for 20-30 min in 0.3 M oxalic acid at 2-5 °C, 40 V, followed by template removal in a mixture of phosphoric acid (6 wt.%) and chromic acid (1.8 wt.%) at 65 ° C for 15-20 min. The second step anodization was performed at the same conditions as the first step and resulted in relatively uniform pore sizes. The alumina barrier layer at the bottom of the template was removed by etching in 5 wt % H_3PO_4 for 30-70 min. The pores are simultaneously widened during this process.

The prepared AAO template on Si substrate and a bare Si substrate were subjected to ALD in a commercial ALD system (f • XALD, Azimuth Technologies Pte. Ltd., Singapore). Two precursors, DEZn and H_2O (deionized water)，were fed into the reaction chamber in turn carried by a flow of 200 sccm N_2 gas, and purged by the N_2 nitrogen gas flow between every ALD cycle. The system background pressure was maintained at 0.6 Torr. The function of N_2 is as carrier gas and purging gas to remove the excess of unreacted precursors and expels gaseous by-products. The cycle of DEZn pulse – purging – H_2O pulse – purging was repeated over and over again. Typical pulse times for DEZn and H_2O feed were both 15 ms, and the times for purging between the precursor exposure were both 10 s. The substrate temperature was kept at 170 °C. The ALD window for ZnO is 130-180° C as reported[6]; and good crystalline quality of ZnO can be obtained around 170° C[7].

In order to remove the AAO template to obtain ZnO nanotube arrays, the ZnO layer on the top surface of AAO template need to be removed first. Inductively coupled plasma-deep reactive ion etching (Unaxis Shuttle Lock ICP system) was employed to remove the ZnO top layer. Consequently, the substrate was dipped into 0.1M NaOH solution to dissolve the template.

The thickness of ZnO films on Si substrate were measured by ellipsometry (WVASE32, J. A. Woollam Co., Inc.) and the morphology and crystallography of a signle ZnO nanotube were checked by high-resolution transmission electron microscope (HRTEM, Philips CM300). A scanning electron microscope (JEOL FESEM, JSM-6700F) was used to examine the morphology of the samples and the crystal structure of the samples was characterized using X-ray diffraction (Bruker D8 GADDS XRD) with copper K_α radiation. Photoluminescence (PL) was recorded using a 325 He-Cd laser as excitation source.

DISCUSSION

ALD is a technique where a binary reaction can be split into two self-limiting chemical reactions. Both reactions are between a gas phase molecular precursor and a surface functional group. In air water is adsorbed on most surfaces, therefore the initial substrate surface is a hydroxyl group in the form of Si-O-H. After placing the substrate in the reaction chamber, DEZn is pulsed into and reacts with the adsorbed hydroxyl groups:

$Zn(CH_2CH_3)_2 + SiOH^* <\text{-}> SiOZn(CH_2CH_3)^* + C_2H_6$

The reaction between the new surface function group and H_2O pulse after purging is:

$2 H_2O + SiOZn(CH_2CH_3)^* <\text{-}> SiOZn(OH)^* + 2C_2H_6$

After the initial reactions, the binary reaction sequence occurred are: [8]

$$ZnOH^* + Zn(CH_2CH_3)_2 <-> ZnOZn(CH_2CH_3)^* + C_2H_6$$
$$Zn(CH_2CH_3) + H_2O <-> ZnOH^* + C_2H_6$$
(the asterisks designate the surface species).

Growth rate

Fig 1 shows film thickness measured by ellipsometry vs. the number of ALD cycles. A linear growth rate of 1.8 Å/cycle was observed over the entire range of films produced. This result is in consistent with the reported growth rate of ~2 Å or below for ALD of ZnO, since one monolayer of ZnO is around 2 Å.[8,9] The same thickness was observed by TEM which was performed on the nanotube. Since the nanotube is grown on the surface of the AAO while the film is grown on (100) Si, it appears that the grow rate is independent of the type substrate. The growth rate is attributed to the chemistry on the growing film rather than the chemistry at the interface.[9]

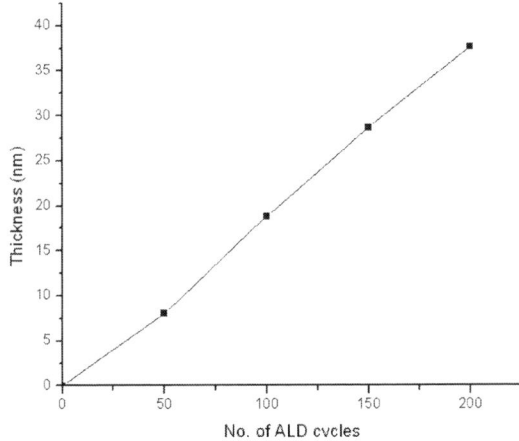

Fig 1. Linear fit of ALD growth rate

Structural properties

Fig 2a shows the AAO nanoporous structure by 2-step anodization. The average pore diameter is 70 nm and inter-pore distance is around 100 nm. The height of the template is 1.5 um. The inset of Fig 2a is the SEM image of ZnO grown onto AAO template after 100 cycles ALD with top layer ZnO removed. The AAO pore size is reduced to 50~55 nm by the embedded ZnO onto the wall of pores. Fig2b is the side view of ordered ZnO nanotube arrays after removal of AAO. The top view is shown in the inset. Conformal thickness and uniform size of ZnO nanotubes well aligned on Si substrate are obtained by ALD technique and AAO template.

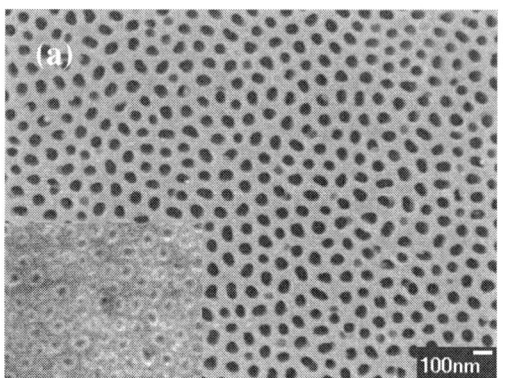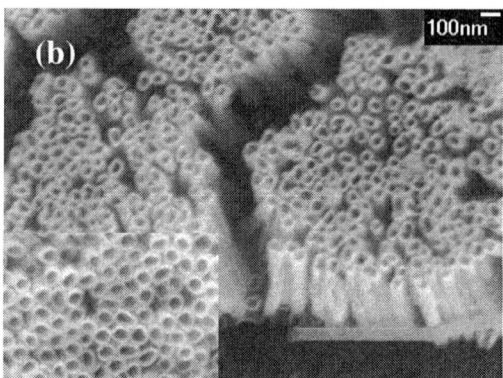

Fig 2. SEM image of (a) AAO template on Si substrate before and after (inset) 100 cycles ALD. (b) Side view and top view (inset) of ZnO nanotube arrays on Si substrate.

The morphologies of the ZnO nanotubes were further examined via TEM by scrapping the nanotubes from the Si substrate. As shown in Fig 3, the wall of nanotube is uniform in thickness and its bottom is closed due to ALD occurring on the underlying substrate also. The orientation and structure of the nanocyrstalline ZnO nanotubes are better illustrated in a HR-TEM image in the inset of Fig 3.

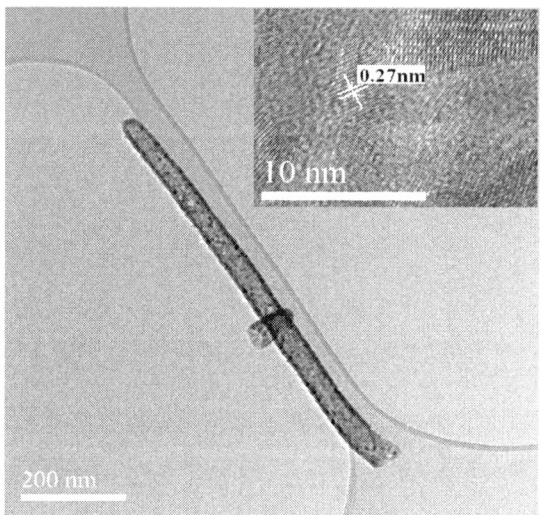

Fig 3. TEM image of a single ZnO nanotube and lattice-resolved TEM image (inset) of the wall.

Typical XRD spectrums ($2\theta = 33° - 66°$) of ZnO nanotubes and films after 100 ALD cycles are presented in Fig4. XRD peaks of wutzite structured ZnO at (100), (002), (101), (102), (110) and (103) can be seen from the spectrum, verifying that the deposited ZnO are polycrystalline structure. The relative intensities of the individual peaks depend on the deposition temperature and other growth process parameters such as pulsing and purging times. [6,10]

66

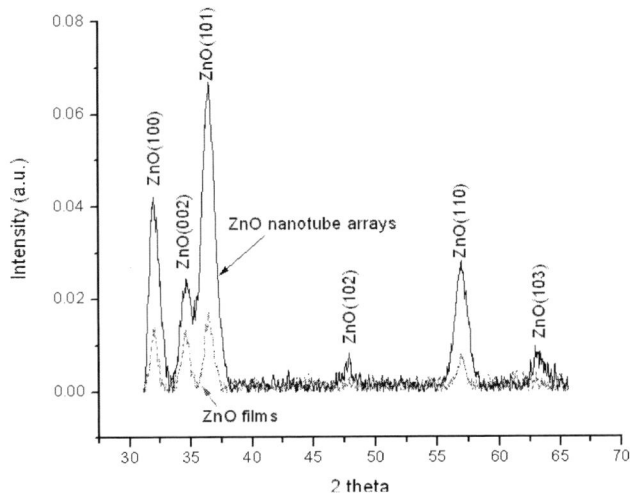

Fig 4. XRD spectrum of ZnO nanotubes and films after 100 cycles ALD

Optical properties

Fig 5 shows the PL spectrum of ZnO films deposited on Si substrate under ALD of 50-200 cycles. The ZnO film thickness increases linearly with the number of ALD cycles, as previously shown in Fig 1. The UV emission begins to be observed from ZnO film with thickness of 20nm after 100 ALD cycles and the intensity was enhanced with increased ZnO film thickness. An important observation from PL spectrum is the UV peak positions show red shift (from 377.3 nm to 381.4 nm) as the number of cycles increased from 100 to 200. The reason for this effect is still controversial.[11] One possible explanation is the changes of ZnO grain size resulting in excitonic confinement.[12] The optical properties and the possibly same size-dependent UV shift effect of ZnO nanotubes are still under investigation.

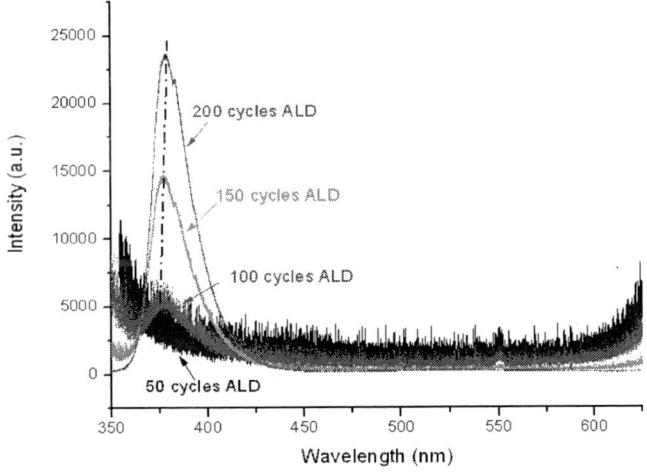

Fig 5. PL spectrum of ZnO films on Si after 50-200 cycles ALD

CONCLUSIONS

ZnO ALD was performed using alternating exposures to DEZn and H_2O. With the aid of AAO template, vertically aligned, uniform dimensions and highly ordered polycrystalline ZnO nanotube arrays can be achieved on Si substrate. The properties of ZnO nanotubes were characterized using Ellipsometry, SEM, TEM and XRD. The dimension of nanotubes can be controlled by AAO anodization conditions and the wall-thickness can be precisely controlled by number of ALD cycles. PL spectrum of the as deposited ZnO films on Si substrate was measured and size-dependent UV shift are observed. This approach provides a well-controlled method for fabricating large-scale self-aligned ZnO nanotubes for many important applications such as photovoltaics and photocatalysis.

ACKNOWLEDGMENTS

This work is supported by Singapore-MIT Alliance. The ALD process was performed at Azimuth Technologies Pte. Ltd., Singapore. The characterizations were carried out at Institute of Materials Research and Engineering, Singapore.

REFERENCES

[1] Wei An, Xiaojun Wu, and X. C. Zeng, J. Phys. Chem. C 2008, 112, 5747

[2] Lori E. Greene, Matt Law, Benjamin D. Yuhas, and Peidong Yang, J. Phys. Chem. C, Vol. 111, No. 50, 2007, 18452

[3] Jiping Cheng, Ruyan Guo and Qing-Ming Wang, Appl. Phys. Lett., Vol. 85, No. 22, (2004) 5140

[4] Sang-Hee Ko Park, Chi-Sun Hwang, Hu Young Jeong, Hye Yong Chu and Kyoung Ik Cho, Electrochemical and Solid-State Letters, 11 (1) H14 (2008) H10

[5] Masuda, Jpn, J. Apll. Phy., Vol 35, (1996)L126

[6] Jongmin Lim, Chongmu Lee, Journal of Alloys and Compounds 449 (2008) 371

[7] Jongmin Lim, Kyoungchul Shin, Hyoun Woo Kim, Chongmu Lee, Materials Science and Engineering B107 (2004) 301

[8] J. D. Ferguson, A. W. Weimer and S. M. George, J. Vac. Sci. Technol. A 23(1), Jan/Feb 2005, 118

[9] A.W. Ott, R.P.H. Chang, Materials Chemistry and Physics 58 (1999) 132

[10] A. Wojcik, M. Godlewski, E. Guziewicz, R. Minikayev and W. Paszkowicz, Journal of Crystal Growth 310 (2008) 284

[11] Lee Kheng Tan, Maria A. S. Chong, and Han Gao, J. Phys. Chem. C, 2008, 112, 69

[12] HyounWoo Kim, Seung Hyun Shim and JongWoo Lee, Nanotechnology 19 (2008) 145601

Scalable Carbon Nanotube Thin Films: Fabrication, Properties and Device Applications

Liangbing Hu,[*] Youngbae Park, David Hecht, Corinne Ladous, Mike O'Connell,
David Thomas, George Gruner, Glen Irvin, Paul Drzaic,
Unidym Inc, 1430 O'Brien Dr, Suite G, Menlo Park, California, CA 94025

ABSTRACT

Carbon nanotubes (CNT) have been under investigation for many years as a material suitable for applications in electronic devices. This paper will focus on the development and production of high quality, high performance, and scalable transparent and conductive CNT thin films using solution based roll to roll coating methods. We demonstrate both additive and subtractive methods for patterning conductive nanotube films. Various types of devices incorporating CNT thin films are demonstrated, including EPD e-paper, touch screen, OLED, flexible OPV, and TFT-LCD. Issues involving the integration of CNT electrodes into various devices are discussed, in particular conformal step coverage. Optical and mechanical properties, environmental stability and large scale uniformity together make Unidym's CNT thin films a viable alternative to transparent conductive oxides in applications requiring transparent, conductive electrodes.

INTRODUCTION

Thin films of carbon nanotubes (CNT) are a promising candidate in the development of alternatives to indium-tin-oxide thin films in applications requiring transparent, conductive films, and have been investigated by a number of groups. Films with high conductivities up to 6000 S/cm have been reported, and CNT integration into various devices such as organic solar cells, OLEDs, and LCD prototypes have been demonstrated.[1-6] However, most of these systems have suffered from one or more limitations. These limitations have included the high cost of CNT materials, film fabrication processes that are not scalable to large volumes, inferior conductivity, and poor lifetimes. Here we report the fabrication of high performance, scalable CNT films useful in applications for transparent conductors, and describe some applications of these materials demonstrating their usefulness in devices.

RESULTS

CNT material is synthesized by Unidym by thermal CVD using a proprietary catalyst and reactor system, and formulated into an aqueous ink with the proper rheology for slot-die roll-to-roll coating. The concentration of the CNT inks range from 0.2 mg/ml to 1.5 mg/ml, depending on the formulation (Figure 1(a)). CNT thin films were coated by using a conventional roll-to-roll slot-die coater with a custom die head. As show in Figure 1(b), the speed of coating can go up to 150 feet per minute. The ink formulation, ink-substrate interaction, drying and encapsulation are critical to obtain uniform and stable CNT thin films. Unidym has demonstrated 2000 ft long CNT coating which meets the specifications of most resistive and capacitive touch panel applications. Figure 1 (c) shows a 30 inch wide and 2000 ft long roll of CNT coating on PET substrate.

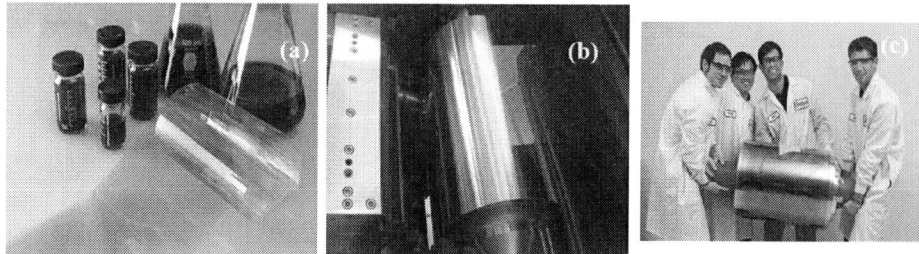

Figure 1. (a) CNT ink; (b) slot die coating set up; (c) a picture of 2000 ft long CNT film on PET substrate;

Figure 2 (a) shows the performance of CNT film coating on PET substrates as a transparent electrode. The sheet resistance and transmittance is controlled by the CNT dry thickness, which in turn is controlled by the CNT concentration and the wet laydown in the coating. For comparison, performance of ITO on PET is also shown in the Figure 2.[7] At the current price of CNT of 400-500$ per gram, the final cost of CNT/PET is close to that of ITO/PET (4-6$/square feet). All transmittance data shown in Figure 2 depict the total film transmittance, which includes both the CNT film and the PET substrate (which is ~ 91% transmissive at 550 nm). Both the production film and the lab scale film are shown in the figure. The calculation of the DC conductivity for CNT thin films is based on R_s and CNT transmittance, with the exclusion of PET substrate transmittance.[4] As shown in Figure 2 (a), the sheet resistance of CNT film on PET substrate is close to that of ITO on PET at a given sheet resistance. It is important to note that the loss of transmission in ITO is primarily due to reflection, while for Unidym's CNT films transmission loss is due primarily to absorbance. This low reflectivity is advantageous in a number of display applications where first-surface reflectivity degrades optical performance.

Figure 2(b) shows an SEM image of a CNT film surface for roll-to-roll coated and encapsulated film on PET substrate. The dense network of overlapping individual CNTs and CNT bundles insures a robust network that is highly resistant to flexure of the underlying substrate, and other mechanical damage. The roughness of CNT film is 8-10 nm and 2-4 nm before and after encapsulation, respectively.

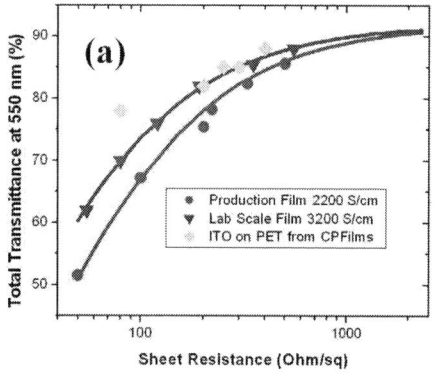
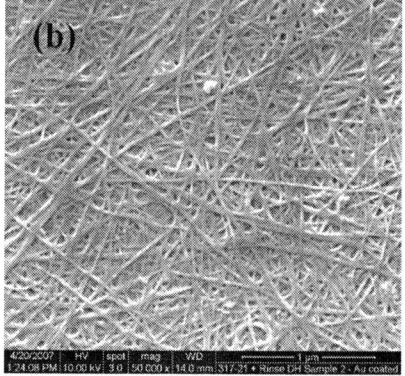

Figure 2. (a) Performance of CNT thin film as a transparent and conductive electrode; (b) SEM image of production CNT thin film coating.

For device integration of transparent CNT thin films, patterning is required. Subtractive methods for patterning CNT thin films includes lithography followed by O_2 plasma etching through a mask, and direct laser etching with or without the use of mask. Figure 3 (a) shows a patterned film in a TFT-LCD device. The detailed parameters are documented in our previous publication.[6] We also demonstrated that other types of plasmas such as Ar and CF_6 plasmas have high etching rate for CNTs. Lithography and plasma etching lead to well-defined patterned films with resolution down to 2 μm. Figure 3(b) shows a patterned film using direct laser patterning without the use of a mask. Direct laser patterning can be applied in a roll-to-roll fashion at high speeds, a method that is currently commercially used for ITO patterning. As opposed to subtractive patterning, CNT films can be patterned additively, using methods such as direct ink jet printing of CNT solution. The additive method is material saving and can be applied at high speeds. Figure 3(c) shows an ink-jet RFID circuit on photography paper printed with the use of an ink jet printer. The formulation concentration is 0.8 mg/mL and filtered through a 5 μm syringe filter. The volume of the jetted droplet is 10 picoliter, which spread on paper to a dimension of 30 μm. During the printing process, the substrate is heated to 40 C to facilitate the drying of the liquid. The patterning resolution achieved in Figure 3 (c) is 20μm. Electrical measurement of the films shows comparable electrical performance with slot die coated CNT films.

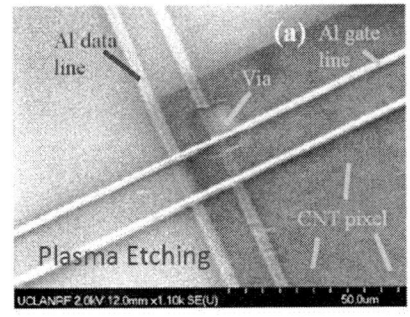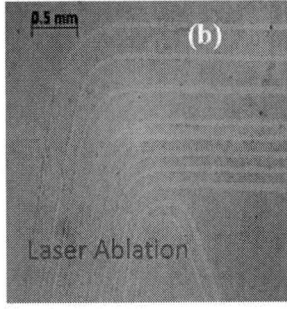

Figure 3. Patterning methods for transparent and conductive CNT thin films. Subtractive methods include (a) Lithography with a mask and plasma etching (b) Direct laser ablation without mask. Additive method includes (c) Ink jet printing.

Compared with traditional ITO electrodes, CNT thin films bring new optical properties and additional mechanical advantages which are beneficial for optoelectronics and flexible electronics. Figure 4 lists the optical properties, mechanical properties and uniformity of Unidym production CNT thin films. In Figure 4(a), the wavelegth-dependent optics of a CNT film and ITO film are compared. The spectra for CNT film are flatter than ITO in the visible range, and the color of the CNT film is neutral while ITO is yellowish. Figure 4(b) shows the reflection of PET, CNT on PET and ITO on PET in the visible and infrared range. After CNT coating on PET, the reflection has little decrease compared to that of PET substrate itself. In the visible range, ITO on PET has much higher reflection than CNT on PET. The analysis is based on the optics for multilayer structures.[8] As such, the transmittance of an ITO electrode is determined by its reflection, while the transmittance of a CNT electrode is determined by its absorption.

Figure 4 (c) shows the reflection, absorption and transmittance of CNT thin films at 550 nm on PET substrates. The thickness of CNT thin films was measured by AFM and optical profiling the edge of patterned films. The thickness was confirmed by comparing the data from the two methods. The absorption of CNT film increases linearly with film thickness while the reflection remains the same. CNT material has a high absorption coefficient of 0.24 at 550 nm.[9] The difference between transmittance of CNT itself and CNT on PET is mainly due to the absorption of the CNT thin films.

Figure 4 (d) illustrates some of the superior mechanical properties of CNT thin films, compared to ITO. CNT films can be bent down to 2 mm without any electrical failure, while ITO on PET begins to fail at a bending radius of 4 mm. The mechanical flexibility and robustness makes CNT electrodes ideal for flexible electronics applications.

Film optical and electrical uniformity is critical for commercial use of the product. Figure 4(e) shows the uniformity of the Unidym production film coated with high speeds up to 300 feet per minute. In this particular sample, the average sheet resistance is 210 Ohm/sq and standard deviation is less than 4% for a 20 inch by 20 inch film. We also found that the standard deviation for optical transmittance is less than 4% for the same CNT film.

Environmental stability is also evaluated by soaking CNT films at 85C and 85 RH for long periods of time and measuring the sheet resistance before and after the soaking. For transparent electrode applications, the requirement for environmental stability varies but generally the resistance change needs to be less than 30% after a 1000 hour soak. Figure 4 (f) shows the stability of two films after 150 hours soaking in 85C and 85 RH environment. The two films included a Unidym proprietary binder at two different thicknesses.

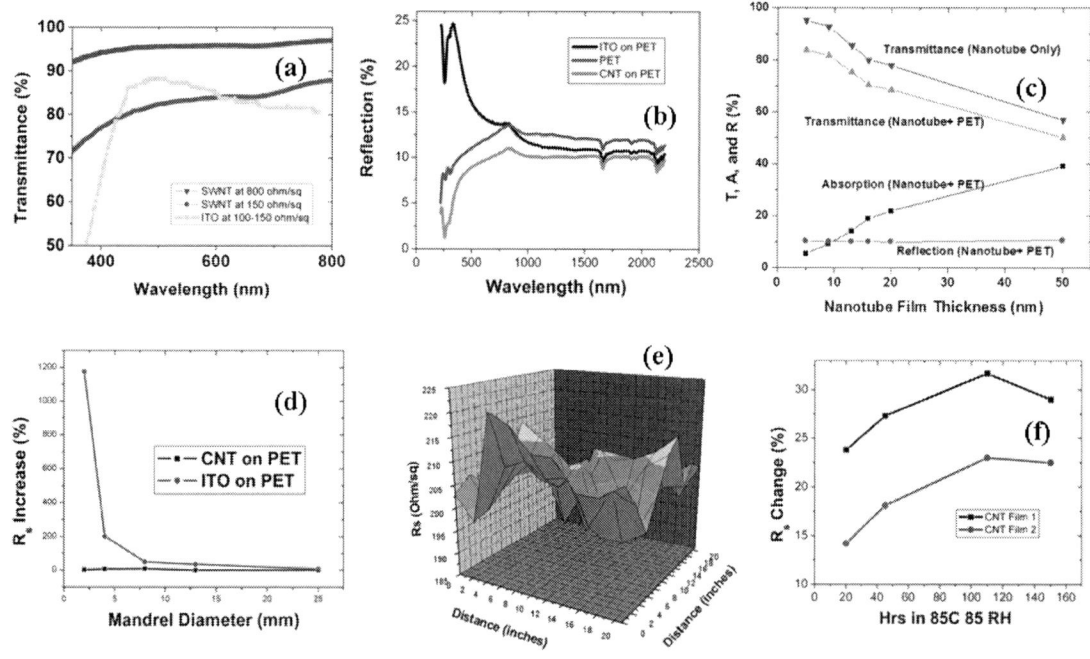

Figure 4. Performance of transparent CNT films: (a) Visible spectra of CNT and ITO films; (b) Reflection in visible and infrared range; (c) Reflection, absorption and transmittance of CNT films with increasing film thickness; (d) Bending test of CNT and ITO; (e) Uniformity of (f) 85C/85 RH stability of production CNT films on PET substrate.

Transparent and conductive CNT films have immediate applications in optoelectronics. The first wave of device application for Unidym transparent electrode is touch panels and liquid crystal displays (LCDs), followed by organic light emitting diode (OLED) displays and solar cells. Functional prototypes including resistive and capacitive touch screens, 14.4" EPD e-paper, 5.5" VGA full color TFT-LCD, OLED and organic photovoltaic devices have been demonstrated.[6] The devices have similar performance with ITO based device but have better mechanical properties. Figure 5 shows a functional resistive touch screen devices. The current performance of CNT electrodes is 600 Ohm/sq sheet resistance and 85% total transmittance. The finished panel has shown much better single point actuation durability, which is due to the CNT mechanical flexibility and robustness.

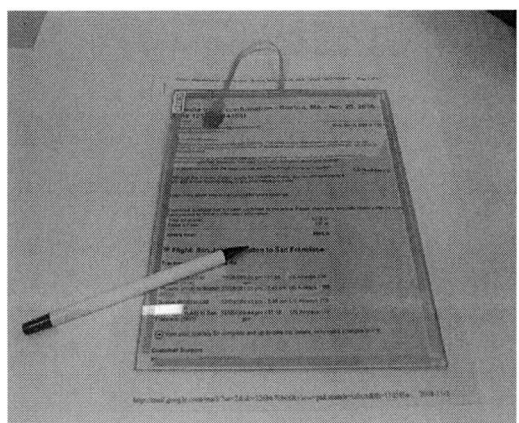

Figure 5. Demonstrated 4 wire resistive touch panel.

For device integration of CNT films in electronic or optoelectronic devices, various challenges exist. Step coverage across different layer heights in a device is a particularly important requirement, one that is not well met in several competitive nanoscale conductive materials. Figure 6 shows an example of CNT coating on a color filter plate of an LCD, with discontinuous edge structure due to different subpixel filter heights. Due to their large aspect ratio, strong nanotube-substrate interaction and their mechanical flexibility, CNTs shows outstanding conformal coating on step structure. Such types of coating ensure good electrical connection for different regions of devices.

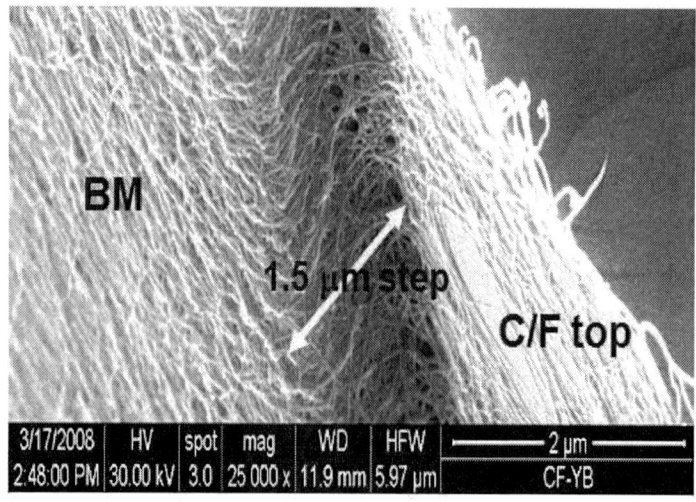

Figure 6 Conformable coating of CNT on color filter (DF) edge to black matrix (BM).

CONCLUSIONS

In conclusion, we demonstrate the achievement of scalable conductive and transparent CNT thin films. These films have shown equivalent or better optical and mechanical properties than ITO-based electrodes, and show dramatic superiority in flexibility. Integration into a number of different device structures has been achieved, demonstrating the potential for practical application of these materials and processes.

* Email: lhu@unidym.com

REFERENCES

1. Z. Wu, Z. Chen, X. Du, J.M. Logan, J. Sippel, M. Nikolou, K. Kamaras, J.R. Reynolds, D. Tanner, A.F.Hebard, A.G.Rinzler, Science, 305, 1273 (2004).
2. L. Hu, D.S. Hecht, G. Grüner, Nano Lett. 4, 2513 (2004).
3. J.Li, L.Hu, L.Wang, Y.Zhou, G. Grüner, T.J. Marks, Nano Lett. 6, 2472 (2006).
4. C. M. Aguirre, S. Auvray, S. Pigeon, R. Izquierdo, P. Desjardins, R. Martel, Appl. Phys. Lett. 88, 183104 (2006).
5. M. W. Rowell, M. A. Topinka, M. D. McGehee, H. Prall, G. Dennler, N. S. Sariciftci, L. Hu, G. Gruner, Appl. Phys. Lett. 88, 233506 (2006).
6. Y-B Park, L Hu, G Gruner, G Irvin and P Drzaic, SID Tech. Dig., 537 (2008).
7. www.CPFilms.com
8. M. Dressel, M, G. Gruner, Electrodynamics of Solids: Optical Properties of Electrons in Matter (Cambridge University Press, 2002)
9. B. Ruzicka, L. Degiorgi, R. Gaal, L. Thien-Nga, R. Bacsa, J. P. Salvetat and L. Forro, Phys. Rev. B, 61, 2468 (2000).

Surface Properties of Polycrystalline Transparent Conducting Oxides

A. Klein, C. Körber, A. Wachau, R. Schafranek, Y. Gassenbauer, F. Säuberlich, G. Venkata Rao
Darmstadt University of Technology, Institute of Materials Science, Surface Science Division,
Petersenstrasse 23, D-64287 Darmstadt, Germany

ABSTRACT

Properties of transparent conducting oxide surfaces have been investigated using photoelectron spectroscopy (XPS, UPS). The TCO films are prepared by magnetron sputtering. Surface analysis is performed in integrated systems, allowing for vacuum transfer between different preparation and analysis chambers. With XPS and UPS, it is possible to assess chemical as well as electronic properties of surfaces and interfaces. The electronic properties include the work function, the Fermi level position with respect to the band edges and barrier heights at interfaces between TCOs and other materials. Results will be presented for magnetron sputtered films of doped and undoped ZnO, In_2O_3, and SnO_2. The variation of surface properties with deposition parameters is described, highlighting the influences affecting the work function of magnetron sputtered films.

INTRODUCTION

Transparent Conducting Oxides are used, e.g., as passive elements such as transparent electrodes in flat-panel displays, electrochromic windows, and solar cells [1-5]. Organic-based optoelectronic devices, which require transparent electrodes include light-emitting diodes (OLEDs) [6] and solar cells (OPVs) [7,8]. For such applications the surface properties including chemical composition, Fermi level and work function, are of great importance [9-11]. Electronic surface properties are also important for flexible/transparent thin film transistors [12,13] or chemical gas sensor applications [14,15]. Chemical and electronic properties of surfaces can be assessed by photoelectron spectroscopy [16]. In-situ sample preparation, i.e. the investigation of clean surfaces, is particularly useful for such experiments. This contribution will give an overview on the various possibilities of this approach applied to transparent conducting oxides.

EXPERIMENTAL

Thin films of ZnO, ZnO:Al, SnO_2 SnO_2:Sb, In_2O_3, In_2O_3:Sn (ITO), and (Zn,Sn) co-doped In_2O_3 (ZITO) were deposited by DC (ZnO) or RF magnetron sputtering (SnO_2 and In_2O_3) using ceramic targets of 2 inch diameter. Sputtering was performed using a power of 5-25 W, a substrate-to-target distances of 5-10 cm, a substrate temperature of 25-500°C, a pressure of 0.5-5 Pa, and an Ar/O_2 gas mixture with oxygen content of 0-50%. Glass substrates, partially coated with conducting F-doped SnO_2 or ITO to avoid binding energy shifts due to charging effects in less conducting films, were used for UPS measurements. Doped films were obtained from 2 wt% Al-doped ZnO, 3 wt% Sb_2O_5-doped SnO_2, 2 wt% and 10 wt% SnO_2-doped In_2O_3, respectively. Targets were purchased from different sources (Lesker, Mateck) with a purity of 99.9% or better. Partial information on the prepared materials has been reported previously [11,17-22].
X-ray and ultraviolet photoelectron spectroscopy (XPS and UPS) were carried out using a Physical Electronics PHI 5700 spectrometer as part of the DArmstadt Integrated System for MATerials research (DAISY-MAT) (Fig. 1). The system combines the multi-technique surface

analytical tool with several thin film preparation chambers with an ultrahigh vacuum sample transfer. All thin film samples were deposited, transferred, and measured without breaking vacuum. Monochromatic Al Kα radiation (hν = 1486.6 eV) was employed as excitation for XPS. At 5.85 eV pass energy, the PHI system provides an overall experimental resolution better than 400 meV for XPS measurements, as determined from the Gaussian broadening of the Fermi edge of a clean Ag sample. The spectrometer was calibrated at regular intervals with sputter cleaned metallic reference samples.

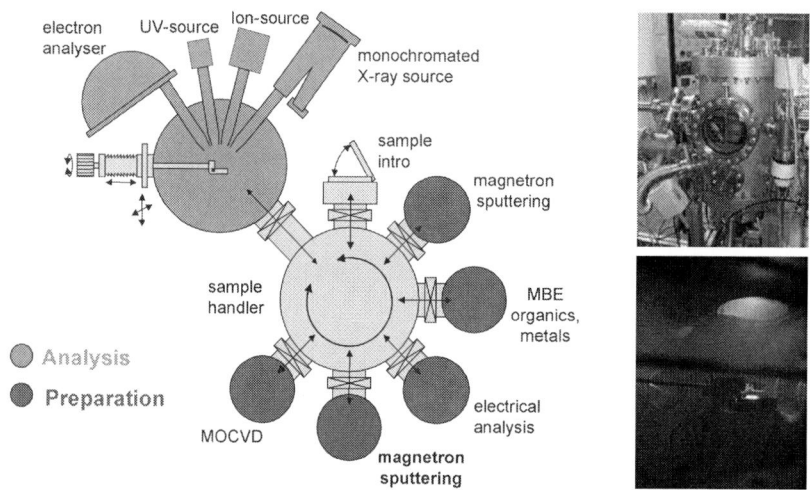

Fig. 1: Layout of the Integrated System (DAISY-MAT) used for the experiments.

RESULTS

The band gap of In$_2$O$_3$

Photoelectron spectroscopy can detect only occupied electronic states. However, for most TCO applications, n-type films are applied, which have a Fermi level close to or above the conduction band minimum of the materials. There are several reports of the observation of conduction band states in highly doped TCO materials by photoelectron spectroscopy [23,24]. However, for lower doped materials the intensity is negligible or absent and only occupied valence band states are detected. The valence band maximum (VBM) can be determined by the leading edge of these emissions (see Fig. 5 below). In order to correlate the VBM data with electrical conductivity, which is determined by the filling of the conduction band states, the difference between the valence band maximum and the conduction band minimum, i.e. the band gap of the materials, needs to be known. The band gaps of ZnO and SnO$_2$ have been well established for a long time [25]. However, the fundamental gap of In$_2$O$_3$, which is important for electrical data and interpretation of photoemission data, has only very recently been established [26].
Early optical measurements of In$_2$O$_3$ single crystals by Weiher and Ley have identified a direct gap of 3.6-3.7 eV and a forbidden indirect gap of 2.6 eV [27]. While there is general agreement about the value for the direct (optically measured) gap, there has been no agreement about the magnitude and even the existence of an indirect gap in In$_2$O$_3$ (see discussion in [21]). Recent band structure calculations gave no evidence for an indirect gap almost ruling out this possibility [26,28,29]. Photoemission data of In$_2$O$_3$ and ITO reproducibly showed valence band maximum

energies with respect to the Fermi energy of <3.5 eV, even for highest doped materials [18,19,21,30]. This is not consistent with a fundamental gap of 3.6-3.7 eV and a Fermi level above the conduction band minimum as expected for degenerately doped material. In our previous work, this discrepancy has been explained by assuming a high density of surface states at the In_2O_3 and ITO surfaces, leading to carrier depletion and band bending at the surface [19,30]. It has further been proposed that oxygen uptake during the cooling of the materials after deposition results in a reduced doping, thus enabling the depletion layer, which are not expected for degenerately doped semiconductors [21,31].

The surface Fermi level position of different conducting In_2O_3 and ITO samples varies by almost 1 eV (even larger variations can be observed for deposition under highly oxidizing conditions). The optical gap of the same films measured by optical transmission varies by the same magnitude (see left graph in Fig. 2 and Ref. [18]). The data suggest that the energy difference between the Fermi level and the valence band maximum measured by XPS and UPS is consistently ~1 eV smaller than the same value measured by optical transmission of thin films. Although this observation is consistent with a surface depletion layer with a constant band bending, a more recently proposed interpretation is more likely. It has been shown by density functional theory electronic structure calculations that optical transitions from the topmost valence bands are not allowed [26]. A similar observation has been reported for $CuInO_2$, which explains that it is easy to achieve both n- and p-type doping with this optically transparent material [32]. The situation for In_2O_3 is sketched in the right graph in Fig. 2. It is, to some extent, surprising that the optical transitions are forbidden for such a wide range of energies, as the energy region between the topmost valence band and those bands giving rise to strong optical transitions contains a large number of bands [26,28,29].

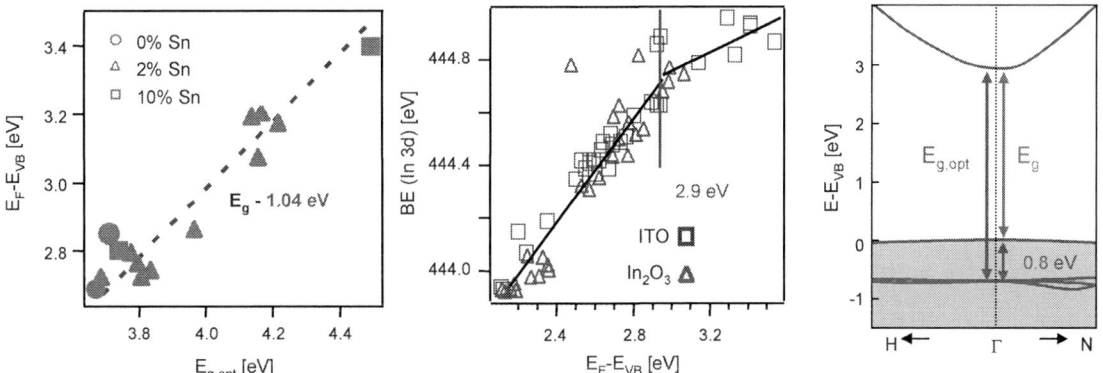

Fig. 2: (left) Surface Fermi level position measured by XPS vs. optical gap measured by transmission at the same films; (middle) Binding energy of the In 3d core level vs. surface Fermi level position [19,21]; (right) Sketch of electronic structure of In_2O_3 with fundamental and optical gap indicated [26]. The conduction bands are rigidly shifted in order to match an optical gap of 3.7 eV

The energy range of the forbidden transitions is given by Walsh et al. as 0.8 eV, i.e. the fundamental gap is 0.8 eV smaller than the optical gap. Taking the optical gap to be 3.6-3.7 eV, this corresponds to a fundamental gap of 2.8-2.9 eV. Such a value is consistent with the observation of the onset of free carrier screening of the photoelectron core hole at E_F-E_{VB} ~2.9 eV (see change of slope in the middle graph in Fig. 2) [19,20]. However, the data in the left graph of

Fig. 2 suggest that the fundamental gap is ~1 eV smaller than the optical gap, resulting in a fundamental gap of 2.6-2.7 eV, as also inferred in recently by Bourlange et al. [33]. After all, some uncertainty remains about the accurate value of the fundamental gap of In_2O_3.

Chemical surface properties

The determination of chemical bonding and composition is the most widely used feature of XPS. Also here, in-situ sample preparation provides much better accuracy for the evaluation of chemical composition. Adsorbate layers, which are typically present at surfaces of ex-situ prepared films, will lead to stronger attenuation of the more surface sensitive species. The surface sensitivity depends on the electron energy and has a minimum at ~50 eV kinetic energy [16]. Photoelectron lines with a high binding energy, as e.g. Zn 2p, are therefore more surface sensitive than lines with lower binding energies. In the presence of adsorbates, the content of an element with high binding energy can therefore be considerably underestimated.

Figure 3 shows the composition of (Zn,Mg)O and Al-doped ZnO films as derived by XPS. In both cases, the Zn content decreases with increasing substrate temperature during deposition. This can be explained by a loss of Zn due to re-evaporation from the substrate. Zinc has a high vapour pressure. Zn atoms adsorbed during the deposition process can therefore re-evaporate if no oxygen is available during the residence of Zn atoms on the surface to form a ZnO compound. Mg and Al are both low vapour pressure metals and have a high sticking probability even at elevated substrate temperature. This means that the higher the temperature during deposition, the higher the enrichment of the less volatile species. In contrast, elevated temperatures during annealing steps after deposition do not have the same effect and hardly change the sample composition as indicated by the two data points at 450°C in Fig. 3 (left). Furthermore, the loss of Zn can be partially suppressed by addition of oxygen to the sputter gas. In the presence of oxygen the Mg content in the film does not change with substrate temperature.

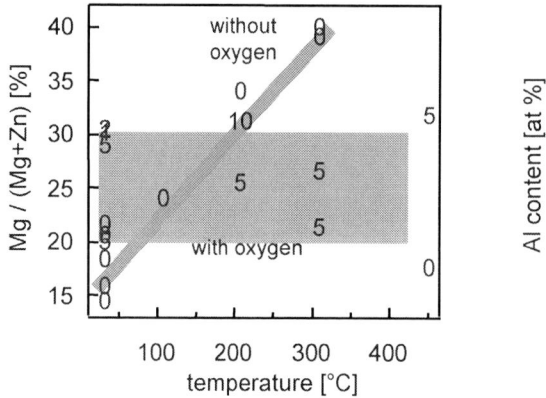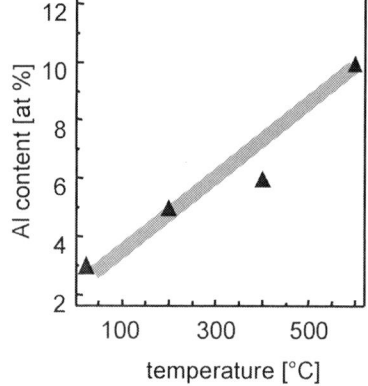

Fig. 3: (left) Composition of (Zn,Mg)O thin films as determined by XPS. The films are prepared by RF magnetron sputtering from a $Zn_{0.85}Mg_{0.15}O$ target. The numbers correspond to the fraction of oxygen in the sputter gas (in %). The data points at 450°C are obtained after deposition at room temperature and subsequent annealing at 450°C. (right) Al content of Al-doped ZnO films determined by XPS. The films are deposited by DC magnetron sputtering from a 2% Al-doped ZnO target. In both cases, higher substrate temperatures lead to a loss of Zn and enrichment of the species with the lower partial pressure. The enrichment can be partially suppressed by addition of oxygen to the sputter gas.

In the case of ITO, segregation of Sn to the surface has been reported for highly conducting films [19]. Such films are prepared under reducing conditions, i.e. with pure Ar as sputter gas. Angle dependent XPS measurements of such a film are shown in left graph of Figure 4. The Sn signal increases with reducing takeoff angle, clearly demonstrating the segregation to the surface. No such segregation is observed for films prepared under more oxidizing conditions, i.e. with addition of oxygen to the sputter gas. Such films show a lower conductivity and correspondingly a lower Fermi level position. As suggested by the right graph of Figure 4, there is a correlation between the Fermi level position and the Sn concentration. The observation has originally been assigned to the presence of the surface depletion layer [19]. However, in the light of the recently established lower fundamental gap of In_2O_3, which rules out the presence of a depletion layer, another interpretation is required.

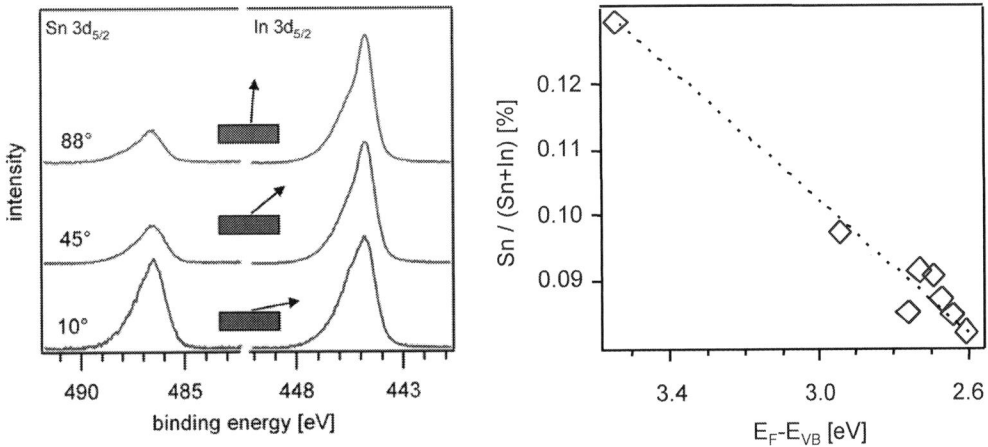

Fig. 4: (left) Sn 3d and In 3d spectra of a highly doped ITO film prepared with pure Ar as sputter gas in dependence on emission angle. The intensities are normalized to the In 3d intensity in order to emphasize the increase of the Sn signal with increasing surface sensitivity. (right) Sn content of ITO samples plotted vs. their Fermi level position. The samples are prepared with different oxygen concentrations in the sputter gas.

The different Fermi level positions of the films prepared with different oxygen contents in the sputter gas are induced by different oxygen interstitial concentrations in the films. A lower oxygen interstitial concentration as a consequence of a lower oxygen fraction in the sputter gas leads to a higher concentration of active Sn_{In} donors (the dominant defect species in ITO are neutral $(2Sn_{In}O_i)^x$ defect clusters [34,35]) and a rise of the Fermi energy. A rise of the Fermi energy generally leads to a decrease of the formation energy of acceptor states and to pinning levels, which correspond to a maximum possible doping level [36]. Basically, the rise of the Fermi level is limited by formation of intrinsic compensating defects. Possible mechanisms, which compensate for donor doping in ITO, are either the formation of acceptors like oxygen interstitials or In vacancies, or the removal of donors from the lattice by segregation of Sn to the surface. It has been shown by high-pressure XPS that the Sn concentration at the surface changes reversibly on a time scale of 1 hour when the oxygen pressure is varied as long the substrate temperature is above ~300°C [19]. Hence, it is proposed that Sn donors are destabilized and segregate to the surface when oxygen is removed from the lattice, and vice versa.

Surface potentials

The determination of surface potentials from photoelectron spectroscopy is illustrated in Figure 5. For the determination of electronic properties, it is important that binding energies are measured with respect to the Fermi energy in XPS. This is achieved by electrical contact of the samples with the spectrometer system [16]. The position of the Fermi energy can be calibrated by metallic reference samples. Then, the onset of the valence band emissions gives a direct measure of the Fermi level position in the gap (E_F-E_{VB}). The low kinetic energy or high binding energy cutoff corresponds to secondary electrons with energy just above the vacuum energy. The work function of the sample (ϕ=E_{vac}-E_F) can thus be directly derived from the energetic position of the secondary electron cutoff. Together with the Fermi level position, the ionization potential (I_P=E_{vac}-E_{VB}) can be calculated. As the work function depends on the Fermi level position, which can vary significantly with the doping of the material, the work function does not provide direct information about the surface termination. The surface termination can change by adsorbates or by the oxygen partial pressure. The latter determines whether a surface is more cation- or anion-terminated [37]. Therefore only the ionization potential, which is independent on the Fermi level position, provides direct access to different surface conditions.

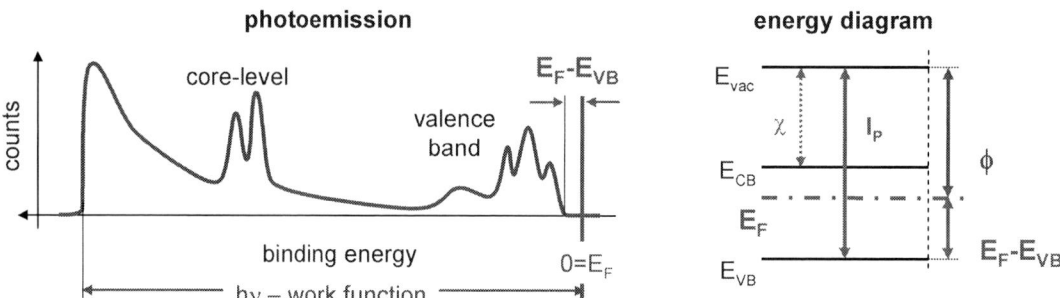

Fig. 5: Surface potentials of the basic TCO materials and their determination by photoelectron spectroscopy. The zero of binding energy corresponds to the Fermi level position, which is calibrated by measurement of a metallic reference sample.

Surface potentials of undoped and doped ZnO and SnO_2, and of ITO films are presented in Fig. 6. Data for undoped In_2O_3 films are not shown for clarity but can be found elsewhere [38]. All films are prepared by magnetron sputtering using the setup presented in Fig. 1. The data points are obtained by variation of deposition conditions, mainly the oxygen content in the sputter gas during deposition. In order to account for the variation of the parameters, the work functions are plotted vs. the Fermi level positions. This has the advantage of providing a general picture of the dependencies of the surface potentials. As long as the surface termination does not change, i.e. the ionization potential remains constant, the data points fall on a line with a slope of -1. Corresponding lines are indicated in Fig. 6.

For all three materials, a large variation of the Fermi level position is observed with higher Fermi level positions found in the doped materials. Except for Sb-doped SnO_2 [39], where the variation is only ~0.4 eV, variations in the Fermi level also correspond to variations in the conductivity of the samples. As indicated by the left graph of Fig. 2, the variation of the Fermi level corresponds well with variation of the optical gap, at least as long as the Fermi level is at or above the conduction band minimum. For lower Fermi level positions, no further decrease of the optical gap is possible. The optical gaps measured for doped SnO_2 and ZnO films correspond well with the measured Fermi level positions. The difference between Fermi level position and optical gap for ITO has been explained above. This indicates that the surface Fermi level positions measured by XPS are close or identical to the bulk Fermi level positions measured by optical spectroscopy. In other words, there is no evidence for strong band bending at the surface for either material.

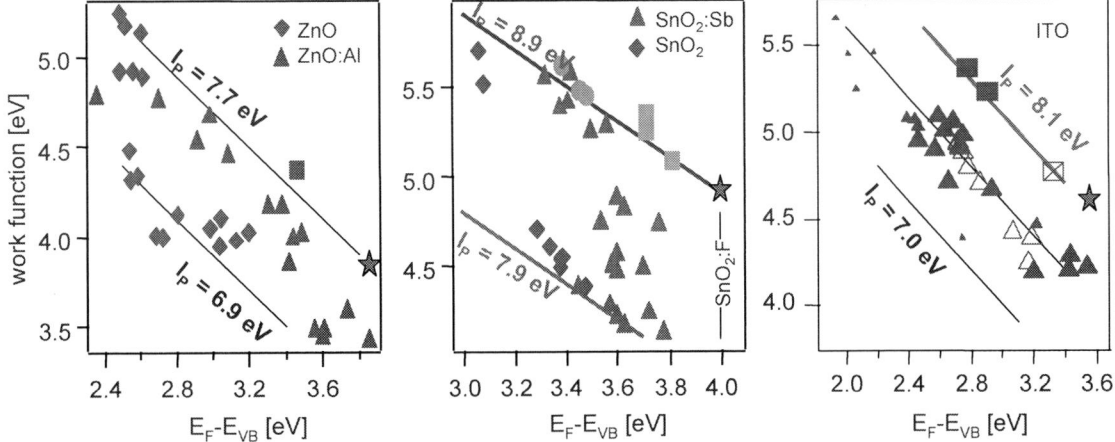

Fig. 6: Surface potentials of the basic TCO materials. Ceramic ZnO:Al, SnO_2:Sb and ITO samples are indicated by filled squares. The crossed square corresponds to an air annealed ITO thin film. For ITO open and filled triangles represent values obtained for 2 and 10wt% SnO_2 doped In_2O_3, respectively. The lines with negative slope represent values of constant ionization potential. Smaller symbols correspond to lower substrate temperatures during deposition The vertical line in the middle graph corresponds to the measured Fermi level position for fluorine doped SnO_2, for which reliable work function data are not yet available. The stars indicate the largest work function to be expected for highest doping level. Additional data are provided and discussed elsewhere [38].

In general, a lower Fermi level position is obtained by deposition with a higher oxygen content in the sputter gas. For all three materials, the addition of oxygen to the sputter gas leads to a lowering of the Fermi level and an associated increase of the work function. In addition, there is also a tendency for an increase of ionization potential for ZnO and SnO_2 with lower Fermi level position. No such trend is observed for ITO. The ionization potential varies between ~6.9 and ~7.7 eV for ZnO, between ~7.9 and ~8.9 eV for SnO_2 and between ~7.7 and ~8.1 eV for ITO. The variation of ionization potential for ZnO and SnO_2 is attributed to a different origin. The limiting values observed for ZnO agree with those reported for different surface orientations of single crystalline ZnO [40,41], where the lower value of 6.9 eV is found for the Zn-terminated (0001) surface and the larger for oxygen terminated $(000\bar{1})$ and non-polar $(10\bar{1}0)$ surfaces. A comparable variation of ionization potential with surface orientation has also been observed for GaAs [42] and for CdS and CdTe [43]. According to these results, a low ionization potential is only observed for the cation-terminated hexagonal (0001) or (111) surfaces, while all other surface orientations show a larger value. The dependency of ionization potential for ZnO suggests, that the lower value and hence (0001) surface orientation, is obtained for higher Fermi level position. This is reasonable, as the higher Fermi level positions are obtained under more reducing deposition conditions, which will also stabilize cationic surface termination. The variation of ionization potential is also consistent with preferred film orientation as determined by X-ray diffraction [44].

A variation of ionization potential has also been reported for SnO_2 (110) single crystal surfaces [45]. No change in surface orientation can be expected in this case. In this case, the changes in ionization potential observed for polycrystalline magnetron sputtered films are assigned to changes in surface termination, in particular to the oxygen content at the surface [45,46]. The reduced surface contains considerably less oxygen but remains electrically neutral due the change of Sn oxidation state from Sn^{4+} to Sn^{2+}. This is consistent with the experimentally observed variation of ionization potential for magnetron sputtered SnO_2. Films prepared under reducing conditions, which are again identified in the middle graph in Fig. 6 by the higher Fermi level positions, exhibit lower ionization potentials as expected for reduced surfaces. Reduced surfaces are also evident in photoemission from the occurrence of emissions above the valence band maximum [22,45,46]. Depositing under more oxidizing conditions by adding oxygen to the sputter gas leads to a lowering of the Fermi energy caused by reduced doping of the film and an increase of the ionization potential due to an oxidation of the surface. The dependence of ionization potential observed for polycrystalline surfaces of magnetron sputtered SnO_2 films is hence comparable to the behaviour observed for SnO_2 (110). Ionization potential or work function data for other SnO_2 surface orientations are not yet available.

No variation of ionization potential with variation of the oxygen content during deposition is observed for ITO surfaces (see right part of Fig. 6 and [11,19,22]). However, while ITO exhibits an ionization potential of ~7.6 eV, smaller values of ~7.0 eV are observed for In_2O_3 and (Zn,Sn) co-doped In_2O_3 films [22,30,38]. The origin of the larger ionization potential of ITO is not clear yet. It is possibly related to the segregation of Sn discussed above. However, so far there is no evidence for segregation of Sn at surfaces of films prepared under oxidizing conditions, which exhibit the same ionization potential. Also, films prepared with only 2 wt% SnO_2 doping exhibit the same ionization potentials as films prepared with standard 10 wt% SnO_2 doping (see Fig. 6). Recently it has been shown that post-deposition treatments performed under strongly oxidizing conditions, e.g. the widely applied UV ozone treatments or a 24 hour annealing in air at 400°C leads to a noticeable increase of the ionization potential of ITO films to ~8.1 eV (see squares in the right graph of Fig. 6). This suggests that the ionization potential measured at as-deposited ITO films corresponds to a "reduced" surface, while the one obtained after the oxidative treatment corresponds to an oxidized surface. Unlike for SnO_2, no microscopic models for the considerably more complex surface structures of In_2O_3 are yet available. The data indicate, however, a pronounced difference between ITO and SnO_2: While the surface termination of SnO_2 can be changed from reduced to oxidized by addition of oxygen to the sputter gas, this is not the case for In_2O_3 and ITO. On the other hand, much larger variations of the Fermi energy in dependence on oxygen content in the sputter gas are observed for ITO. This difference is most likely related to the different defect structures of the materials. The bixbyite lattice of ITO allows for an easy incorporation of oxygen [35], while oxygen interstitials have rather high formation energies in the rutile structure [47].

The plots in Fig. 6 allow for a direct comparison of surface potentials of the three different TCO materials. For most applications, highly doped films are required. Due to the direct correspondence of doping and Fermi level position, highest doped films should have the highest possible Fermi level position. Taking the maximum possible ionization potentials of 7.7 eV for ZnO, 8.9 eV for SnO_2, and 8.1 eV for ITO, the highest possible work functions for the highest doping levels are ~3.9 eV for ZnO, ~5.0 eV for SnO_2, and 4.7 eV for ITO, respectively. The values are indicated in Fig. 6 by stars. Larger work functions are possible for any of the materials if the Fermi level is lowered.

SUMMARY AND CONCLUSIONS

We have presented the application of photoelectron spectroscopy for the study of chemical and electronic properties of magnetron sputtered polycrystalline transparent conducting oxides. Due to the use of in-situ sample preparation as implemented in integrated UHV surface analysis and preparation systems well-defined surface properties are obtained allowing for systematic studies of the dependencies of TCO surface properties on preparation conditions. Important advantages of photoelectron spectroscopy are (i) the simultaneous measurement of chemical composition and electronic surface properties and their interdependence and (ii) the simultaneous measurement of Fermi level position and work function, which provide detailed insights into the different processes affecting the surface potentials.

The electronic surface potentials vary considerably for all studied TCO materials. The variation corresponds well with changes in optical gaps which is caused by filling of conduction band states (Burstein-Moss shift), thus suggesting that the surface Fermi level position measured by photoelectron spectroscopy represents also the bulk Fermi level position. Photoelectron spectroscopy therefore provides a unique and direct way to determine the Fermi level positions of TCO materials. The striking difference between Fermi level positions measured by photoelectron spectroscopy and the optical gaps derived from thin film transmission experiments for In_2O_3 and ITO has contributed to unravel the band gap of this important TCO material.

The variation of surface potential is mainly caused by changing the oxygen content in the sputter gas. The variations of Fermi level position are caused by changes in defect concentration, e.g. the variation of oxygen interstitial concentration in ITO. The change of oxygen concentration during sputter deposition also changes the ionization potential of ZnO and SnO_2, which correspond to a variation of work function independent on Fermi level changes. For ZnO, the variation is attributed to changes in surface orientation, while for SnO_2 different surface terminations (oxidized and reduced) are most likely responsible for the variation. For ITO, oxidation of the surface does not occur during deposition under oxidizing conditions, but can be induced by post-deposition oxidation treatments.

ACKNOWLEDGEMENT

The authors acknowledge close collaborations with T.O. Mason and S.P. Harvey from Northwestern University on defect chemistry and complex composition TCOs and with K. Albe, P. Erhart, and P. Agoston from TU Darmstadt on electronic structure and defect calculations. The contributions of A. Knop-Gericke, M. Hävecker, and R. Schlögl from Fritz-Haber-Institute on high-pressure photoemission, D. Payne and R.G. Egdell on high-energy photoemission, A. Walsh and S.-H. Wei from NREL on calculations of optical transitions in In_2O_3 are highly appreciated. Our work has been supported by the German Science Foundation in the framework of the collaborative research center SFB 595 and in the Materials World Network program (project number KL1225/4) as well as by the German Bundesministerium für Bildung und Forschung, project No. 01SF0034.

REFERENCES

1. H.L. Hartnagel, A.L. Dawar, A.K. Jain, and C. Jagadish: *Semiconducting Transparent Thin Films*, (Institute of Physics Publishing, Bristol, 1995).

2. D.S. Ginley, and C. Bright, MRS Bulletin **25 [8]** (2000), 15.

3. C. Jagadish, and S.J. Pearton (ed): *Zinc Oxide: Bulk, Thin Films and Nanostructures*, (Elsevier, Oxford, 2006).

4. C.G. Granqvist, Sol. Energy Mat. Sol. Cells **91** (2007), 1529.

5. K. Ellmer, A. Klein, and B. Rech (ed): *Transparent Conductive Zinc Oxide: Basics and Applications in Thin Film Solar Cells*, (Springer-Verlag, 2008).

6. L.S. Hung, and C.H. Chen, Mater. Sci. Eng. R **39** (2002), 143.

7. C.J. Brabec, N.S. Sariciftci, and J.C. Hummelen, Adv. Funct. Mat. **11** (2001), 15.

8. P. Peumans, A. Yakimov, and S.R. Forrest, J. Appl. Phys. **93** (2003), 3693.

9. H. Ishii, K. Sugiyama, E. Ito, and K. Seki, Adv. Mater. **11** (1999), 605.

10. T. Kugler, W.R. Salaneck, H. Rost, and A.B. Holmes, Chem. Phys. Lett. **310** (1999), 391.

11. Y. Gassenbauer, and A. Klein, J. Phys. Chem. B **110** (2006), 4793.

12. K. Nomura, H. Ohta, A. Takagi, T. Kamiya, M. Hirano, and H. Hosono, Nature **432** (2004), 488.

13. T. Kamiya, and H. Hosono, International Journal of Applied Ceramic Technology **2** (2005), 285.

14. N. Yamazoe, G. Sakai, and K. Shimanoe, Catalysis Surveys from Asia **7** (2003), 63.

15. G. Korotcenkov, Sensors and Actuators B **107** (2005), 209.

16. A. Klein, T. Mayer, A. Thissen, and W. Jaegermann, Bunsenmagazin **10** (2008), 124.

17. A. Klein, and F. Säuberlich, in: *Transparent Conductive Zinc Oxide: Basics and Applications in Thin Film Solar Cells*, Eds. K. Ellmer, A. Klein and B. Rech, (Springer-Verlag, Berlin, 2008).

18. Y. Gassenbauer, and A. Klein, Solid State Ionics **173** (2004), 141.

19. Y. Gassenbauer, R. Schafranek, A. Klein, S. Zafeiratos, M. Hävecker, A. Knop-Gericke, and R. Schlögl, Phys. Rev. B **73** (2006), 245312.

20. Y. Gassenbauer, R. Schafranek, A. Klein, S. Zafeiratos, M. Hävecker, A. Knop-Gericke, and R. Schlögl, Solid State Ionics **177** (2006), 3123.

21. S.P. Harvey, T.O. Mason, Y. Gassenbauer, R. Schafranek, and A. Klein, Journal of Physics D: Applied Physics **39** (2006), 3959.

22. S.P. Harvey, T.O. Mason, C. Körber, Y. Gassenbauer, and A. Klein, Appl. Phys. Lett. **92** (2008), 252106.

23. Y. Dou, T. Fishlock, R.G. Egdell, D.S.L. Law, and G. Beamson, Phys. Rev. B **55** (1997), R13381.

24. Y. Dou, and R.G. Egdell, Surf. Sci. **372** (1997), 289.

25. O. Madelung (ed): *Semiconductors Basic Data (2nd ed.)*, (Springer Verlag, Berlin, 1996).

26. A. Walsh, J.L.F. Da Silva, S.-H. Wei, C. Körber, A. Klein, L.F.J. Piper, A. DeMasi, K.E. Smith, G. Panaccione, P. Torelli, D.J. Payne, A. Bourlange, and R.G. Egdell, Phys. Rev. Lett. **100** (2008), 167402.

27. R.L. Weiher, and R.P. Ley, J. Appl. Phys. **37** (1966), 299.

28. P. Erhart, A. Klein, R.G. Egdell, and K. Albe, Phys. Rev. B **75** (2007), 153205.

29. F. Fuchs, and F. Bechsted, Phys. Rev. B **77** (2008), 155107.

30. A. Klein, Appl. Phys. Lett. **77** (2000), 2009.

31. A. Klein, Mater. Res. Soc. Symp. Proc. **666** (2001), F1.10.

32. X. Nie, S.-H. Wei, and S.B. Zhang, Phys. Rev. Lett. **88** (2002), 066405.

33. A. Bourlange, D.J. Payne, R.G. Egdell, J.S. Foord, P.P. Edwards, M.O. Jones, A. Schertel, P.J. Dobson, and J.L. Hutchison, Appl. Phys. Lett. **92** (2008), 092117.

34. G. Frank, and H. Köstlin, Appl. Phys. A **27** (1982), 197.

35. G.B. González, T.O. Mason, J.P. Quintana, O. Warschkow, D.E. Ellis, J.-H. Hwang, and J.P. Hodges, J. Appl. Phys. **96** (2004), 3912.

36. A. Zunger, Appl. Phys. Lett. **83** (2003), 57.

37. M. Batzill, and U. Diebold, Progr. Surf. Sci. **79** (2005), 47.

38. A. Klein, C. Körber, A. Wachau, F. Säuberlich, Y. Gassenbauer, R. Schafranek, S.P. Harvey, and T.O. Mason, Thin Solid Films (submitted),

39. C. Körber, and A. Klein, (unpublished results),

40. R.K. Swank, Phys. Rev. **153** (1967), 844.

41. K. Jacobi, G. Zwicker, and A. Gutmann, Surf. Sci. **141** (1984), 109.

42. W. Ranke, Phys. Rev. B **27** (1983), 7807.

43. J. Fritsche, D. Kraft, A. Thissen, T. Mayer, A. Klein, and W. Jaegermann, Mater. Res. Soc. Symp. Proc. **668** (2001), H6.6.

44. F. Säuberlich, *Oberflächen und Grenzflächen polykristalliner kathodenzerstäubter Zinkoxid-Dünnschichten*, Thesis, (Technische Universität Darmstadt, 2006).

45. D.F. Cox, T.B. Fryberger, and S. Semancik, Phys. Rev. B **38** (1988), 2072.

46. M. Batzill, K. Katsiev, J.M. Burst, U. Diebold, A.M. Chaka, and B. Delley, Phys. Rev. B **72** (2005), 165414.

47. C. Kilic, and A. Zunger, Phys. Rev. Lett. **88** (2002), 095501.

Mater. Res. Soc. Symp. Proc. Vol. 1109 © 2009 Materials Research Society

Large electron mass anisotropy in anatase $Ti_{1-x}Nb_xO_2$ transparent conductor

Yasushi Hirose[1,2], Naoomi Yamada[2], Shoichiro Nakao[2], Taro Hitosugi[2,3], Toshihiro Shimada[1,2], Seiji Konuma[2], and Tetsuya Hasegawa[1,2]

[1] Department of Chemistry, The University of Tokyo, 7-3-1 Hongo, Bunkyo-ku, Tokyo 113-0033, Japan

[2] Kanagawa Academy of Science and Technology, Kawasaki 213-0012, Japan

[3] WPI-Advanced Institute for Materials Research, Tohoku University, Sendai 980-8577, Japan.

ABSTRACT

(012)-oriented anatase $Ti_{1-x}Nb_xO_2$ (TNO) epitaxial films were grown on $LaAlO_3$ (110) substrates by pulsed laser deposition technique, and electron mass (m^*) of the TNO along the c-axis as well as along the a-axis were determined by polarized optical measurements with Drude analysis. The m^* value of TNO (x = 0.01-0.06) along the a-axis was ~0.6 m_0, which was comparable to the electron masses of other transparent conducting materials, while m^* along the c-axis was much larger, ~2-4 m_0. These results clearly indicate that TNO is more conductive along the a-axis than along the c-axis, and therefore, control of crystallographic orientation is needed for improvement in the conductivity of polycrystalline TNO films.

INTRODUCTION

In the past years, titanium dioxide (TiO_2) has been studied extensively as photo-catalytic and dielectric materials. Recently, we found that Nb-doped anatase TiO_2 (TNO) shows excellent conductivity ($\rho > 1 \times 10^3 \ \Omega^{-1}cm^{-1}$) and transparency for visible light (>75%) in the form of both epitaxial [1, 2] and polycrystalline [3, 4] film.

In contrast to conventional TCOs, such as Sn-doped In_2O_3 (ITO), Al-doped ZnO (AZO), and F-doped SnO_2 (FTO), which possess conduction bands composed of isotropic *s* orbitals, TNO has a conduction band with an anisotropic Ti 3*d* character. Therefore, it is anticipated that electrical transport in TNO is highly anisotropic compared with those of conventional TCOs, and that the conductivity of polycrystalline films are significantly affected by the crystallographic orientation of grains. However, due to the difficulty in growing bulk single crystals or epitaxial films with an orientation other than (001), anisotropic carrier transport in TNO has not been measured so far.

In this study, we have investigated the anisotropy in effective electron mass of TNO with optical measurements. We have first grown (012)-oriented TNO epitaxial films with high conductivity by pulsed laser deposition (PLD) technique [5, 6]. The obtained films were subjected to polarized infrared (PIR) spectroscopy, to derive anisotropic electron mass with Drude model. As a result, we found that electron mass of TNO along the c-axis, $m^*_{<001>}$, were 4-6 times larger than that along the a-axis, $m^*_{<100>}$. This tells us that it is of crucial importance to control the crystallographic orientation for improving conductivity of TNO polycrystalline films.

EPITAXIAL GROWTH OF (012)-ORIENTED TNO FILMS

Ti$_{1-x}$Nb$_x$O$_2$ (x = 0, 0.01, 0.03, and 0.06) films were grown on LaAlO$_3$ (LAO) (011) single-crystalline substrates by PLD. TNO targets were made by sintering the pellets composed of TiO$_2$ and Nb$_2$O$_5$. A Kr:F excimer laser (248 nm, 2 J cm^{-2}, 2 Hz) was used for ablating the targets, and the deposition rate was ~0.1 Å pulse^{-1}. The substrate temperature was set to 600 °C, and oxygen partial pressure (P_{O2}) was varied from 1x10^{-3} Torr to 1x10^{-5} Torr.

Figure 1(a) is θ-2θ X-ray diffraction (XRD) patterns of TNO films grown under different P_{O2}. The XRD spectra of TNO films grown under the $P_{O2} \geq$ 1x10^{-4} Torr show (024) peaks of anatase TiO$_2$ without any secondary phases, such as rutile TiO$_2$, Ti$_n$O$_{2n-1}$ and Nb$_2$O$_5$. On the other hand, a rutile TiO$_2$ (110) peak appears, together with anatase (024) peak, for the P_{O2} = 1x10^{-5} Torr film. It is known that TNO films grown under oxidative conditions ($P_{O2} \geq$ 1x10^{-4} Torr) tend to show low conductivity. In order to prevent the growth of rutile phase, we attempted to use a TNO homo-epitaxial buffer layer deposited at P_{O2} of 1x10^{-3} Torr. Figure 1(b) compared the XRD patterns of the films grown under P_{O2} = 1x10^{-5} Torr with and without buffer layer of ~3 nm thickness. As seen from the figure, the homo-epitaxial buffer layer effectively suppressed the growth of rutile phase. The TNO films obtained by the homo-epitaxial buffer layer technique were used for further optical characterizations.

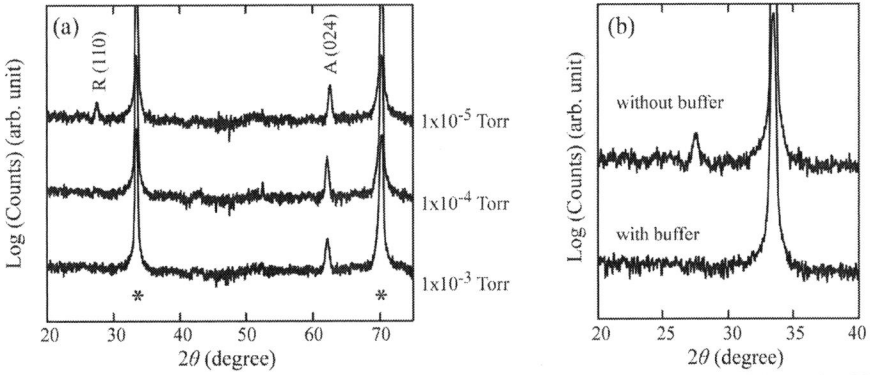

Figure 1. θ-2θ XRD patterns of TNO films (a) grown under different P_{O2} without buffer layer, and (b) grown under P_{O2} = 1x10^{-5} Torr with and without homo-epitaxial buffer layer. Labels A and R denote anatase and rutile TiO$_2$, respectively. Asterisks are the diffraction peaks from LAO substrates.

Figure 2(a) shows φ-scan XRD patterns of the TNO films, clearly indicating that the (102)-oriented TNO films consist of two equivalent epitaxial domains with 180° different in-plane orientations (fig. 2(b)), being consistent with the reports on the undoped anatase TiO$_2$ films grown under more oxidative conditions [5, 6]. Figure 3 shows high-resolution transmission electron microscope (TEM) images of Ti$_{0.97}$Nb$_{0.03}$O$_2$ (012)/LAO (011), which also confirm epitaxial growth of (012)-oriented TNO films.

PORALIZED INFRARED MESUREMENTS

PIR reflection and transmission spectra were measured by an FT-IR spectrometer (PerkinElmer, Spectrum100) with a grid polarizer (Shimadzu, GPR-8000). The configuration of the measurements is illustrated in fig. 2(b): The incident angle of IR light was set at 0°, and the electric polarization vector **E** was either parallel ($\mathbf{E}_{//}$) or perpendicular ($\mathbf{E}\perp$) to the [100] direction of the TNO films. All measurements were performed at room temperature. The thicknesses of the TNO films, ~500 nm, were evaluated with a stylus profiler (Veeco, Dektak 6M).

Figure 4 compared PIR spectra of TNO films with different Nb concentrations. Although the undoped ($x = 0$) film was highly transparent in the IR region, the Nb-doped films ($x \geq 0.01$) showed a finite absorption in the IR region. In addition, the spectral shapes of the Nb-doped films were strongly dependent on the polarization condition, that is, frequency of the absorption maxima in $\mathbf{E}\perp$ shifted to lower frequency side than in $\mathbf{E}_{//}$. In our previous study on (001)-oriented TNO films [2, 4], it was revealed that the optical absorption in the IR region is explained well by free-carriers in the framework of Drude model. In Drude model, the plasma frequency ω_p, which represents the frequency at absorption maxima, is defined as

$$\omega_p{}^2 = (n_e \cdot q^2)/(\varepsilon_\infty \cdot m^*) \tag{1}$$

where ε_∞ is the permittivity at the high-frequency limit, n_e is the carrier concentration, and q is the charge of electron. Thus, the fact the PIR spectra observed here show remarkable **E** dependence suggests large anisotropy in m^* of TNO.

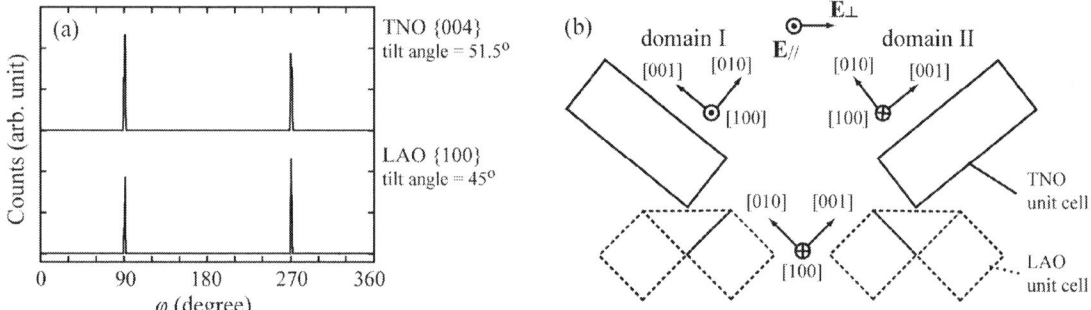

Figure 2. (a) Typical φ-scan patterns of the (012)-oriented TNO film. (b) Schematic illustration of crystallographic structure and polarization conditions in the PIR measurements.

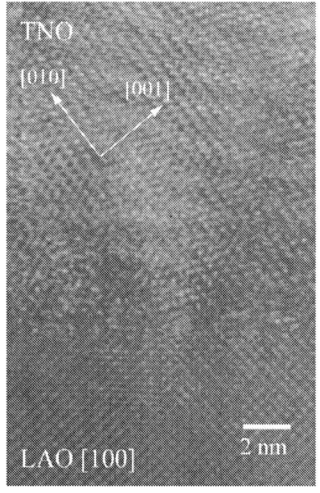 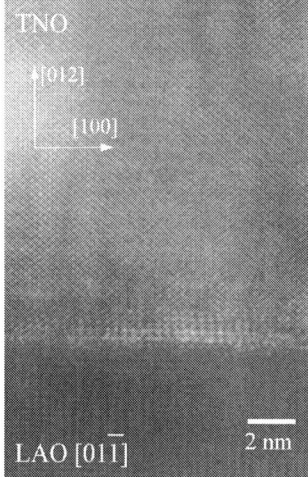

Figure 3. High-resolution cross-sectional TEM images of (012)-oriented $Ti_{0.97}Nb_{0.03}O_2$ film on LAO (110) substrate.

88

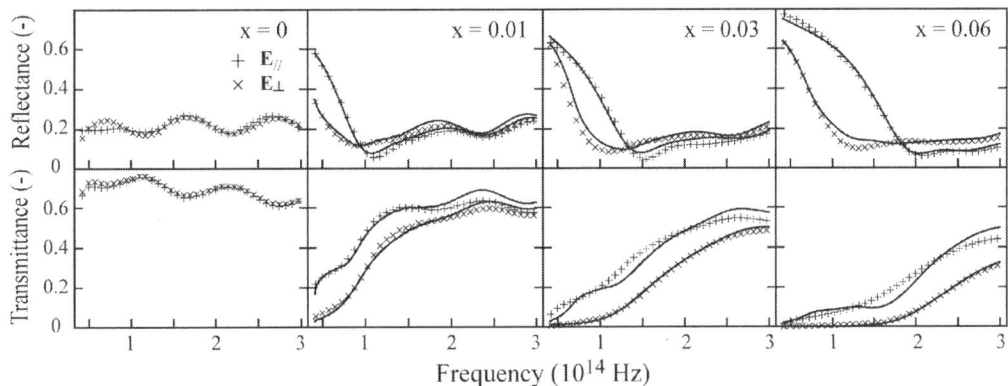

Figure 4. PIR reflection and transmission spectra of TNO (x = 0-0.06). Plus (+) and cross (x) symbols represent the spectra under $E_{//}$ and $E_\perp$, respectively. Solid lines are the simulated spectra based on Drude model with $m^*_{<001>}$ and $m^*_{<100>}$ in Table I.

DRUDE ANALYSIS OF THE PIR SPECTRA

We have evaluated anisotropic m^* values of TNO by Drude analysis of the PIR spectra, as follows. The detailed analytical procedure will be given elsewhere [7].

As mentioned above, our (102)-oriented TNO films consist of two epitaxial domains (domain I and II, see fig. 2(b)). These two domains are equivalent under the $E_{//}$ condition, where the electric field is parallel to <100> principal axis of TNO. Under the $E_\perp$ condition, on the other hand, the two domains are not equivalent with respect to $E_\perp$. In addition, the electric filed is not parallel to any of the principal axis of TNO. To treat such a complicated system, we used effective medium approximation (EMA).

The EMA approach is valid for the present TNO films, because the domain size (~50-200 nm, evaluated by TEM and atomic force microscope) is much smaller than the wavelength of IR light (> 1 µm). We assume the followings, for simplification: (1) Each domain has a cylindrical shape (i.e. depolarization factors are 1/2) and (2) the volume fractions of domain I and II are the same. Under these assumptions, Bruggeman-type effective permittivity is determined by the following equation, which is applicable to the anisotropic system by dyadic description [8].

$$(\bar{\bar{\varepsilon}}_I + \bar{\bar{\varepsilon}}_{eff})^{-1}(\bar{\bar{\varepsilon}}_I - \bar{\bar{\varepsilon}}_{eff}) + (\bar{\bar{\varepsilon}}_{II} + \bar{\bar{\varepsilon}}_{eff})^{-1}(\bar{\bar{\varepsilon}}_{II} - \bar{\bar{\varepsilon}}_{eff}) = 0 \tag{2}$$

where $\bar{\bar{\varepsilon}}_{eff}$, $\bar{\bar{\varepsilon}}_I$ and $\bar{\bar{\varepsilon}}_{II}$ are the permittivity dyadics of the homogenized effective medium and those of two domains in experimental co-ordinate, respectively. The superscript "-1" represents inverse of corresponding dyadic.

Co-ordinate transformation from crystal to experimental systems, yields $\bar{\bar{\varepsilon}}_I$ and $\bar{\bar{\varepsilon}}_{II}$, as,

$$\bar{\bar{\varepsilon}}_{I(II)} = \varepsilon_{\langle 100 \rangle}\mathbf{u}_x\mathbf{u}_x + \left(\varepsilon_{\langle 100 \rangle}\cos^2\theta + \varepsilon_{\langle 001 \rangle}\sin^2\theta\right)\mathbf{u}_y\mathbf{u}_y + \left(\varepsilon_{\langle 100 \rangle}\cos^2\theta + \varepsilon_{\langle 001 \rangle}\sin^2\theta\right)\mathbf{u}_z\mathbf{u}_z$$
$$+ c\left(\varepsilon_{\langle 001 \rangle} - \varepsilon_{\langle 100 \rangle}\right)\sin\theta\cos\theta\mathbf{u}_y\mathbf{u}_z + c\left(\varepsilon_{\langle 001 \rangle} - \varepsilon_{\langle 100 \rangle}\right)\sin\theta\cos\theta\mathbf{u}_z\mathbf{u}_y \tag{3}$$

where $\mathbf{u}_x$, $\mathbf{u}_y$, and $\mathbf{u}_z$, are the unit vectors in experimental co-ordinate, $\varepsilon_{<100>}$ and $\varepsilon_{<001>}$ are the permittivity of TNO along the <100> and <001> principal axis, respectively, and θ is the angle between surface normal (z-direction) and the c-axis of TNO (51.5°), and constant c is 1 for $\bar{\bar{\varepsilon}}_I$ and -1 for $\bar{\bar{\varepsilon}}_{II}$. By assuming Drude model, $\varepsilon_{<100>}$ and $\varepsilon_{<001>}$ were obtained, as

$$\varepsilon_{(hkl)}(\omega) = \varepsilon_{\infty(hkl)}\{1 - \omega_{p(hkl)}{}^2/(\omega^2 - i\tau^{-1}\omega)\} \tag{4}$$

where $\varepsilon_{\infty<hkl>}$ is the permittivity at the high-frequency limit, $\omega_{p<hkl>}$ is the plasma frequency for the $<hkl>$ direction. The homogenized effective medium composed of the domains I and II has $\bar{\bar{\varepsilon}}_{eff}$ with C_{2V} symmetry in micrometer scale, where the C_2 axis is parallel to z-axis and two mirror planes include x- and y- axis. In this case, $\bar{\bar{\varepsilon}}_{eff}$ can be written in the following diagonal form.

$$\bar{\bar{\varepsilon}}_{eff}(\omega) = \varepsilon_{eff,x}(\omega)\,\mathbf{u}_x\mathbf{u}_x + \varepsilon_{eff,y}(\omega)\,\mathbf{u}_y\mathbf{u}_y + \varepsilon_{eff,z}(\omega)\,\mathbf{u}_z\mathbf{u}_z \tag{5}$$

Finally, from eqs. (2), (3) and (5), $\varepsilon_{eff,\,x}(\omega)$ and $\varepsilon_{eff,\,y}(\omega)$ are obtained as

$$\begin{aligned}
\varepsilon_{eff,x}(\omega) &= \varepsilon_{(100)}(\omega) \\
\varepsilon_{eff,y}(\omega) &= \left(\varepsilon_{(100)}(\omega)\cos^2\theta + \varepsilon_{(001)}(\omega)\sin^2\theta\right) \\
&\quad \cdot \sqrt{1 - \frac{\left(\varepsilon_{(001)}(\omega) - \varepsilon_{(100)}(\omega)\right)^2 \sin^2\theta\cos^2\theta}{\left(\varepsilon_{(100)}(\omega)\cos^2\theta + \varepsilon_{(001)}(\omega)\sin^2\theta\right)\left(\varepsilon_{(100)}(\omega)\sin^2\theta + \varepsilon_{(001)}(\omega)\cos^2\theta\right)}}
\end{aligned} \tag{6}$$

Because PIR spectra under the polarization conditions $\mathbf{E}_{//}$ and $\mathbf{E}\perp$ are given as functions of $\varepsilon_{eff,\,x}(\omega)$ and $\varepsilon_{eff,\,y}(\omega)$, respectively, we can determine $m^*{}_{<001>}$ and $m^*{}_{<100>}$ through eqs. (1), (4) and (6) by nonlinear least-squares fitting of the spectra. In the fitting procedure, the multiple reflection interference in the films [9] was taken into account, and for carrier concentration n_e and film thickness d, experimental values determined by Hall and stylus profiler measurements were used. The $\varepsilon_{<hkl>\infty}$ of TNO was assumed to be the same as that of pure anatase TiO_2, i.e., 5.9 for $\mathbf{E}_{//}$ and 5.8 for $\mathbf{E}\perp$ [10].

ELCTRON MASS ANISOTROPY IN TNO

The m^* values of TNO obtained from the above-mentioned analysis, together with those of conventional s-electron-based TCOs, were shown in Table I. The PIR spectra reproduced with these m^* values (solid lines in fig. 4) corresponded well to the experimentally obtained spectra, proving the validity of the present Drude analysis. As seen from Table I, the $m^*{}_{<100>}$ values of TNO, 0.5-0.6 m_0, are comparable to the m^* of conventional TCOs, while the $m^*{}_{<100>}$ values, ~2-4 m_0, are 4-6 times larger than $m^*{}_{<100>}$ in the whole Nb concentration range examined here. These imply that TNO is more conductive along the a-axis than along the c-axis. At a rough estimate, conductivity of a (001)- oriented TNO film is ~1.5 times higher that of a randomly orientated polycrystalline film.

Finally, we briefly discuss on the origin of this large mass anisotropy. Recent first-principle band calculation of TNO ($Ti_{0.9375}Nb_{0.0625}O_2$) based on density functional theory (DFT) showed that the bottom of the conduction bands are mainly composed of isolated Ti $3d_{xy}$ orbitals [11], as well as the case of undoped anatase TiO_2 [12]. The Ti $3d_{xy}$ orbitals spread in the a-b plane direction, and this directional nature of the d_{xy} orbitals results in larger $m^*{}_{<001>}$ compared with $m^*{}_{<100>}$.

Table I. Effective electron mass of TNO and other conventional TCOs.

	$Ti_{1-x}Nb_xO_2$			In_2O_3[13]	ZnO[14]*	SnO_2[15]
	$x = 0.01$	$x = 0.03$	$x = 0.06$			
$m^*_{<100>}$	$0.5m_0$	$0.6m_0$	$0.6m_0$	$0.3m_0$	$0.33m_0$	$0.30m_0$
$m^*_{<001>}$	$2.0m_0$	$2.4m_0$	$3.9m_0$		$0.31m_0$	$0.23m_0$

(*) obtained from LCAO calculation

SUMMARY

The $m^*_{<100>}$ and $m^*_{<001>}$ values of TNO were determined by PIR measurements of (012)-oriented epitaxial TNO films on the basis of anisotropic Drude model. The $m^*_{<100>}$ values, 0.5-0.6 m_0, were much larger than $m^*_{<001>}$, ~2-4 m_0. This large mass anisotropy is attributable to the d-orbital-dominated conduction band, which is a unique feature of TNO. From these results, we concluded that control of crystallographic orientation is crucial for further improving conductivity of polycrystalline TNO films.

ACKNOWLEDGMENTS

We thank Dr. K. Itaka and Dr. S. Yaginuma of the University of Tokyo and Mr. Y. Yukioka of KAST for their kind assistance with PIR measurements. This research was partially supported by MEXT Elements Science and Technology Project.

REFERENCES

1. Y. Furubayashi, T. Hitosugi, Y. Yamamoto, K. Inaba, G. Kinoda, Y. Hirose, T. Shimada, and T. Hasegawa, App. Phys. Lett. **86**, 252101 (2005).
2. Y. Furubayashi, N. Yamada, Y. Hirose, Y. Yamamoto, M. Otani, T. Hitosugi, T. Shimada, and T. Hasegawa, J. App. Phys. **101**, 093705 (2007).
3. T. Hitosugi, A. Ueda, S. Nakao, N. Yamada, Y. Furubayashi, Y. Hirose, T. Shimada, and T. Hasegawa, App. Phys. Lett. **90**, 212106 (2007).
4. N. Yamada, T. Hitosugi, N. L. H. Hoang, Y. Furubayashi, Y. Hirose, T. Shimada, and T. Hasegawa, Jpn. J. App. Phys. Part 1 **46**, 5275 (2007).
5. R. J. Kennedy and P. A. Stampe, J. Cryst. Growth 252,333 (2003).
6. W. Gao, R. Klie, and E. I. Altman, Thin Solid Films **485**, 115 (2005).
7. Y. Hirose, N. Yamada, S. Nakao, T. Hitosugi, T. Shimada, and T. Hasegawa, Phys. Rev. B submitted.
8. A. Sihvola, Electromagnetic mixing formulas and applications (The Institution of Electrical Engineers, London, 1999) Chap. 3-5.
9. O. S. Heavens, Optical properties of thin solid films (Dover, New York, 1991) Chap. 5.
10. N. Hosaka, T. Sekiya, C. Satoko, and S. Kurita, J. Phys. Soc. Jpn. **66**, 877 (1997).
11. H. Kamisaka, T. Hitosugi, T. Suenaga, T. Hasegawa, and K. Yamashita, Comput. Mater. Sci., submitted.
12. R. Asahi, Y. Taga, W. Mannstadt, and A. J. Freeman, Phys. Rev. B 61, 7459 (2000).
13. Z. M. Jarzebski, Phys. Stat. Sol. A **71**, 13 (1982).
14. M-Z. Huang and W. Y. Ching, J. Phys. Chem. Solids **46**, 977 (1985).
15. K. J. Button, D. G. Fonstad, and W. Dreybradt, Phys. Rev. B **4**, 4539 (1971).

Mater. Res. Soc. Symp. Proc. Vol. 1109 © 2009 Materials Research Society

Characterization of amorphous indium-gallium-zinc-oxide (a-IGZO) films deposited by dc magnetron sputtering with H_2O introduction

T. Aoi*[1], N. Oka[1], Y. Sato[1], R. Hayashi[2], H. Kumomi[2], Y. Shigesato[1]

[1]Graduate School of Science and Engineering, Aoyama Gakuin University, 5-10-11 Fuchinobe, Sagamihara, Kanagawa 229-8558, Japan
[2]Canon Inc., 3-30-2 Shimomaruko, Ohta-ku, Tokyo 146-8501, Japan

ABSTRACT

Amorphous indium-gallium-zinc-oxide (a-IGZO) films were deposited by dc magnetron sputtering with H_2O introduction and how the H_2O partial pressure (P_{H2O}) during the deposition affects the electrical properties of the films was investigated. The total pressure of the Ar and H_2O gas mixture was maintained at 0.5 Pa during deposition. P_{H2O} was controlled precisely using a needle valve with monitoring P_{H2O} through quadrupole mass spectrometry (QMS). The depth profile of the hydrogen concentration was analyzed quantitatively by secondary ion mass spectroscopy (SIMS). It was found that the electrical properties of a-IGZO films could be controlled precisely by adjusting P_{H2O} during deposition. Furthermore, TFTs that used a-IGZO films deposited by dc magnetron sputtering with H_2O introduction as channel layers were fabricated. The TFT characteristics indicated the potential for controlling device characteristics by varying the H_2O partial pressure.

I. INTRODUCTION

Transparent amorphous oxide semiconductors (TAOSs) have attracted much attention as high performance channel materials for thin film transistors (TFTs). TAOSs can be fabricated on plastic substrates at low temperature by physical vapor deposition methods such as the conventional dc sputtering method. In particular, the carrier density of amorphous indium gallium zinc oxide (a-IGZO) can be precisely controlled at very low levels, which is stable under various conditions [1-2]. Thus, a-IGZO has been the strongest candidate for the semiconductor layer in TFTs. Yabuta et al. have demonstrated high-performance TFTs using a-IGZO films deposited by rf sputtering on flexible plastic substrates [3]. But it is known that the electrical properties of ZnO films deposited by sputtering are strongly affected by the H_2O partial pressure (P_{H2O}) of the residual gas in the vacuum chamber [4]. There have been attempts to use industrially applicable H_2O as a reactive gas to improve the electrical properties or etching rate of indium tin oxide (ITO) films [5-7]. Therefore, introducing H_2O into the deposition chamber should yield a-IGZO films with the electrical properties appropriate for TFT applications. In this study, we investigated the effects of P_{H2O} on the electrical properties of a-IGZO films and on the TFT's device characteristics.

II. EXPERIMENTAL DETAILS

A. Deposition of the a-IGZO films by dc sputtering with H_2O introduction

200-nm-thick a-IGZO films were deposited on fused silica glass and alkali-free glass (Corning #1737) substrates by dc magnetron sputtering using a polycrystalline $InGaZnO_4$

ceramic sputtering target. The sputtering gas was a mixture of Ar and H_2O vapor, whose total pressure was maintained at 0.5 Pa. These depositions were carried out under various P_{H2O} from 1.0×10^{-3} to 6.0×10^{-2} Pa. As illustrated in Fig. 1, the P_{H2O} during sputter deposition of a-IGZO films was precisely controlled using a precision needle valve and quadrupole mass spectrometer (QMS), where P_{H2O} can be monitored quantitatively. The distance between the substrate and target was 60 mm, and the sputtering power was kept at 50 W.

B. Fabricating TFTs using a-IGZO films

Bottom gate, bottom contact-type TFT devices were fabricated on alkali-free glass (Corning #1737) substrates. As shown in Fig. 2, the TFT is composed of a gate electrode of Mo film (thickness: 100 nm), a gate insulator of SiO_2 film (thickness: 200 nm), a channel layer of a-IGZO film (thickness: 30 nm), source and drain electrodes of a ITO film (thickness: 30 nm) and a channel protection layer of SiO_2 film (thickness: 200 nm). The channel width and length were 180 μm and 30 μm, respectively. The a-IGZO channel layer was deposited by dc magnetron sputtering with H_2O introduction, where the P_{H2O} was maintained at 1.6×10^{-2}, 6.2×10^{-2}, and 8.6×10^{-2} Pa during the sputter-deposition runs.

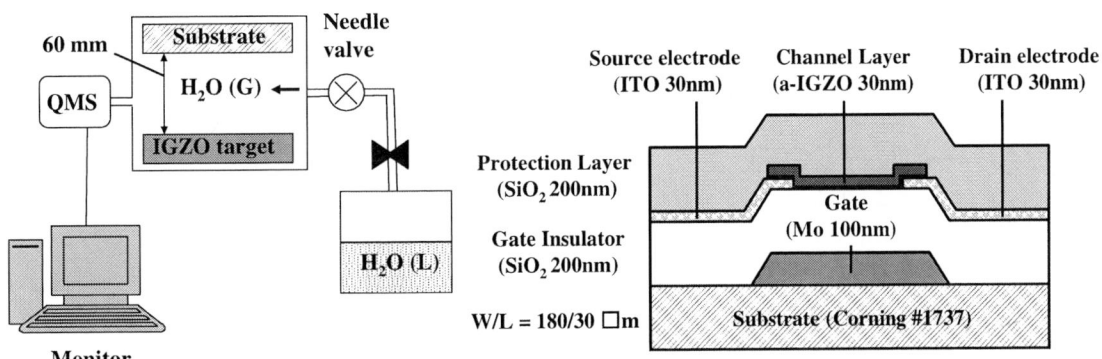

Fig. 1 Schematic of precisely controlled system of H_2O introduction attached to the IGZO sputter deposition system.

Fig. 2 Cross-sectional structure of TFT using the a-IGZO film as a channel layer.

III. RESULTS AND DISCUSSION

A. Structural and optical properties of a-IGZO films

Figures 3 and 4 show X-ray diffraction patterns and optical properties (optical transmittance and reflectance) of a-IGZO films deposited under various P_{H2O}. Regardless of P_{H2O}, all films exhibited an amorphous structure and optical transparency in the entire visible and near-infrared region. In addition, x-ray photoelectron spectroscopy (XPS) revealed that the chemical composition was In:Ga:Zn = 1.3:1.0:1.0 in atomic ratio, regardless of P_{H2O}.

Figure 5 shows FTIR spectra of a-IGZO films deposited under various P_{H2O}. Broad absorption peaks at around 3400 cm^{-1} due to the stretching vibration of OH groups clearly increase for the films deposited under higher P_{H2O} values, indicating that hydrogen is incorporated into the a-IGZO films by the introduction of H_2O. The hydrogen concentration in the films was estimated quantitatively by secondary ion mass spectrometry (SIMS). Figure 6 shows depth profiles of hydrogen concentration in the a-IGZO films. It was found that there was a considerable amount of incorporated hydrogen (more than 10^{21} atoms/cm^3) in the a-IGZO films deposited with H_2O introduction, where the hydrogen concentration increased with increasing P_{H2O}. Although the chemical bonding state of hydrogen in the a-IGZO films remains unclear, a certain portion of the hydrogen atoms is thought to be incorporated as indium hydroxide because H_2O molecules dissociate into H and OH in the plasma, as we have previously reported [8].

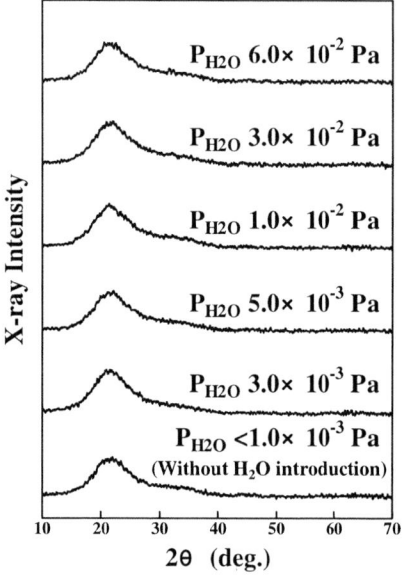

Fig. 3 XRD patterns of a-IGZO films deposited under various P_{H2O}.

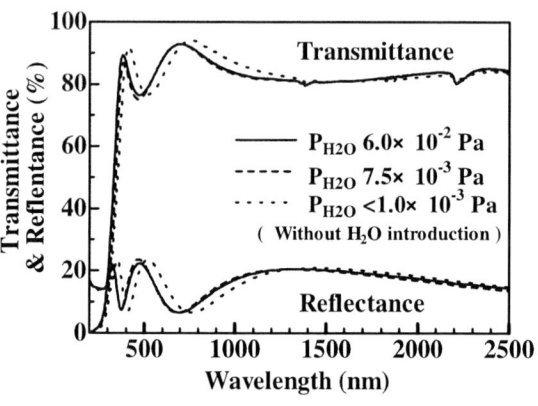

Fig. 4 Transmittance and reflectance of a-IGZO films deposited under various P_{H2O}.

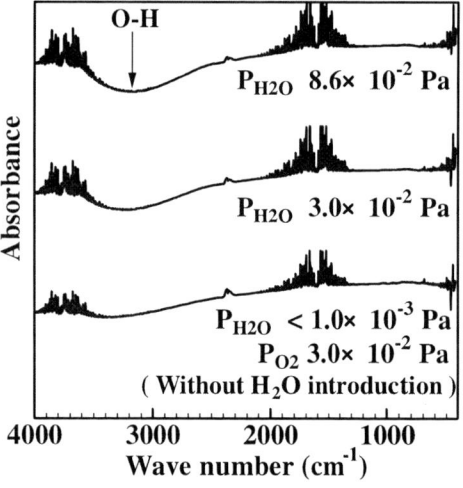

Fig. 5 FTIR spectra of the a-IGZO films deposited under various P_{H2O}.

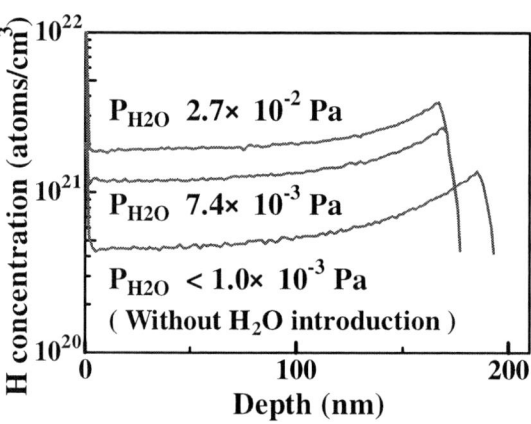

Fig. 6 Depth profiles of hydrogen in the a-IGZO films deposited under various P_{H2O}.

B. Electrical properties of the a-IGZO films

Figure 7 shows the electrical properties (carrier density, Hall mobility, and resistivity) of a-IGZO films as a function of the P_{H2O} during the sputter deposition. The resistivity increased with increasing P_{H2O}. It should be noted that the carrier density of the a-IGZO films was dramatically reduced at a P_{H2O} above 1.0×10^{-2} Pa. This behavior is similar to that observed in the case of the introduction of O_2 reactive gas during the sputter-deposition, as seen in Fig. 7. Thus, the introduction of H_2O during sputter-deposition could be effective in reducing oxygen vacancies which can release donors, and hence decrease the carrier density of the a-IGZO films. Figure 8 shows the temperature dependence of electrical conductivity for a-IGZO films with a carrier density controlled by the introduction of H_2O or O_2. The temperature of the films was regulated by a combination of liquid nitrogen cooling and electrical heating. The electrical conductivity showed thermally activated behavior, and $\sigma = \sigma_0 \exp[-A/T^{1/4}]$, suggesting variable range hopping conduction, when the carrier density was below 10^{15} cm^{-3}. The behavior is similar to that of a-IGZO with the introduction of O_2, which has already been explained in terms of percolation conduction through potential barriers in the vicinity of the conduction band edge [9]. The change in carrier transport mechanism in a-IGZO with H_2O introduction will presumably be similar to the case of O_2 introduction.

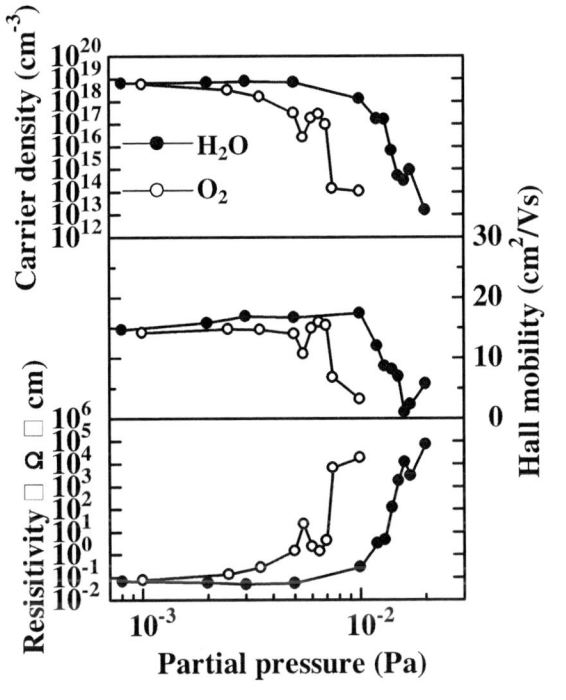

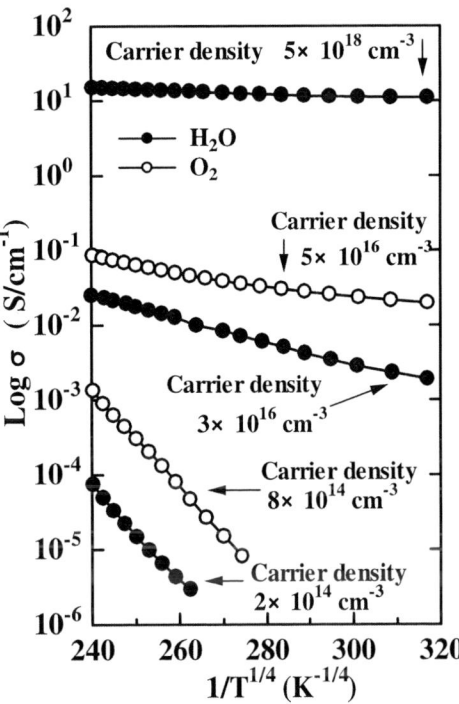

Fig. 7 Electrical properties of a-IGZO films as a function of H_2O or O_2 partial pressure during sputter-deposition.

Fig. 8 Temperature dependence of the electrical conductivity of a-IGZO films with H_2O or O_2 introduction.

C. Device Characteristics of TFTs using a-IGZO films with H_2O introduction

Figure 9 shows transfer characteristics of the TFTs fabricated using a-IGZO deposited with the introduction of H_2O or O_2. All TFTs had the same channel dimension and operated at V_{DS} = 12 V in the saturated region. While the performances of a-IGZO TFTs with H_2O introduction are lower than those of a-IGZO with O_2 introduction, Fig. 9 indicates that the TFTs operate as enhanced-mode n-channel field-effect transistors exhibiting sufficient performance. Device parameters and the hydrogen concentration of a-IGZO channel layers are summarized in Table I. The hydrogen concentration was estimated by the least-squares method from the observed values for 200-nm-thick a-IGZO films by SIMS. The results indicate that the device characteristics of a-IGZO TFTs can be controlled by varying P_{H2O} during sputter-deposition.

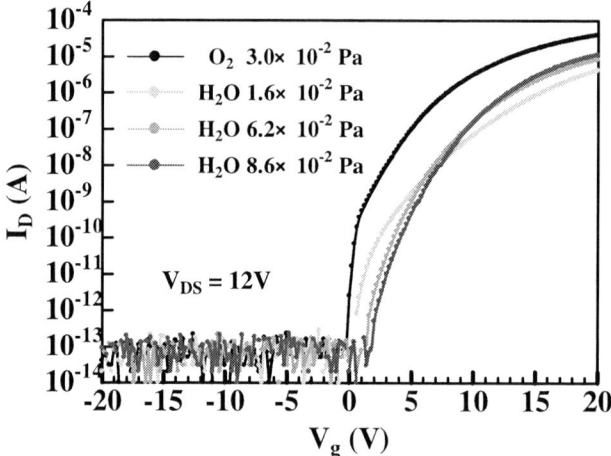

Fig. 9 Transfer characteristics of TFT devices fabricated using a-IGZO films deposited under various P_{H2O}.

Table I. Various properties of the TFTs

H_2O partial pressure	**(Pa)**	$< 1.0 \times 10^{-3}$	1.6×10^{-2}	6.2×10^{-2}	8.6×10^{-2}
O_2 partial pressure	**(Pa)**	3.0×10^{-2}	$< 1.0 \times 10^{-4}$	$< 1.0 \times 10^{-4}$	$< 1.0 \times 10^{-4}$
Hydrogen content	**(atoms/cm^3)**	1.0×10^{21}	2.5×10^{21}	3.8×10^{21}	4.1×10^{21}
Threshold voltage	**(V)**	6.78	12.04	10.92	9.75
On voltage	**(V)**	0.46	2.53	4.24	3.64
Subthreshold swing	**(V/decade)**	0.50	1.59	1.28	1.03
Field effect mobility	**(cm^2/Vs)**	5.27	1.37	1.90	2.96
On-off current ratio		2.2×10^8	3.9×10^7	7.2×10^7	1.0×10^8

IV. CONCLUSION

In this study, a-IGZO films were deposited by dc magnetron sputtering with H_2O introduction and the relationship between H_2O partial pressure during sputter-deposition and the structural/electrical properties of the films was investigated in detail. It was found that the hydrogen concentrations incorporated in the a-IGZO films were over 10^{21} cm^{-3}, and the electrical properties can be precisely controlled by varying the H_2O partial pressure. The behavior of electrical properties is similar to that observed in the case of using the O_2 reactive gas which is effective in reducing oxygen vacancies to induce free electrons. In addition, H_2O incorporated a-IGZO TFTs operated and exhibited sufficient device characteristics. These results demonstrate the potential for controlling device characteristics of a-IGZO films by the introduction of H_2O vapor into the sputter-deposition process.

ACKNOWLEDGEMENTS

This work was partially supported by a High-Tech Research Center project for private universities with the matching fund subsidy from the Ministry of Education, Culture, Sports, Science and Technology (MEXIT) of the Japanese Government.

REFERENCES

[1] K. Nomura, H. Ohta, A. Takagi, T. Kamiya, M. Hirano, and H. Hosono, *Nature* **432**, 488 (2004).

[2] K. Nomura, A. Takagi, T. Kamiya, H. Ohta, M. Hirano, and H. Hosono, *Jpn. J. Apl. Phys.* **45** 4303 (2006).

[3] H. Yabuta, M. Sano, K. Abe, T. Aiba, T. Den, H. Kumomi K. Nomura, T. Kamiya, and H. Hosono, *Appl. Phys. Lett.* **89**, 112123 (2006).

[4] P. K. Song, M. Watanabe, M. Kon, A. Mitsui, and Y. Shigesato, *Thin Solid Films* **411**, 82 (2002).

[5] S. Ishibashi, Y. Higuchi, Y. Ota, and K. Nakamura, *J. Vac. Sci. Technol. A*, **8**, 1399 (1990).

[6] M. Ando, M. Takabatake, E. Nishimura, F. Leblanc, K. Onisawa, and T. Minemura, *J. Non-Cryst. Solids* **198-200** 28 (1996).

[7] E. Nishimura, H. Ohkawa, P. K. Song, and Y. Shigesato, *Thin Solid Films*, **445**, 235 (2003).

[8] Y. Shigesato, I. Yasui, Y. Hayashi, S. Takai, T. Oyama, and M. Kamei, *J. Vac. Sci. Technol. A*, **13**, 268 (1995).

[9] A. Takagi, K. Nomura, H. Ohta, H. Yanagi, T. Kamiya, M. Hirano, and H. Hosono, *Thin Solid Films* **486**, 38 (2005).

Mater. Res. Soc. Symp. Proc. Vol. 1109 © 2009 Materials Research Society

Layer-by-layer Self-assembly of Unilamellar Nanosheet Crystallites of Ruthenium Oxides

Katsutoshi Fukuda,[1] Hisato Kato,[2] Wataru Sugimoto,[1,2] Yoshio Takasu[2]

[1]Collaborative Innovation Center for Nanotech Fiber, Shinshu University, 3-15-1 Tokida, Ueda, Nagano 386-8567, Japan

[2]Faculty of Textile Science and Technology, Shinshu University, 3-15-1 Tokida, Ueda, Nagano 386-8567, Japan

ABSTRACT

Unilamellar nanosheet crystallites of ruthenium oxide obtained via delamination of layered potassium ruthenate and polyvinylamine-polyvinylalcohol diblock copolymer were alternately deposited onto various substrates by electrostatic self-assembly reactions, resulting in almost transparent nanosheet/polymer films. Ultraviolet and visible light absorption spectroscope was used to monitor the effect of thickness in multilayer films grown via repeated self-assembly. Atomic force microscope was utilized to visualize the surface coverage of the nanosheets adsorbed on the substrate. In-plane diffraction revealed that the ruthenate nanosheet possessed two-dimensional crystallinity. A monolayer film with thickness of about 1 nm, in which the nanosheets were densely packed and partially overlapped, exhibited good electronic conductivity.

INTRODUCTION

Indium tin oxide (ITO) possessing high transparency and conductivity at room temperature has been widely used as a commercial transparent electrode for optical applications. A variety of alternative electrodes including fluorine doped tin oxide,[1] zinc oxide,[2] conductive polymers[3] and graphene[4,5] have also been proposed. From the viewpoint of the limited abundance of indium as well as poor corrosion resistance against acidic/basic media, the development of an indium-free, anti-corrosive, transparent conducting system is an important research subject for many applications.

Ruthenium dioxide is known for its versatile properties, e.g. metallic conductivity at room temperature,[6] large pseudo-capacitance,[7] and electrocatalytic activity for chlorine evolution.[8] Recently, we succeeded in the preparation of a layered potassium ruthenate and its delamination into conductive nanosheet crystallites of ruthenium oxide through a soft-chemical route.[9] The obtained nanosheet crystallites, as the elementary fragment of the layered compound, are characterized by thickness in the range of molecular scale and lateral size in bulk dimension. These nanostructures are useful as functional building blocks for electrophoretic deposition (EPD)[10,11] and electrostatic self-assembly[12,13] owing to the polyelectrolytic nature of the nanosheets. A transparent electrode has been fabricated by EPD of ruthenate nanosheet onto ITO substrate for use as transparent electrochemical capacitor.[11] However, the EPD method in principle requires the use of conductive substrate, which is a drawback for many applications. On the other hand, the electrostatic self-assembly of two-dimensional anisotropic crystallites has advantages to construct more sophisticated architectures regardless of the substrate. Oppositely charged organic polymers are generally used as an adhesive binder in this film growth.

Here, we report successful film growth via the electrostatic layer-by-layer self-assembly of ruthenate nanosheets and oppositely charged diblock copolymers. Unique transparent electrodes composed of the conductive ruthenate nanosheets would be realized based on this build-up fabrication.

EXPERIMENT

Layered potassium ruthenate was obtained by heating of a mixture of K_2CO_3 and RuO_2 (5 : 8 molar ratio) at 1123 K for 12 h under Ar atmosphere.[9] The greenish black product was washed with ultra-pure water and then treated with 1 mol dm^{-3} HCl solution for 3 days at room temperature to promote proton exchange. The protonated derivative, $H_{0.2}RuO_{2.1} \cdot 0.9H_2O$, (0.1 g) was mixed with a 25 cm^3 aqueous solution of 0.126 mmol dm^{-3} tetrabutylammonium hydroxide, yielding an auburn colloidal suspension including dispersed nanosheet crystallites with a chemical composition of $RuO_{2.1}{}^{0.2-}$ and a small amount of un-exfoliated residue. The black residue could be readily separated by centrifugation at 2000 rpm as sediment.

Si wafer and SiO_2 glass substrates were cleaned by dipping in 12 mol dm^{-3} HCl + 24 mol dm^{-3} CH_3OH (1:1 by volume) solution and then in 18 mol dm^{-3} H_2SO_4 solution to obtain hydrophilic surfaces. The substrates were immersed in an aqueous solution containing 10 g dm^{-3} diblock copolymer composed of ~14% polyvinylamine and ~86% polyvinylalcohol (Mitsubishi Chemical Corporation, hereafter abbreviated as PVA copolymer) to precoat the surface with positive charge. The PVA copolymer precoated substrates were then dipped in a colloidal suspension containing negatively charged nanosheets (pH ~11.3 as-prepared, 0.22 g dm^{-3}) to fabricate a "monolayer film" of $RuO_{2.1}$ nanosheet/PVA copolymer pair. Subsequent dipping into the copolymer solution and the nanosheet suspension yielded a 2-layer film. Through this deposition cycle, multilayer films up to 10 layers were fabricated. Sheet resistance of the film was measured by two gold current collectors with a gap of 2.1 × 4.2 mm^2, which were formed on the monolayer film by means of vacuum vapor deposition.

X-ray diffraction (XRD) data for nanosheet films were collected by Bragg-Brentano-type diffractometers (Rigaku Rint 2000) with Cu Kα radiation (λ = 0.15405 nm). In-plane XRD analysis on the nanosheet monolayer deposited on Si substrate was carried out with a four-axis diffractometer installed on the BL-6C at the Photon Factory in the High Energy Accelerator Research Organization. A tapping-mode atomic force microscope (AFM) with Si-tip cantilever (Seiko Instruments SPA400) was used to evaluate the dependence of the dipping-time and concentration of the suspension with regard to the nanosheet coverage. Ultraviolet and visible light (UV-Vis) absorption spectra for the self-assembled films were recorded on a Hitachi U-4100 spectrophotometer to monitor the multilayer film growth.

DISCUSSION

$RuO_{2.1}$ nanosheet/PVA copolymer monolayer film

Restacking of the nanosheets, which consequently disturbed film growth, occurred when the pH of the ruthenate nanosheet suspension was reduced to below 10. This made the use of the commonly used cationic polymers such as polydiallyldimethyl-ammonium (PDDA) and polyethylenimine (PEI)[12,13] difficult and less favorable as counter polycations for electrostatic self-assembly. A monolayer film of $RuO_{2.1}$ nanosheet/polycation was successfully fabricated by employing a cationic PVA copolymer as the counter polycation for the electrostatic self-assembly process. Compared to PDDA and PEI, the PVA copolymer should have lower charge density owing to the large amount of neutral polyvinylalcohol within the polymer. This seems to be important for the successful electrostatic self-assembly in the present ruthenate nanosheet system.

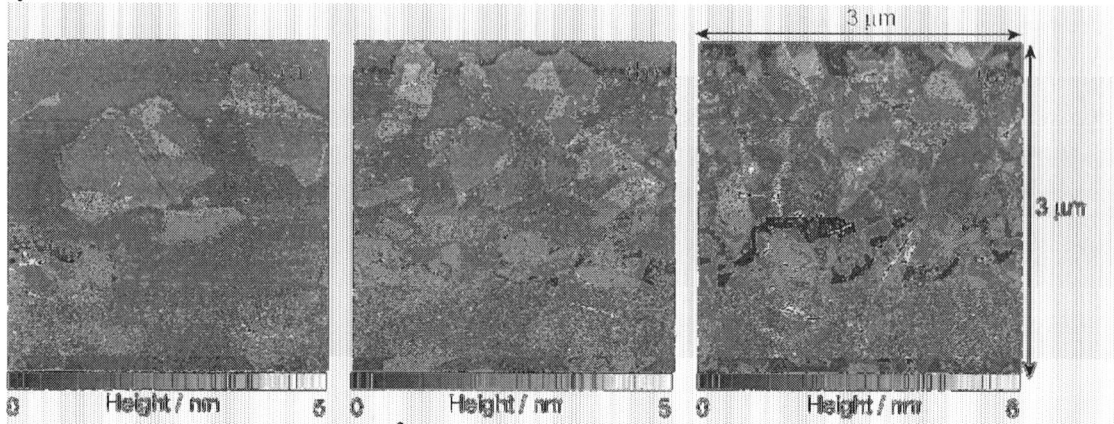

Figure 1. AMF images (3 × 3 μm²) for nanosheet monolayer films adsorbed on Si substrate at different nanosheet concentration and duration of adsorption; (a) 0.08 g dm⁻³ for 1 min, (b) 0.08 g dm⁻³ for 5 min and (c) 0.22 g dm⁻³ for 10 min.

AFM images revealed that nanosheets with thickness of about 1 nm and lateral size ranging from sub-micrometer to micrometers were adsorbed onto the substrate (fig. 1). The nanosheet coverage increased with the increase in dipping time or concentration of the nanosheet suspension. Approximately 90% coverage with about 40% overlapping of the nanosheet could be obtained, yielding a fairly transparent film (fig. 1c). It should be noted that most of the nanosheets in the high coverage film were connected with the neighboring nanosheets via overlapping and therefore the film appears like a "patchwork" of the nanosheets.

Figure 2 shows a synchrotron radiation in-plane XRD pattern for the monolayer film grown on a Si substrate. The ruthenate nanosheets exhibited many sharp peaks in the $1/d$ region from 2.2 to 8.2 nm⁻¹. All of the observed peaks are indexed to hk reflections of a 2D oblique cell, and its refined cell parameters were $a = 0.5610(8)$ nm, $b = 0.5121(6)$ nm and $\gamma = 109.4(2)°$. Clearly, the ruthenate nanosheet monolayer film has high two-dimensional crystallinity presumably inherited from the parent material. The crystallographic parameters of the parent material are presently unclear due to considerable difficulty in its structure analysis.

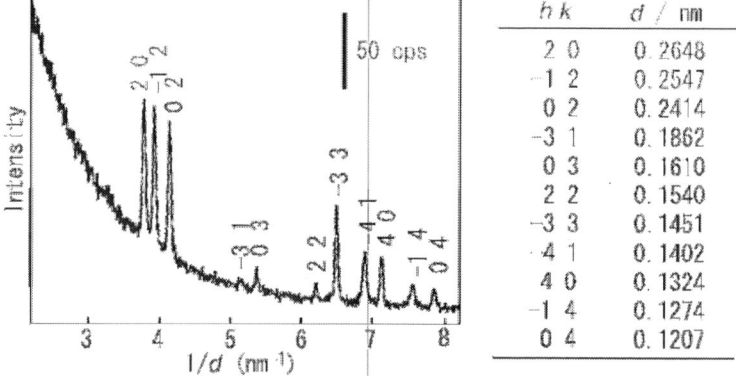

h k	d / nm
2 0	0.2648
-1 2	0.2547
0 2	0.2414
-3 1	0.1862
0 3	0.1610
2 2	0.1540
-3 3	0.1451
-4 1	0.1402
4 0	0.1324
-1 4	0.1274
0 4	0.1207

Figure 2. Synchrotron radiation in-plane XRD pattern for the monolayer film of the ruthenate nanosheets. Right panel lists the index and d value of each reflection.

RuO$_{2.1}$ nanosheet/PVA copolymer multilayered film

A thick, translucent RuO$_{2.1}$/PVA copolymer multilayered film was fabricated by repeated dipping cycles. The absorption spectra acquired after each nanosheet deposition step are shown in fig. 3. The ruthenate nanosheet has a sharp absorption peak at 220 nm and two broader peaks at 360 and 500 nm, which accounts for the auburn color in the resultant thick film. The peak intensities increased proportionally with the number of deposition cycle of RuO$_{2.1}$/PVA copolymer pair (fig. 3 inset), indicating that quantitative growth was accomplished during the fabrication.

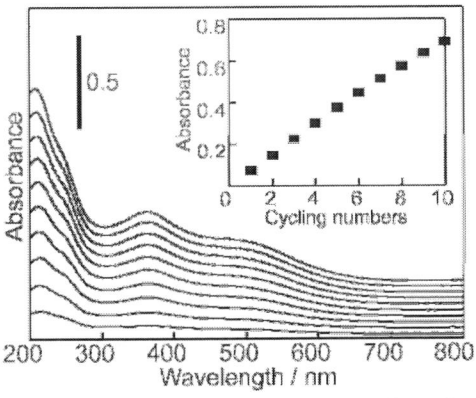 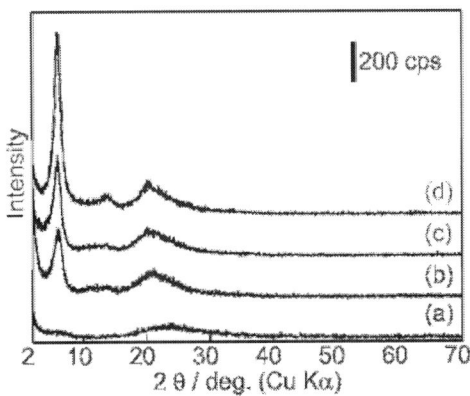

Figure 3. UV-vis absorption spectra of RuO$_{2.1}$/PVA copolymer multilayer films grown on both sides of quartz. The observed absorbance at 360 nm is plotted against the number of deposition cycle in the inset.

Figure 4. XRD patterns of (a) 1, (b) 3, (c) 6, and (d) 10-layer films of RuO$_{2.1}$/PVA copolymer pairs grown on the quartz substrate.

XRD patterns illustrated in fig. 4 depict the evolution of Bragg peaks attributable to a superlattice reflection. The intensity of the prominent peak at 6° enhanced with the increase in the cycle number, in accordance with the UV-vis spectra showing the successful built-up of the 2D ordered RuO$_{2.1}$ nanosheet/PVA copolymer multilayer. Note that even the monolayer film exhibited a small broad reflection at the same position, which can be attributed to the presence of the overlapped regions of ruthenate nanosheets as mentioned above. The interlayer distance between the deposited nanosheets, i.e. the sum of nanosheet and polymer thickness, was calculated as about 1.5 nm from the peak position. This value is in agreement with the topographical thickness of approximately 1 nm for the ruthenate nanosheet observed in the AFM image.

From these results, we can conclude that ultathin superlattice films composed of RuO$_{2.1}$ nanosheet/PVA copolymer were obtained by the solution-based electrostatic self-assembly on an insulating substrate. This approach allows the precise control of film thickness, which is critical to the optical and physicochemical properties, and may offer a rational design for various nanostructured materials using conductive ruthenium oxide.

Conductivity of the nanosheet patchwork film

Rutile-structured RuO$_2$ is well known for its metallic conductivity[6] as well as its chemical and electrochemical stability.[14] The layered ruthenate has also been shown to exhibit chemical and physical properties comparable to rutile RuO$_2$.[15] Thus, it is intriguing to investigate the conductivity of the self-assembled ruthenate nanosheet films obtained in this study. The conductivity of transparent monolayer films with high and low coverage of the nanosheets on the SiO_2 glass substrate was investigated.

The dense monolayer film with about 90 % nanosheet coverage was electro-conductive, showing a sheet resistance of approximately 11 kΩ / $\square$ on the basis of measurement by the two-probe method. Providing that the nanosheet thickness is ~1 nm, the conductivity of this film can be roughly evaluated as 900 S cm^{-1}. This value is comparable to those of other transparent conductors. The suitable conductivity of the nanofilm, combined with the chemical and electrochemical stability of ruthenate nanosheet, may pave the way for new surface modification directed toward electrochemical material synthesis on various substrates.

In contrast to the dense case, the sparse film was non-conducting. Clearly, the conductivity of the self-assembled monolayer film depends on the coverage and overlapping of the nanosheets. This dependence may be understood by a two-dimensional percolation path for electron flow, which should be formed by the overlapping of the nanosheets. Taking into account the high conductivity of the rutile-structured RuO_2, the observed conductivity is believed to be mostly governed by the resistance between the nanosheets. Therefore, a careful design of the sheet gap and overlapping may be important for improving the film conductivity in this case.

CONCLUSIONS

Ruthenate nanosheets were successfully deposited onto hydrophilic substrates via electrostatic self-assembly using diblock copolymer with low charge density. In-plane diffraction study of monolayer film revealed the presence of well-defined two-dimensional crystallites deposited on the substrate. A transparent film, in which the nanosheets were densely packed, exhibited relatively good conductivity for a 1 nm thin film. The present study shows the possibility of constructing novel nanoscopic electrodes by stacking the conductive ruthenate nanosheets with unique properties, e.g. high chemical and electrochemical stability.

ACKNOWLEDGMENTS

This work was supported in part by a "Creation of Innovation Centers for Advanced Interdisciplinary Research Areas" Project in Special Coordination Funds for Promoting Science and Technology of the Ministry of Education, Culture, Sports, Science and Technology, Japan and CREST of the Japan Science and Technology Agency (JST). The in-plane XRD measurement was performed with the approval of the Photon Factory Program Advisory Committee (2005G159). The authors wish to thank Professors Mutsumi Kimura and Yoko Tatewaki (Shinshu University) for fruitful help and advice in conductivity measurements.

REFERENCES

1. Fantini, M.; Torriani, I. *Thin Solid Films* **1986**, *138*, 255-265.
2. Minami, T.; Sato, H.; Nanto, H.; Takata, S. *Jpn. J. Appl. Phys.* **1985**, *24*, 781-784.
3. Ouyang, J.; Xu, Q.; Chu, C.; Yang, Y.; Li, G.; Shinar, J. *Polymer* **2004**, *45*, 8443-8450.
4. Wang, X.; Zhi, L.; Müllen, K. *Nano Lett.* **2008**, *8*, 323-327.
5. Becerril, H. A.; Mao, J.; Liu, Z.; Stoltenberg, R. M.; Bao, Z.; Chen, Y. *ACS Nano* **2008**, *2*, 463-470.

6. Ryden, W. D.; Lawson, A. W. *Phys. Rev. B* **1970**, *1*, 1494-1500.

7. Zheng, J. P.; Cygan, P. J.; Jow, T. R. *J. Electrochem. Soc.* **1995**, *142*, 2699-2703.

8. Trasatti, S. *Electrochim. Acta*, **1991**, *36*, 225-241.

9. Sugimoto, W.; Iwata, H.; Yasunaga, Y.; Murakami, Y.; Takasu, Y. *Angew. Chem. Int. Ed.* **2003**, *42*, 4092-4096.

10. Yui, T.; Mori, Y.; Tsuchino, T.; Itoh, T.; Hattori, T.; Fukushima, Y.; Takagi K. *Chem. Mater.* **2005**, *17*, 206-211.

11. Sugimoto, W.; Yokoshima, K.; Ohuchi, K.; Murakami, Y.; Takasu, Y. *J. Electrochem. Soc.* **2006**, *153*, 255-260.

12. Lvov, Y.; Ariga K.; Ichinose, I.; Kunitake, T. *Langmuir* **1996**, *12*, 3038-3044.

13. Sasaki, T.; Ebina, Y.; Watanabe, M.; Decher, G. *Chem. Commun.* **2000**, 2163-2164.

14. Trasatti, S. in Electrodes of conducting metal oxides (Trasatti, S. Ed.), Elsevier, Amsterdam, Netherland **1980**, 301-358.

15. Sugimoto, W.; Omoto, M.; Yokoshima, K.; Murakami, Y.; Takasu, Y. *J. Solid State Chem.* **2004**, *177*, 4542-4545.

Mater. Res. Soc. Symp. Proc. Vol. 1109 © 2009 Materials Research Society   1109-B03-28

PMP-MOCVD Grown $Zn_xCd_{1-x}Se$ cladded $Zn_yCd_{1-y}Se$ Quantum Dot Structures Exhibiting Diode Like Characteristics for Electroluminescent Application

F. Al-Amoody[1], A. Rodriguez[2], E. Suarez[1], W. Huang[3], F. Papadimitrakopoulos, and F. Jain[1]

[1]Electrical and Computer Engineering Department, University of Connecticut, 371 Fairfield Road, Storrs, CT 06269-2157, U.S.A.

[2]Intel Corp., Rio Rancho, NM, U.S.A.

[3]US Military Academy, West Point, NY, U.S.A.

[4]Chemistry Department and Institute of Materials Science, University of Connecticut, Storrs, CT, U.S.A.

ABSTRACT

This paper reports the fabrication of devices that exhibit diode like characteristics using pseudomorphic $Zn_xCd_{1-x}Se$ (core) quantum dots (QDs) cladded with $Zn_yCd_{1-y}Se$ (x < y) layers. The cladded dots were grown using Photoassisted Microwave Plasma (PMP) Metalorganic Chemical Vapor Phase Deposition (PMP-MOCVD) technique. The devices were fabricated by growing QDs on ITO coated glass substrates with CsF as the thin barrier layer between the dots and the aluminum top contact cathode. Current-voltage characteristics are presented with and without CsF layers.

INTRODUCTION

Earlier we reported growing 3-8 nm CdSe and pseudomorophic $Zn_xCd_{1-x}Se$ (core) quantum dots (QDs) cladded with $Zn_yCd_{1-y}Se$ (x < y) in a Photoassisted Microwave Plasma Metalorganic Chemical Vapor Phase Deposition (PMP-MOCVD) reactor [1, 2]. Influence of growth parameters including microwave power, ultraviolet intensity, gas phase II/VI [Zn&Cd/Se] molar ratio, temperature of growth, and post-growth processing was investigated.

EXPERIMENT

Growth

The ITO coated glass substrate was cleaned using warm trichloroethylene, acetone, and methanol and then left in a boiling propanol before it was loaded inside the PMP-MOCVD reactor. The substrate was heated in the reactor while the carrier gas was passing. The heating of the substrate was to remove any excess water and chemicals left on the surface from cleaning procedure. In addition to this, the heating helps in the nucleation of cladded dots. With the plasma turned on, the deposition was seen nucleating on the surface of the substrate. The UV lamp assisted in the incorporation of Zn hence forming different composition of ZnCdSe used as core and cladding.

Characterization

The dots grown, using 3-step method involving substrate heating prior to nucleation, are shown in Fig. 1, which presents a high-resolution transmission electron micrograph (HR-TEM). These dots can be compared with those reported before [1]. They seem to be comparable to dots prepared by other methods [2, 3].

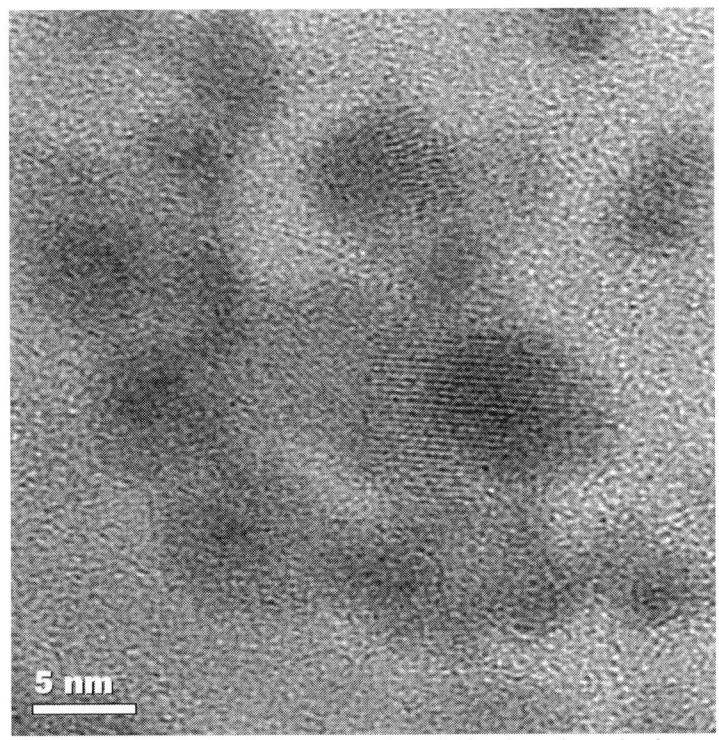

Fig. 1 Transmission electron micrograph of ZnCdSe dots deposited on grids.

Figure 2 shows the X-ray diffraction plot of grown dots on ITO/glass substrate. The recently obtained photoluminescence spectrum on our cladded dots deposited on glass substrate is shown in Fig. 3. The exciton peaks are distinguishable. Two different dot sizes were observed.

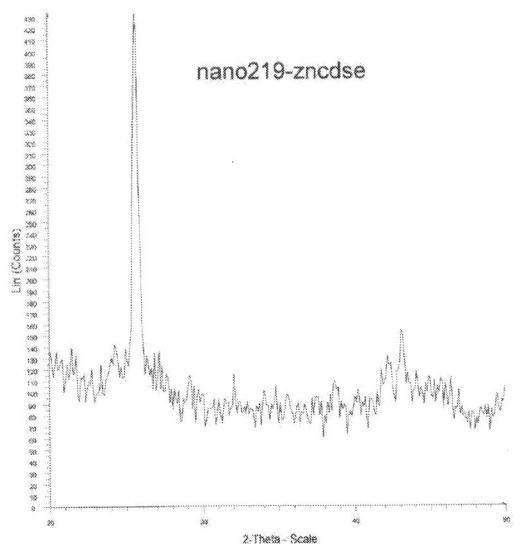

Fig. 2. X-ray diffraction plot of grown dots on ITO/glass substrate.

It has been shown [1] that location of photoluminescence (PL) peaks and the value of full width at half maximum (FWHM), and X-ray diffraction (XRD) data can be used to calculate the core/shell size.

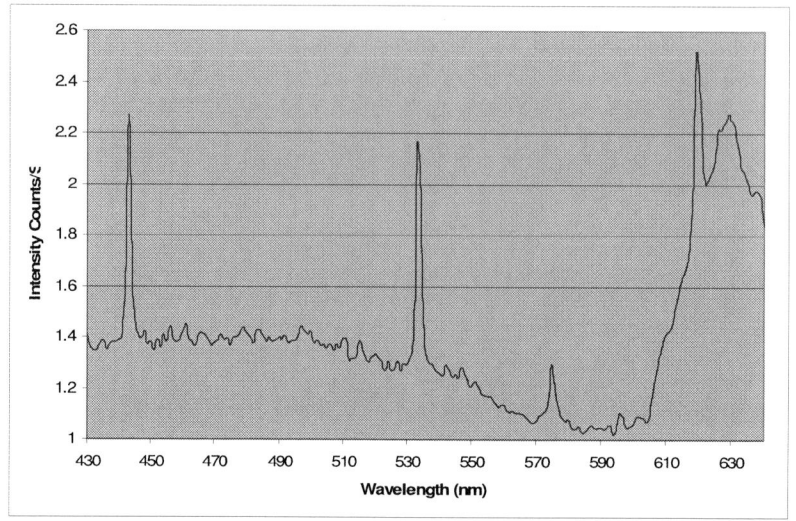

Fig. 3. Fluorescence intensity spectra from QDs on ITO on glass substrates.

DISCUSSION AND RESULTS

Figures 4(a) and 4(b) show the current-voltage characteristic of fabricated ZnCdSe QDot diodes grown on ITO coated glass substrate. Here, the Fig. 4(b) shows improved reverse biased behavior. The forward-biased current is also higher by a factor of 10. The dot diameters were ~10 mils.

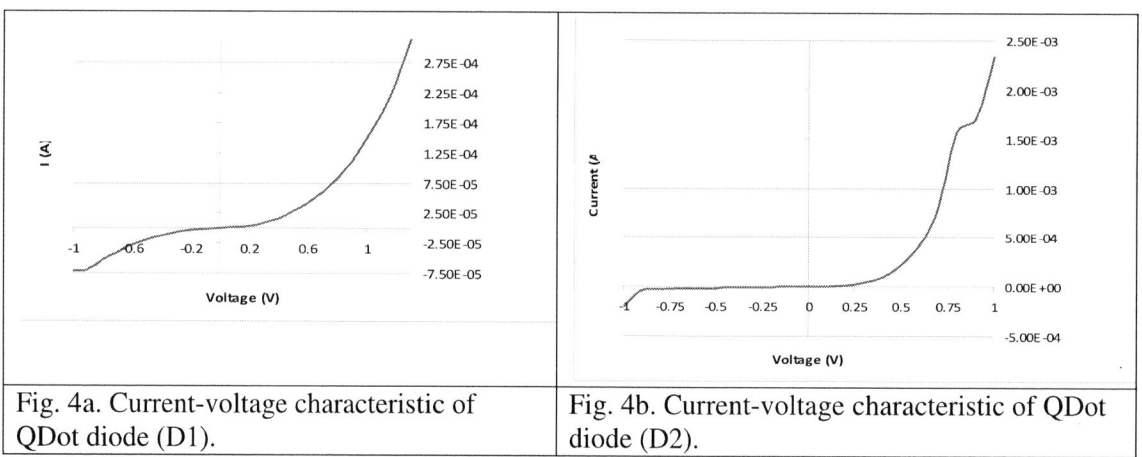

Fig. 4a. Current-voltage characteristic of QDot diode (D1).	Fig. 4b. Current-voltage characteristic of QDot diode (D2).

The recently obtained photoluminescence spectrum on our cladded dots deposited on glass substrate shows two exciton peaks at 445nm and 535nm. This indicates existence of at least two different dot sizes. It is hard to predict their influence of I-V characteristics of the diodes. We have also simulated the optical gain of cladded quantum dots including the effect of strain in the cladding for different composition of cladding layer. Simulation is based on excitonic model reported by Jain and Huang [4] with some modification.

CONCLUSIONS

We have presented growth and characterization of ZnCdSe quantum dots using PMP-MOCVD vapor phase nucleation methodology. The PL data shows two exciton peaks indicating two different dot sizes. The fabricated diodes exhibit good reverse characteristics. The forward characteristics do not show higher built-in voltage as we expected. Mesa etching (using reactive ion etching via methane/hydrogen chemistry) around the top contacts will be employed to improve this behavior. Work is in progress for growth on matching substrates.

ACKNOWLEDGMENTS

This work is supported by ONR Contracts N00014-02-1-0883 and N00014-06-1-0016, and NSF-Grant ECS 0622068. Discussions with Dr. D. Purdy (ONR) and Dr. R. Khosla (NSF), and technical assistance in processing by S. Vaddiraju are gratefully acknowledged.

REFERENCES
1. A. Rodriguez, R. Li, P. Yarlagadda, F. Papadimitrakopoulos, W. Huang, J. Ayers, and F. Jain, NSTI, Boston Conference, May 8, 2006.
2. X. Peng, M.C. Schlamp, A.V. Kadavanich and A. P. Alivisatos, J. Am. Chem. Soc., **119**, 7019-7019.(1997).
3. B.O. Dabbousi, J. Rodriguez-Viejo, F.V. Mikulec, J.R. Heine, H. Mattoussi, R. Ober, K.F. Jensen, M.G. Bawendi. J. Phys. Chem. B, **101**, 9463-9475(1997).
4. F. Jain, W Huang, J. Appl. Phys., **85**, 2706-2712 (1999).

Mater. Res. Soc. Symp. Proc. Vol. 1109 © 2009 Materials Research Society 1109-B03-32

Effect of Annealing on Rectifying Contacts on ZnO Thin Films Grown using Pulsed Laser Deposition

A. Bhattacharya, R.K. Gupta, P.K. Kahol, K. Ghosh
Department of Physics, Astronomy, and Materials Science, Missouri State University,
Springfield, MO-65897

ABSTRACT

Zinc oxide (ZnO) is probably the most explored oxide based semiconductors in this century, for its versatile characteristics in number of domains. ZnO has drawn a strong attention in optoelectronics for being transparent as well as highly conducting in nature, making it suitable to be used for transparent electrodes. Having wide optical band gap ~3.34 eV, ZnO offers promise to be utilized as UV detectors. Also ZnO has tasted some success in the domain of spintronics as well. But in-spite of this fame, the thirst of understanding the basic transport phenomena in ZnO is still in search, as the situation is enough challenging to make good contact on ZnO. Based upon the growth technique a metal-ZnO contact can be ohmic or rectifying in nature. The ultimate hurdle therefore lies to make a stable rectifying contact with metal (Au, Ag, and Pt) and ZnO. In this article we therefore aim to make some systematic study towards the metal (Au, Ag) and ZnO contact, and optimization of the growth parameters. It has been found that inherent oxygen vacancies act as source of conducting nature of ZnO. Here we grow ZnO films in ambient oxygen pressure (5.0×10^{-2} mbar) at 400°C using pulsed laser deposition technique. The films with (100) preferential growth, are annealed for different time scales in higher oxygen pressure. This makes the carrier concentration of ZnO to reduce from 10^{20} cm^{-3} to 10^{17} cm^{-3} or order below that. Thickness of the films were limited to 100 nm, with rms surface roughness ranging 1~5 nm. The ZnO surface was ablated with energy pulses 5 mJ/cm^2 to reduce some surface defects. The transparency of the films grown in ambient oxygen pressure is nearly 85%. Later, Ag and Au electrodes were grown using same technique on the ZnO film. Detailed observations of I-V characteristics with different annealing time, surface roughness, and XRD data will be presented.

INTRODUCTION

Research in ZnO has found renewed interest for its numerous applications, especially in the field of Optoelectronic devices. A large band-gap value of ~3.34 eV, inherent high carrier concentration, high mobility promotes it for usage in photodetectors, light emitting diodes and transparent conducting oxides [1,2]. The device properties of ZnO substantially depend on the nature of the contact formed with any metal electrodes. In general the contact shows ohmic behavior, but subjected to certain treatments the metal-ZnO junction could be rectifying type as well [3,4]. The second type of contact, familiar as Schottky contact forms the basis of some optoelectronics devices. But the formation of thermally stable Schottky barrier remains remarkably challenging even after almost a decade from the very first the attempt of Schottky & Mott (1931).The major hurdle for achieving a rectifying contact is the presence of inherent high carrier concentration in ZnO, which has its source to oxygen vacancies and surface defects.

Successful growth of rectifying Au, Ag, Pd, Pt contacts on n-ZnO with barrier height of 0.6-0.8 eV has been reported [5,6,7]. But none of these observations support totally the theory of

Schottky-Mott. According to Schottky-Mott theory, the effective barrier height (Φ_B) gets determined by the difference between metal work function (Φ_M) and electron affinity of the semiconductor (Φ_S). This deviation accounts for the incorporation of the influence of defect states originating from vacancies [8,9,10], affecting the formation of barrier. Strong presence of defect states leads to pinning of Fermi level. Also in most of the occasions the ideality factor, an important parameter of the junction, stays greater than unity. In recent days, surface treatments methods using chemical etching of ZnO by HCl, H_2O_2 or oxygen plasma cleaning aided to improve contact properties from ohmic to rectifying one [11]. Occasionally, non-destructive treatments like etching through pulsed laser has also shown enhancement of contact performance [12]. But this type of techniques asks for added caution for applying in ZnO films. In this article carrier concentration of vacancies are controlled through annealing the films at different time scales, under ambient oxygen pressure. The gradual change in time scale, assist to growth of single crystalline ZnO film with betterment in stoichometry. In this communication, we report effect of annealing on rectifying contact of ZnO films grown by pulsed laser deposition (PLD) technique.

EXPERIMENT

Thin films of ZnO were deposited using a sintered ceramic ZnO target with a purity of 99.99% (Alfa-Aesar, USA). For making the target, the ZnO powder was cold pressed using a hydraulic press of 15 tons loads and sintered in air at 800 °C for 10 hours. Thin films were deposited on quartz substrate using KrF excimer laser (Lambda Physik COMPex, wavelength 248 nm) at 400 °C under oxygen pressures of 5.0×10^{-3} mbar. The laser was operated at a pulse rate of 10 Hz, with energy of 300 mJ/pulse. The laser beam was focused onto a rotating target at a 45° angle of incidence. The deposition chamber was initially evacuated to 4×10^{-5} mbar and during deposition oxygen gas was introduced into the chamber to obtain the pressures mentioned above. The thickness of the films was approximately 80 nm. The deposited ZnO films were annealed under oxygen pressure of 3.0×10^{-2} mbar. After annealing, Au and Ag circular contacts were deposited using PLD technique. The thicknesses of the contacts were approximately 100 nm. The effective area of the device was 0.3 cm^2.

RESULTS AND DISCUSSION

Figure 1(a) shows the effect of annealing on x-ray diffraction patterns of ZnO films grown on glass substrate. It is evident from the figure that all the films are highly oriented along (002) direction. It is observed that the annealing improves the crystallinity of the films. A shift in peak towards higher 2θ value is observed with increase in annealing time. This may correspond to contraction of interplanar spacing due to relaxation of compressive strain present in the structure. Annealing in oxygen ambient pressure aids oxygen atoms to move at proper lattice sites with the available kinetics. Strain in the films were calculated using Williamson-Hall equation and observed that the strain in the films reduce with annealing in oxygen ambient condition [13]. Figure 1(b) shows the effect of oxygen annealing time on full width half maximum (FWHM) of ZnO films. FWHM decreases with increase in oxygen annealing time, indicating improvement in film crystallinity.

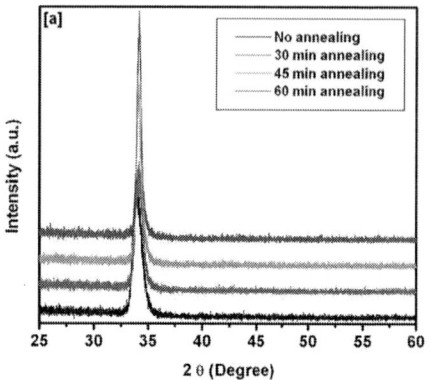

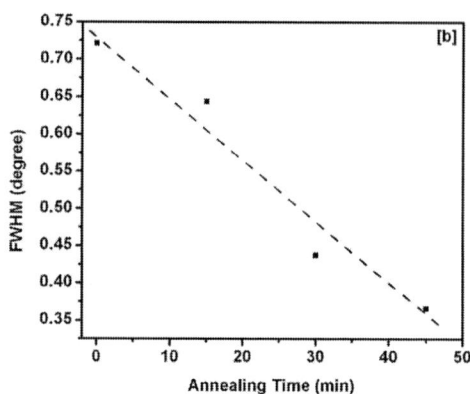

Fig. 1(a) XRD patterns of ZnO films.

Fig. 1(b) Variation of FWHM with annealing time (line is guide to eye).

The effect of annealing on optical transmittance of ZnO films is discussed next. Transmittance spectra of ZnO films were recorded in the range of 350-900 nm at room temperature (Figure 2a). All the films are highly transparent in visible region of solar spectrum with average percentage transmittance more than 85%. It is evident from the figure that oxygen annealing improves the optical transmittance of ZnO films. The effect of annealing on band gap energy of ZnO is estimated using wavelength versus transmittance data. The absorption coefficient (α) of the ZnO films were calculated using equation [14], $T = A exp(-\alpha d)$, where A is a constant, T is the transmittance, and d is the film thickness. Having the value of absorption coefficient, the band gap E_g is estimated using the expression, $(\alpha h v)^2 = (h v - E_g)$, where $h v$ is the photon energy. By extrapolation of the linear region of $(\alpha h v)^2$ vs. $h v$ plot to the photon energy axis, band gap is determined. . Inset of Figure 2(b) shows $(\alpha h v)^2$ vs. $h v$ plot for as grown ZnO(without annealing) film. The effect of oxygen annealing time on bandgap energy of ZnO films are shown in Figure 2(b). The optical bandgap is found to contract from 3.60 eV to 3.24 eV from sample with high carrier concentration to low carrier concentration, i.e. the films go through redshift in optical spectra.

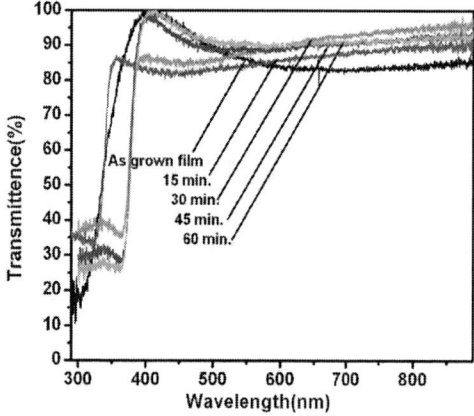

Fig. 2(a) UV-visible spectra of ZnO films.

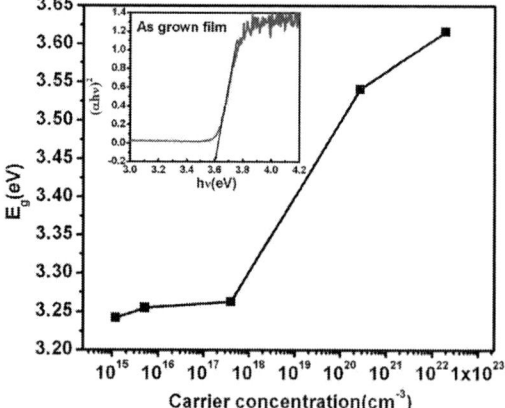

Fig. 2(b) Effect of annealing on band gap of ZnO films.

110

Morphological characterization of ZnO films was done using atomic force microscopy (AFM). AFM image of ZnO film annealed for 60 min is shown in Figure 3. The scan was carried out in tapping mode. The spring constant of the cantilever was ~42 N/m. The cantilevered tip was oscillated close to the mechanical resonance frequency of the cantilever (typically, 200–300 kHz) with amplitudes ranging from 10 to 30 nm. It is observed that surface roughness of the films depends on annealing. The root mean square (rms) values of surface roughness for the ZnO films were found to vary from 1-5 nm for various annealing time.

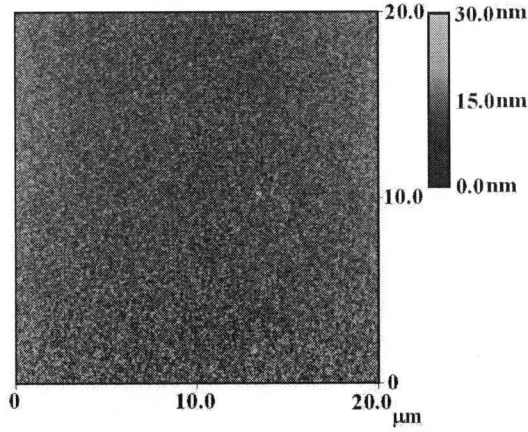

Fig. 3. AFM image of ZnO film annealed for 60 min.

It is observed that oxygen annealing could affect structural and optical properties of ZnO films. Next, we studied the effect of annealing on electrical properties of ZnO films. It is seen that oxygen annealing strongly affects the electrical properties. The resistivity of the films increases continuously with increase in annealing time. The resistivity of the film rises from 6.5 $\times$ 10^{-3} Ωcm to 8.2 $\times$ 10^{1} Ωcm from no annealing to 60 min annealing in ambient oxygen respectively (Figure 4). On the other hand, the carrier density is observed to decrease with annealing time. The carrier concentration drops from 10^{20}cm^{-3} to 10^{15} cm^{-3}, as the samples pass through annealing. On the other hand, mobility of an annealed sample climbs to 0.92 cm^2 v^{-1} s^{-1} from 0.064cm2 v^{-1} s^{-1}, mobility of an as grown sample.

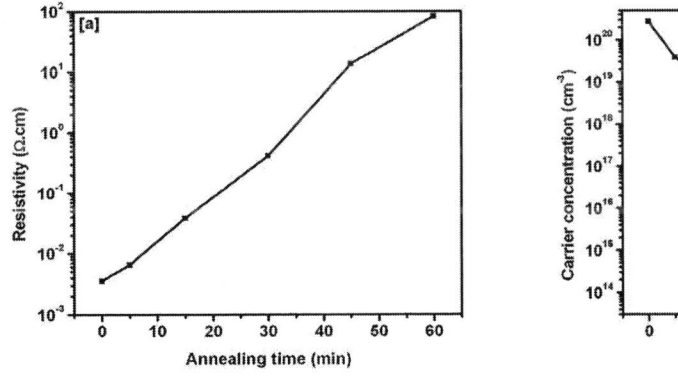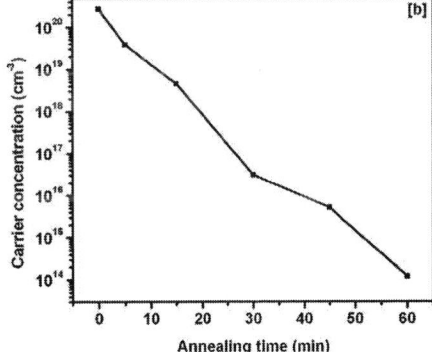

Fig. 4 Effect of annealing time on (a) resistivity and (b) carrier concentration of ZnO film.

The final outcome of the optimizations done so far shows the gradual change from an ohmic to rectifying contact (Figure 5). All the way from Figure 5(a) to Figure 5(d) annealing duration is prolonged, which reduces the carrier concentration. By careful observation of the applied bias and corresponding rise in current, it has been found that the Ag-ZnO junction acts as

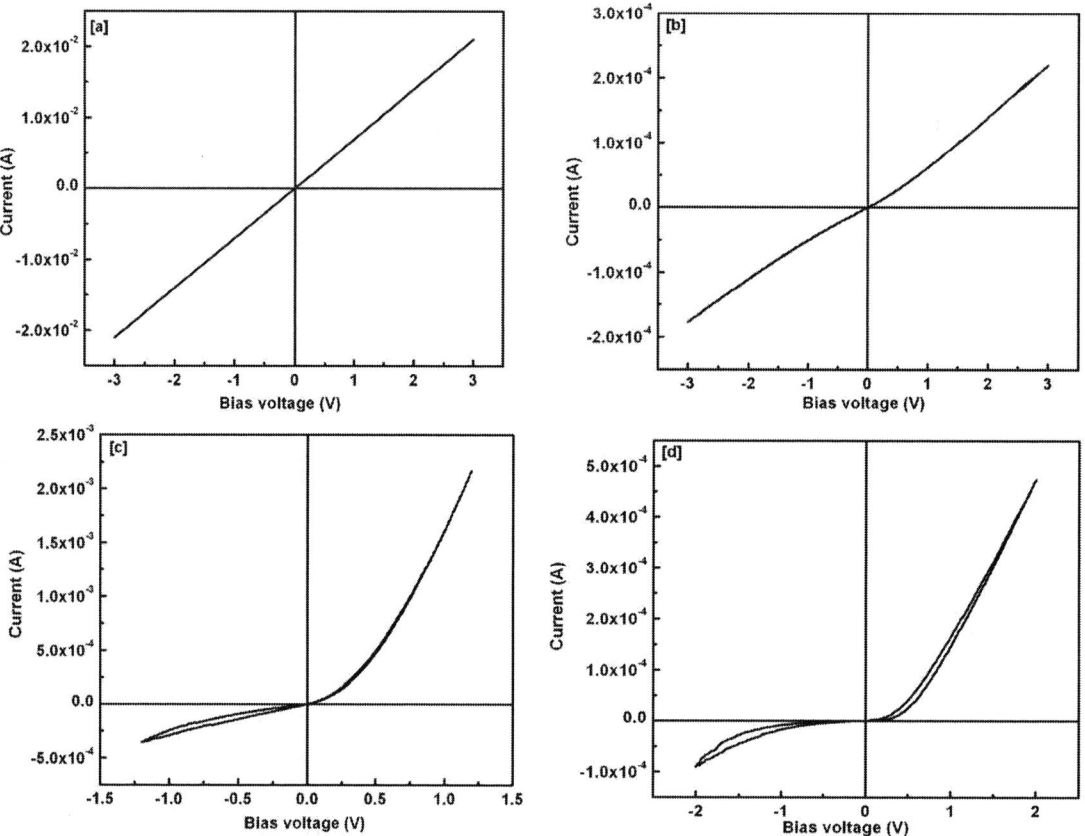

Figure 5. Effect of annealing times (a) 15 min, (b) 30 min, (c) 45 min, and (d) 60 min on current-voltage characteristics of Au/ZnO/Ag junctions.

a rectifying one in our structure. Ag even in oxide state also forms good rectifying contact. The Au contact on ZnO is used as an Ohmic back contact. The nature of carrier transport in the rectifying junction, guided by thermionic emission follows the expression below [15]

$$I = I_s \left[exp \left(q \left(V - IR_s \right) / KT \right) - 1 \right]$$

where I_s is reverse saturation current, V is applied voltage, and R_s is the series resistance. The reverse saturation current is give as

$$I_s = A^* T^2 \, exp \left(-q \phi_b / KT \right)$$

where A^* is the effective Richardson constant, which has a value of 32 A cm^{-2} K^{-2} for ZnO. The potential barrier developed at the junction is denoted as ϕ_b. The reverse saturation current lies nearly to 2.5×10^{-4} A at -1.5 V for the sample annealed for 45 min, as shown in Figure 5(c) which has some effect of leakage current. The saturation current further improves to 4.5×10^{-5} A for the 60 min annealed sample (Figure 5(d)). But still there lies ample room for improvement in saturation current density, as the junction suffers by significant contribution from leakage current. Extensive annealing can improve this problem but at the cost of additional series resistance [12]. The barrier voltage developed, is found to be nearly 0.40 V.

CONCLUSIONS

Thin films of ZnO were grown using pulsed laser deposition technique. The effect of annealing on structural, optical, and electrical properties were discussed in details. The film crystallinity and optical transparency increases with increase in annealing time. The electrical properties of the ZnO films strongly depend on annealing duration in oxygen atmosphere. This process effectively brings the carrier concentration moderately low to an order of 10^{17} cm^{-3} or lower which aids in avoiding the pinning of Fermi level. The reduction of defects, present inherently in ZnO films reduces the contribution of tunneling current component over thermionic emission current component. In other words, the control over carrier concentration substantially guides the formation of Schottky barrier diode

REFERENCES

[1] G. Newman, Phys. Status Solidi B 105 (1981) 605.

[2] R.K. Gupta, K. Ghosh, S.R. Mishra, P.K. Kahol, Mater. Res. Soc. Symp. Proc. 1035E (2007) 1035-L-11-17.

[3] A.Y. Polyakov, N.B. Smirnov, E.A. Kozhukhova, V.I. Vdovin, K. Ip, Y.W. Heo, D.P. Norton, S.J. Pearton, Appl. Phys. Lett 83 (2003) 8.

[4] H.L. Mosbacker , S.E. Hage, M. Gonzalez, A. Ringel, M. Hetzer, D.C. Look, G. Cantwell, J. Zhang, J.J. Song, L.J. Brillson, J. Vac. Sci. Technol. B 25 (2007) 421.

[5] K. Ip, Y.W. Heo, K.H. Baik, D.P. Norton, S.J. Pearton , S. Kim, J.R. LaRoche, F. Ren, Appl. Phys. Lett. 84 (2004) 15.

[6] Q.L. Gu, C.K. Cheung, C.C. Ling, A.M.C. Ng, A.B. Djurisic, L.W. Lu, X.D. Chen, S. Fung, C.D. Beling, H.C. Ong, J. Appl. Phys. 103 (2008) 093706.

[7] M.W. Allen, S.M. Durbin, J.B. Metson, Appl. Phys. Lett. 91 (2007) 053512.

[8] H.L. Mosbacker, Y.M. Strzhemechny, B.D. White, P.E. Smith, D.C. Look, D.C. Reynolds, C.W. Litton, L.J. Brillson, Appl. Phys. Lett. 87 (2005) 012102.

[9] H.L. Mosbacker, S.E. Hage, M. Gonzalez, S.A. Ringelb, M. Hetzer, D.C. Look, G. Cantwell, J. Zhang, J.J. Songd, L.J. Brillson, J. Vac. Sci. Technol. B 25 (2007) 4.

[10] L.J. Brillson, H.L. Mosbacker, M.J. Hetzer, Y. Strzhemechny, G.H. Jessen, G. Cantwell, J. Zhang, J.J. Song, Appl. Phys. Lett. 91 (2007) 102116.

[11] S.H. Kim, H.K. Kim, T.Y. Seong, Appl. Phys. Lett. 86 (2005) 112101.

[12] M.N. Oh, D.K. Hwang, J.H. Lim,,Y.S. Choi, S.J. Park, Appl. Phys. Lett. 91 (2007) 042109.

[13] X.D. Zhoua, W. Huebner, Appl. Phys. Lett. 79 (2001) 21.

[14] H.C Pan, M.H. Shiao, C.Y Su, C.N. Hsiao, J. Vac. Sci. Technol. A 23 (2005) 4.

[15] S.M. Sze, Physics of Semiconductor Devices, 2nd ed. Wiley, New York, 1981.

Mater. Res. Soc. Symp. Proc. Vol. 1109 © 2009 Materials Research Society

Luminescence of ZnO Thin Films Grown on Glass by Radio-Frequency Magnetron Sputtering

K. Liu and M. Shur

Department of ECE and CIE, Rensselaer Polytechnic Institute, Troy, NY 12180, USA

G. Tamulaitis

Semiconductor Physics Department and Institute of Materials Science and Applied Research, Vilnius University, Saultekio 9-III, 10222 Vilnius, Lithuania

S. Cho

Department of Electronic Materials Engineering, Silla University, Busan 617-736, Korea

ABSTRACT

We investigate polycrystalline ZnO thin films grown on glass by rf magnetron sputtering with different O_2 to $Ar + O_2$ ratio. XRD patterns and AFM images showed that the strongest (0002) orientation and the flattest surface of the ZnO layer was achieved in the sample grown with the highest flow ratio (50%). At low temperature, bound exciton was prevailing in the photoluminescence (PL) spectra under high excitation intensity, while a broad band peaked at 3.10 eV was dominant at low excitation. Under short pulse excitation at room temperature, PL band blue shifted and became much narrower with increasing the O_2 / $Ar+O_2$ ratio. However, PL decay kinetics had no significant dependence on the flow ratio. The PL kinetics had very fast initial decay with effective decay time below the time resolution in our experiment (~30 ps), which was followed by a slow decay. The lifetime of nonequilibrium carriers was found to be ~500 ps by using light-induced transient grating technique.

INTRODUCTION

The recent shift in technology and applications of optoelectronic devices to shorter wavelengths enhances interest in science and technology of zinc oxide, a wide bandgap semiconductor ($E_g \approx 3.37$ eV at 300 K), which has previously been used mainly as an optically transparent material for conducting window layers in solar cells and flat-panel displays. A large exciton binding energy (60 meV) introduces interesting features in optical properties of ZnO. ZnO has some advantages over GaN, which currently takes a lead in short wavelength optoelectronics, especially for its inexpensive growth technologies of epilayers and bulk crystals. In addition to such techniques for growth of ZnO epilayers as molecular beam epitaxy and metalorganic chemical vapor deposition, which ensure especially high quality of ZnO epilayers, more simple and less expensive ZnO layer deposition techniques as rf-sputtering [1] and pulsed laser deposition (PLD)[2] are of considerable interest in view of possible applications.[3] However, the deposition technology for these techniques is still in the process of optimization. The most common substrate for deposition of ZnO layers is sapphire, though ZnO growth on GaAs[4], Si[5], SiC[6], and CaF_2[7] is also reported. Meanwhile, glass is the cheapest substrate for

ZnO and has been extensively used for transparent conducting, photoconducting, and other functional ZnO films[8].

We report on deposition and study of a set on ZnO epilayers grown on glass by rf magnetron sputtering technique under different flow ratios of O_2 and $Ar+O_2$. Photoluminescence (PL) and time-resolved photoluminescence, as well as light-induced transient grating technique were used for characterization. XRD and AFM were used for structural analysis.

EXPERIMENTAL

ZnO films under study were grown on glass substrates (Corning 7059) by a rf sputtering technique at a working pressure of 35 mtorr and at substrate temperature of 400°C. A two inch-diameter ZnO target was synthesized by a conventional solid-state reaction using ZnO powders with a purity of 99.999%. The ZnO target was sputtered in gas mixture at O_2 to $Ar+O_2$ ratios of 0, 10%, 30%, and 50%. The rf power was fixed at 70 W and the distance between the target and the substrate was 60 mm.

In quasi-steady-state PL study, ZnO films were excited by the 4th harmonic (266 nm) of Q-switched Nd: YAG laser radiation (pulse duration 4.4 ns) in back scattering geometry. The spectra were recorded by using a UV-enhanced intensified charge coupled device (ICCD) camera and a TRIAX 550 spectrometer. A closed cycle helium cryo-system was used to vary and maintain the sample temperature in the range from 12 to 300 K. In time-resolved PL measurements, ZnO films were excited by the 4th harmonic (266 nm) of the mode-locked YAG:Nd laser radiation (pulse duration 30 ps). A *Hamamatsu* streak camera (time resolution of 2 ps) was used to record the PL decay in time.

In light-induced transient grating (LITG) experiments, two beams of the 4th harmonic (266 nm) of the mode-locked YAG:Nd laser radiation overlapped at the sample surface and created a thin grating via band-to-band carrier excitation. Variably delayed pulses of the 2nd harmonic (532 nm) probed the decay of the grating in the transparency region of the structure. The decay of the transient grating is caused by carrier recombination and diffusion. The grating decay time, τ_G, can be expressed via the carrier lifetime τ_R, the diffusion coefficient D, and the grating spacing Λ as[9]:

$$\frac{1}{\tau_G} = \frac{1}{\tau_R} + \frac{4\pi^2 D_a}{\Lambda^2}$$

(1)

We selected the grating spacing large enough for the diffusion term to be neglected, thus, τ_G was a good approximation for τ_R.

RESULTS AND DISCUSSIONS

The structural properties of ZnO films were characterized by XRD and AFM measurements. Figure 1 shows XRD patterns for the ZnO thin films grown at four different $O_2/Ar + O_2$ ratios. Only two diffraction peaks were observed in XRD patterns of all the ZnO films under this study. The main peak occurs at 34.3° in correspondence with the diffraction from the ZnO (0002) plane. The second peak centered at 72.4° represents diffraction from the ZnO (0004) plane. The XRD patterns indicate preferential growth of ZnO grains oriented along the *c*-axis, i.e. perpendicular to the substrate surface. However, as $O_2 / Ar+O_2$ flow ratio was decreased down to zero, the relative intensities and FWHM of the ZnO (0002) diffraction peak gradually decreased, indicating degradation of film crystallinity. The $O_2 / Ar+O_2$ flow ratio has no significant effect on the crystal orientation, but affects the crystallinity of ZnO films. AFM images showed that the root mean square (RMS) value decreased from 31 nm to 0.3 nm for the samples grown at $O_2 / Ar+O_2$ ratio of 0 and 50%, respectively. The large surface roughness of the samples grown under pure Ar might be an explanation of the dramatically decreased intensity of the (0002) diffraction peak in XRD pattern[10].

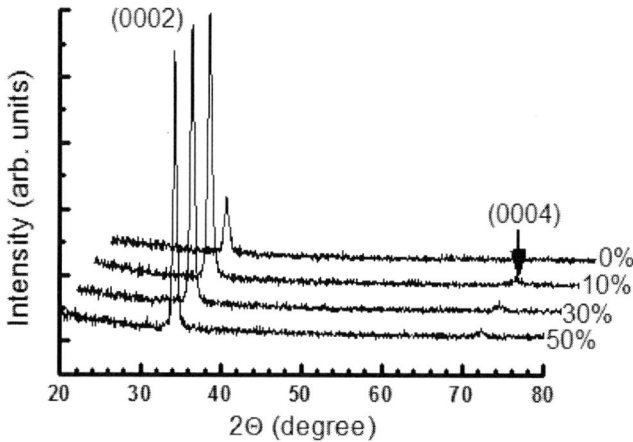

Figure 1. XRD patterns of ZnO thin films grown at different O_2 flow ratios (indicated).

The PL spectrum under quasi-steady-state excitation at low temperatures strongly depends on excitation power density (see Figure 2a). At low excitation, a broad band peaked at 3.10 eV is prevailing in the spectrum. Saturation of the broad band, as well as its spectral position indicate that defect-related levels are involved into the recombination responsible for this band. With increasing pump intensity, two narrow bands on the high energy slope of the broad band become dominant. Position of the first narrow band peaked at 3.375 eV is close to the position of the band, which has been observed in the spectra of high-quality bulk ZnO crystal at 3.378 eV and was interpreted as being caused by free A exciton (FX_A) recombination[11]. Another, more intense band, which is located by 40 meV to the low energy side from the free exciton band and has FWHM of 12 meV, can be attributed to recombination of a neutral donor-bound exciton (D^0X), since binding energies of up to 50 meV are reported [12]. The inhomogeneous broadening of the two PL bands is probably caused by the presence of defect levels related to point defects, such as oxygen vacancies and zinc interstitials.

Spectra at room temperature do not show significant dependence on excitation power density (see Figure 2b). A single broad band is peaked at 3.24 eV, as observed in many previous publications [13, 14].Though the FWHM value of the PL band at room temperature is quite broad (~260 meV), we do not have any additional PL bands related to deep levels, as it has been observed by Jeong *et al* [15] in the PL spectra of ZnO films grown by rf magnetron sputtering under different $O_2 / Ar+O_2$ ratio on Si (100). This is an indication of good quality of our ZnO films. Evolution of the PL spectra with increasing temperature is demonstrated in Fig. 2c. The D^0X peak can be tracked up to 150 K and redshifts slightly with elevated temperature. The band at 3.26 eV appears with increasing temperature, becoming a dominant band above 100 K. This band can be assigned to phonon relica of the D^0X band. The energy of LO-phonon in ZnO is 72 meV [16], which equals the energy split between the D^0X peak and the 3.26 eV peak in our sample (~75 meV). The lines broaden and the PL spectrum transforms into one broad band as the temperature is elevated above ~200 K.

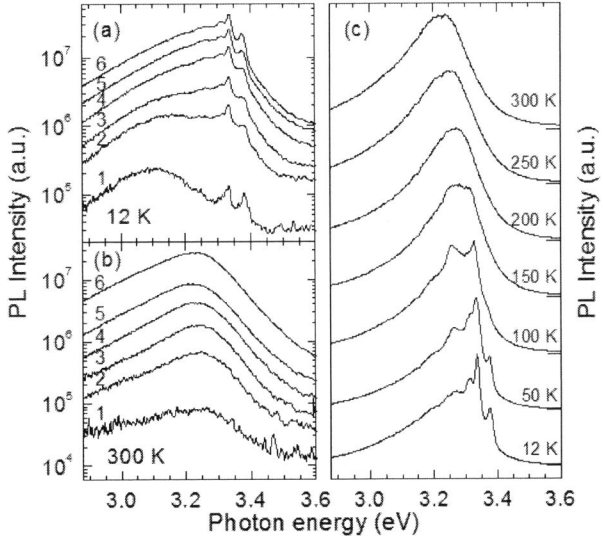

Figure 2. PL spectra under different excitation power densities: 4 (1), 13 (2), 26 (3), 52 (4), 100 (5), and 210 MW/cm^2 (6) at room temperature (a) and 12 K (b), and PL spectra at 210 MW/cm^2 and different temperatures (c) for ZnO sample deposited under O_2 / $Ar+O_2$ flow ratio of 30%. The spectra in (c) are vertically shifted for clarity.

Spectra of three samples grown under different O_2 / $Ar+O_2$ flow ratios are presented in Fig. 3a. These spectra are recorded under intense photoexcitation by pulsed picosecond laser, so the defect-related recombination channels have to be saturated. In this case, a notable blue shift with increasing O_2 / $Ar+O_2$ ratio was observed, accompanied with increasing PL intensity and decreasing FWHM. The dependence of the PL spectra on the O_2 / $Ar+O_2$ ratio indicates that the PL spectrum is related to the stoichiometry of the ZnO film. The ZnO film grown in pure Ar ambient shows very weak PL emission, presumably due to nonstoichiometric composition. The oxygen vacancy is a shallow intrinsic donor in ZnO. With increasing oxygen flow ratio, the stoichiometry is improved and the density of point defects such as oxygen vacancy and Zn interstitials is reduced. As a result, the PL band blueshifts and FWHM of the band decreases.

Carrier dynamics of ZnO thin films were studied by time-resolved PL and LITG techniques. PL decay kinetics is very similar in all three samples (see PL kinetics for one sample presented in Figure 3b). A fast decay with effective decay time below the time resolution in our experiment (~40 ps) is observed after the excitation. After the fast initial decay, the PL intensity decreases at a considerably smaller rate. If approximated by an exponent, the decay has a characteristic time of approximately 220 ps. Though the spectrum obtained in our experiments at the second decay stage is quite noisy, we can conclude that the spectrum does not change considerably in time. This indicates the dominance of the same recombination mechanism througout the entire PL decay process.

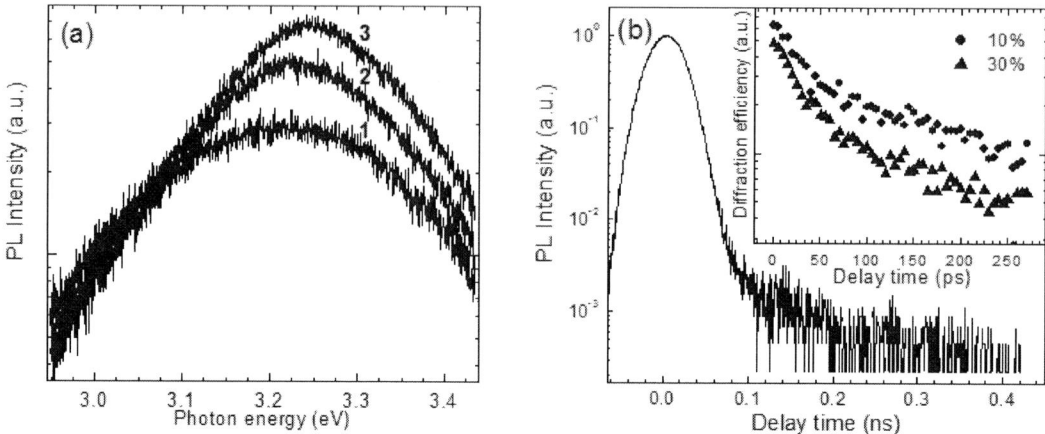

Figure 3. (a) Time-integrated PL spectra of ZnO samples deposited under O_2 / $Ar+O_2$ flow ratios of 10% (1), 20% (2) and 30% (3) and (b) typical spectrally integrated PL kinetics of the samples. Insert: decay kinetics of the light induced transient gratings for samples deposited under the flow ratios of 10% (circle) and 30% (triangle).

Results obtained by using LITG technique are presented in the insert of Figure 3b for two samples. The results for the third sample are not reliable due to a high level of light scattered in this sample and its surface. The LITG experiment was carried out at grating spacing large enough for the diffusion influence to be neglected. Thus, the decay is determined by the carrier lifetime. The decay proceeds faster at the initial decay stage and approaches approximately an exponential decay afterwards. The characteristic PL decay time at the late decay stage equals ~200 ps. Since PL intensity for band-to-band recombination is proportional to carrier density squared, the carrier lifetime should be approximately 400 ps. LITG experiments provided a higher accuracy. The diffraction efficiency decay (see inset in Fig. 3b) also has two stages: a fast initial decay and a slower decay corresponding to carrier lifetime of ~500 ps. This carrier lifetime is comparable to the longest carrier lifetimes in ZnO layers grown on sapphire by PLD under different growth temperature [17] and is longer than that of ZnO layer grown on Si by MOCVD [18]. Note, however, that the initial fast decay both in PL intensity and LITG efficiency might reflect fast capturing of nonequilibrium carriers to the centers of nonradiative recombination, while the subsequent slow decay might be determined by activation of trapped carriers. In this case, a long lifetime does not necessarily mean that radiative recombination is strong in respect to nonradiative one.

CONCLUSIONS

Study of ZnO epilayers deposited on glass by rf magnetron sputtering technique under different $O_2/Ar+O_2$ flow ratio shows the importance of the flow ratio for quality of the layers. Increase in ratio of oxygen flux up to 50% results in increasing structural quality and light emission efficiency, as evidenced by XRD patterns, surface roughness and PL properties of the ZnO layers under study. The observed optical properties of ZnO films also demonstrate the potential of rf magnetron sputtering on glass as a low cost technology for the deposition of comparatively high quality polycrystalline ZnO films.

REFERENCES

[1] K.-K. Kim, J. H. Song, H. J. Jung, S. J. Park, J. H. Song, and J. Y. Lee, *J. Vac. Sci. Technol.* **A18**, 2864 (2000).

[2] V. Craciun, J. Elders, J.G.E. Gardeniers, and I.W. Boyd, *Appl. Phys. Lett.* **65**, 2963 (1994).

[3] U. Ozgur, Ya.I. Alivov, C. Liu, A. Teke, M.A. Reshchikov, S. Dogan, V. Avrutin, S.-J. Cho, and H. Morkoc, *J. Appl. Phys.* **98**, 041301 (2005).

[4] Y. Ma, G.T. Du, J. Z. Yin, T.P. Yang and Y. T. Zhang, *Semicond. Sci. Technol.* **20**, 1198-1202 (2005).

[5] J.Y. Huang, Z.Z. Ye, H.M. Lu, L.Wang, B.H. Zhao and X.H. Li, *J. Phys. D: Appl. Phys.* **40**, 4882–4886 (2007).

[6] A. B. M. A. Ashrafi, N. T. Binh, B. P. Zhang, and Y. Segawa, *J. Appl. Phys.* **95**, 7738-7741 (2004).

[7] Y. Ma, G.T. Du, X. Wang, W.C. Li, J.Z. Yin, D.L. Qiu, B. Song, X. Zhang, Y.T. Zhang and D.L. Liu, *Appl. Surf. Sci.* **243**, 24-29, (2005).

[8] X. Jiang, F. L. Wong, M. K. Fung, and S. T. Lee, *Appl. Phys. Lett.* **83**, 1875 (2003)

[9] A. Miller, in: Nonlinear Optics in Semiconductors II, edited by E. Garmire and A. Kost, Semiconductors and Semimetals, Vol. 59 (Academic, New York, 1999).

[10] E. Fortunato, P. Nunes, D. Costa, D. Brida, I. Ferreira and R. Martins, *Vacuum* **64** 233–236 (2002).

[11] D. W. Hamby, D. A. Lucca, M. J. klopfstein, and G. Cantwell, *J. Appl. Phys.* **93**, 3214 (2003).

[12] B. K. Meyer H. Alver, D.M. Hofmann, W. Kriegseis, D. Forster, F. Bertram, J, Christen, A. Hoffmann, M. Straburg, M. Dworzak, U. Haboeck, A.V. Rodina, *phys. stat. sol. (b)* **241**, 231 (2004).

[13] F.K. Shan, G.X. Liu, W.J. Lee, G.H. Lee, I.S. Kim, B.C. Shin, Y.C. Kim, *J. Crys. Grow.* **277**, 284–292 (2005).

[14] P. Sagar, M. Kumar, R. M. Mehra, H. Okada, A. Wakahara, A. Yoshida, *Thin Soli Films*, **515**, 3330 (2007).

[15] S. Jeong, B. Kim, and B. Lee, *Appl. Phys. Lett.*, **82**, 2625 (2003).

[16] D. C. Reynolds, D. C. Look, B. Jogai, R. L. Jones, C. W. Litton, W. Harsch, and G. Cantwell, *J. Lumin.* **82**, 173 (1999).

[17] S. Cho, S. I. Kim, Y. H. Kim, J. Micevičius, G. Tamulaitis, and M. S. Shur, *phys. Sta. Sol (a)*, **203**, 3699 (2006).

[18] B. Guo, Z. R. Qiu, K. S. Wong, *Appl. Phys. Lett.*, **82**, 2290 (2003).

First-principles calculation for effect of impurities on electronic states
of amorphous In-Ga-Zn-O

[1]Hideyuki Omura, [2]Tatsuya Iwasaki, [2]Hideya Kumomi, [3]Kenji Nomura, [3,4]Toshio Kamiya, [3,5]Masahiro Hirano, and [3,4,5]Hideo Hosono

[1]Analysis Technology Development Center, Canon Inc., 3-30-2 Shimomaruko, Ohta-ku, Tokyo 146-8501, Japan

[2]Materials Technology Development Center, Canon Inc., 3-30-2 Shimomaruko, Ohta-ku, Tokyo 146-8501, Japan

[3]ERATO-SORST, JST, in Frontier Research Center, Tokyo Institute of Technology, 4259, Nagatsuta, Midori-ku, Yokohama 226-8503, Japan

[4]Materials and Structures Laboratory, Tokyo Institute of Technology, 4259, Nagatsuta, Midori-ku, Yokohama 226-8503, Japan

[5]Frontier Research Center, Tokyo Institute of Technology, 4259, Nagatsuta, Midori-ku, Yokohama 226-8503, Japan

ABSTRACT

This paper presents theoretical investigation on the effect of hydrogen impurity in amorphous In-Ga-Zn-O (a-IGZO) based on the density functional theory (DFT). After a structure model of pure a-IGZO was modeled by classical molecular dynamics (MD) simulations, the model was relaxed by DFT calculations using the projector-augmented wave method at the generalized-gradient-approximation level. In all the a-IGZO:H models examined, the hydrogen atoms form O-H bonds with oxygen ions. The addition of hydrogen raises the Fermi level above the conduction band minimum and does not form an in-gap state. This result suggests that hydrogen acts as a donor in a-IGZO and increases the carrier density in a-IGZO, which have actually observed in a-IGZO TFTs.

INTRODUCTION

Since the first demonstration [1], amorphous indium-gallium-zinc oxide (a-IGZO) has attracted keen attention as a new channel material for high-performance and low-temperature thin-film transistors (TFTs). Amorpuous IGZO thin films are transparent for visible light [2] and are formed by sputtering deposition at room temperature [3,4]. The TFTs exhibit large electron mobilities over $10 \text{ cm}^2\text{V}^{-1}\text{s}^{-1}$, small subthreshold voltage swing [3,4], good short-range uniformity because of the amorphous nature [5], and electrical stability better than hydrogenated amorphous Si (a-Si:H) TFTs [6]. These features make a-IGZO TFTs a promising candidate for TFTs in active-matrix back planes for future flat-panel displays. Actually, organic light-emitting diode displays [7-10], liquid-crystal displays [11], and electronic papers [4, 12] have been demonstrated using a-IGZO TFTs.

To apply a-IGZO TFTs to commercial products, intrinsic properties and limiting factors of the material, such as behaviors of defects and impurities, should be understood well. It would be possible a-IGZO films contain more impurities than conventional semiconductor materials do because of impurities in a polycrystalline sputtering target and residual hydrogen-containing molecules in a deposition chamber. Stability test against an ambient atmosphere revealed that water vapors adsorbed on an exposed a-IGZO surface affected the transistor properties [13]. Also it is recently reported in the first demonstration of coplanar homojunction TFTs [14] that conductivity of an a-IGZO thin film is significantly increased by plasma-enhanced chemical-vapor-deposition of a SiN_X:H film onto an a-IGZO layer and subsequent thermal annealing, where the hydrogen plasma attacks the a-IGZO and the hydrogen atoms diffuse into the a-IGZO. In other transparent semiconductors such as ZnO [15] and amorphous $2CdO$-GeO_2 [16], hydrogen ion implantation into these films enhances their conductivity as if the implanted hydrogens worked as shallow donors. These facts suggest that incorporation of hydrogen related species would influence threshold voltages, on-to-off current rations, and stability of TFTs using these oxides including a-IGZO for channels. However, there have been few theoretical approaches to the effect of impurities in the a-IGZO.

This paper presents a theoretical investigation on the effect of hydrogen impurity in the a-IGZO based on density functional theory (DFT). The structure relaxation calculations revealed that the hydrogen atoms always make O-H bonds to network oxygen ions in a-IGZO. The electronic structure calculations revealed that the addition of a hydrogen atom raises the Fermi level above the conduction band maximum (CBM). These results suggest that incorporation of hydrogen influences the threshold voltage and the on-to-off current ratio in a-IGZO TFTs, which would be a reason why electrical characteristics of a-IGZO TFTs are sensitive to their exporsure to hydrogen-related species such as H_2.

THEORY AND CALCULATION MODELS

At first, an a-IGZO structure model was built by classical molecular dynamics simulations (MD), which will be used as an initial model for subsequent first-principles structure relaxation calculations. We employed a crystalline $InGaZnO_4$ MD cell containing 2016 atoms and performed constant temperature, constant pressure MD simulations with 1 fs times steps using the module OFF of Cerious2 MSI software package. Empirical two-body interionic potentials of the Buckingham type were employed. The empirical parameters were adjusted to reproduce the crystal structure of $InGaZnO_4$ and the density of the a-IGZO films (6.1 g/cm^3). The 2016 atoms cell was first melted at 8000 K and then cooled to 12.5 K at a rate of 125 $K \cdot ps^{-1}$. Another 100 ps MD steps were performed at 1 K to obtain a relaxed amorphous structure. Next, the cell size was reduced to 84 atoms cell to fit the cell size to our computational capability for the first-principles calculations. The 84 atoms MD cell was again melted and cooled to obtain a MD-relaxed amorphous structure. We confirmed that the density (5.993 g/cm^3) and local coordination structure of this model were essentially the same as that obtained on the amorphous structure of the 2016 atoms MD cell. We also confirmed that the distances of the oxygen - metal bonds were consistent with those measured by extended X-ray absorption fine structure on K edges of metals: In-O = 0.214 nm, Ga-O = 0.187 nm and Zn-O = 0.201 nm.

Then variable-cell relaxation for the relaxed amorphous structure of the 84 atoms MD cell was performed to obtain a quantum-mechanically stable structure by the DFT calculations based on the projector augmented wave (PAW) method at the generalized-gradient-approximation level using the PBE functional [17] with a code VASP [18-20]. Based on the 84 atoms DFT-relaxed cell of a-$InGaZnO_4$ (pure a-IGZO) (Fig. 1(a)), several initial structures of hydrogen-doped a-IGZO (a-IGZO:H) were built by adding a hydrogen atom to the a-IGZO model at different positions such as anti-bonding sites, bond center sites (Fig. 1(b)), and interstitial sites (Fig. 1(c)). The initial hydrogen position for the anti-bonding site model was chosen to be 0.15 nm apart from an ion forming an oxygen – metal bond on the opposite side of the bond. For the a-IGZO:H models, only atomic positions were relaxed with the fixed cell parameters.

In all the DFT calculations, the PAW method with the frozen-core approximation [21,

22] was used for the ion-electron interactions. The PAW potentials explicitly treat 13 valence electron configurations for In ($4d^{10}$, $5s^2$, $5p^1$) and Ga ($3d^{10}$, $4s^2$, $4p^1$), 12 for Zn ($3d^{10}$, $4s^2$), and 6 for O ($2s^2$, $2p^4$). The atoms are steadily relaxed toward equilibrium until the Hellmann–Feynman forces have become less than 0.001 eV/nm during all the DFT relaxations. After non-spin-polarized calculations were performed for the first trial DFT relaxation to suppress the calculation cost, spin-polarized DFT relaxations were carried out. The conditions for the plane-wave cutoff energy and Monkhost-Pack special k points were 400 eV and $2\times2\times3$, respectively, in the relaxation calculations. Then, ground-state SCF calculations were carried out using the cutoff energy of 400 eV and $3\times3\times4$ special k points. The density of states (DOS) were calculated using $4\times5\times6$ special k points. In both the calculations, spin-polarized calculations were employed.

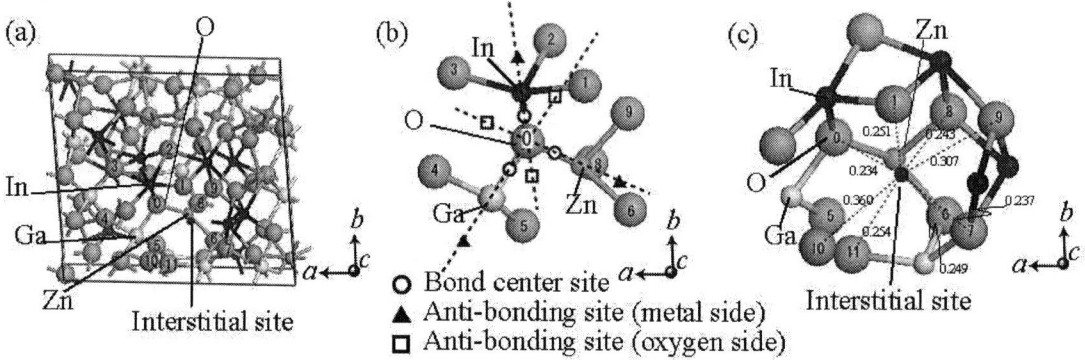

Figure 1.

(a) The 84 atoms DFT relaxed cell of pure a-IGZO. The numbers in the gray circles represent the index numbers of the oxygen ions in Figs. 1 (b) and (c).

(b) A local atomic structure of a bond center site and an anti-bonding site of an oxygen ion extracted from Fig. 1 (a). The open circles represent the bond center sites. The open squares and closed triangles represent the anti-bonding sites at oxygen and metal sides, respectively. Hydrogen atoms were located at anti-bonding sites which are 0.150 nm apart from these metal and oxygen atoms in the initial a-IGZO:H structures.

(c) A local atomic structure of an interstitial site extracted from Fig. 1(a). The numbers along the dotted lines represent distances between oxygen atoms and the interstitial site used for the initial models.

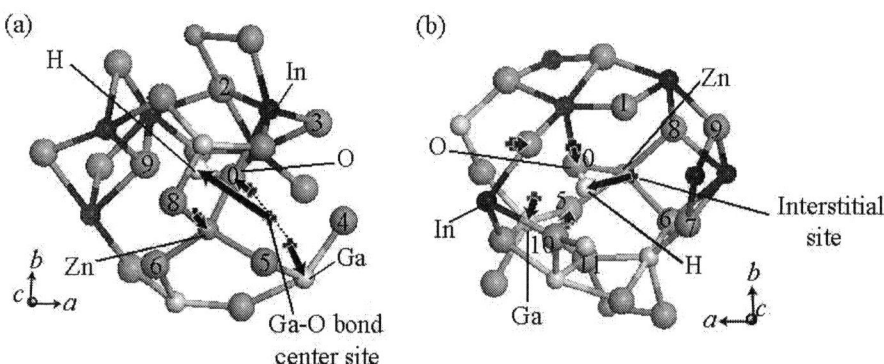

Figure 2.
Two examples of the relaxed local atomic structures of the *a*-IGZO:H models. The initial structure of (a) and (b) are the Ga-O bond center site and the interstitial site, respectively. In these figures, the arrows and the crosses represent the displacement vectors and the initial positions of atoms, respectively. The numbers in the gray circles corresponds to those in Figs. 1 (b) and (c).

RESLTS AND DISCUSSION
Atomic structure

For all the relaxed *a*-IGZO:H models, the hydrogen atoms move toward oxygen ions and finally form O-H bonds with the bond lengths ranging from 0.098 nm to 0.105 nm. Figure 2 (a) and (b) show the local atomic structures of the relaxed *a*-IGZO:H models starting from the Ga-O bond center site and the interstitial site, respectively. The formation of the O-H bond was observed even when the hydrogen is located sufficiently away from oxygen ions in the initial structure. After relaxation calculations, the hydrogen shows the largest displacement and bonds to an oxygen ion positioned at the nearest distance of the hydrogen in the initial structures. Cations positioned near the relaxed hydrogen move outward from the hydrogen. On the other hand, the oxygen ion positioned near but not bonded with the hydrogen moves toward for the hydrogen inward (see Figs. 2(a) and (b)). In Fig. 2(a) the Ga ion of the Ga-O bond shows large displacement because the Ga-O bond is broken and the Ga ion bonds to another oxygen ion (not shown in the figure).

Electronic structure

Figure 3(a) shows total DOSs (TDOSs) and projected DOSs (PDOSs) of the pure *a*-IGZO. The vertical dashed line at 0.80 eV represents conduction band minimum (CBM) of pure *a*-IGZO. Figure 3(b) shows an expanded view of Fig. 3(a) around 1.0 eV. The conduction band is composed of *s*-states of all the metal ions and *s*- and *p*-states of the oxygen ion where the contributions from In *s*- and O *s*- and *p*- states are dominant. This is well consistent with the previous results obtained by Nomura *et al.* [23].

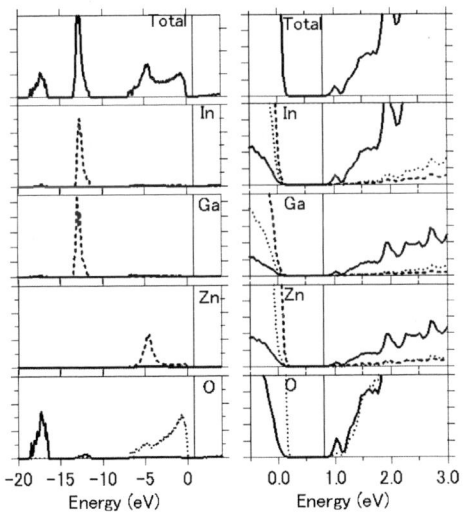

Figure 3.
(a) Total and projected density of states of pure *a*-IGZO. (b) Expanded view of (a). The eigenvalues were broadened by a Gaussian function with a 0.05 eV full width at half maximum. In these figures, the energies are measured from the valence band maximums (VBM), and the vertical solid lines indicate the eigenvalues of conduction band minimums (CBM).

For all of DOSs of the relaxed a-IGZO:H models, the Fermi levels lie above the CBM (here, all of these Fermi levels correspond to the highest occupied states). Figure 4 shows TDOSs [(a) and (c)] and PDOSs of hydrogen [(b)] of the relaxed a-IGZO:H models in the Ga-O bond center site and interstitial site,where Fig. 4 (c) shows the expanded views of Figs. 4 (a). In the TDOSs in Fig. 4(a), it seems that hydrogen does not change the TDOSs from those of the pure a-IGZO (Fig. 3(a)). However, as seen in the PDOSs of hydrogen of Fig.4 (b) the peaks observed around –19 eV and –7 eV correspond to O $2s$ and O $2p$ bands of the pure a-IGZO (Fig. 3(a)), which indicate that hydrogen s orbital is hybridized with O $2s$ and O $2p$ orbital. As shown in Fig. 4 (c), for all the relaxed a-IGZO:H structures, the Fermi levels lie above the CBM and do not form any in-gap state. It is suggested that the added hydrogen raises the Fermi level above the CBM, acts as a donor in a-IGZO and increases the carrier density of the a-IGZO.

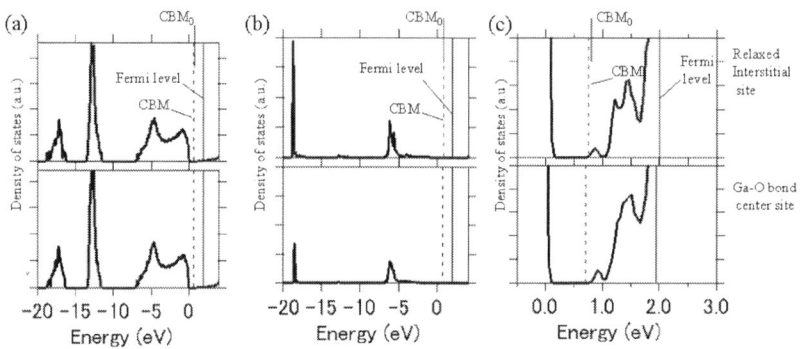

Figure 4.
(a) Total density of states of a-IGZO:H models. (b) Projected density of states of hydrogen atom in the a-IGZO:H models. (c) Expanded view of (a). The eigenvalues were broadened by a Gaussian function with a 0.05 eV full width at half maximum. The energies of VBM of a-IGZO:H are aligned to and measured from VBM of the pure a-IGZO moidel. Vertical doted lines indicate CBM. CBM$_0$ represents CBM of the pure a-IGZO. The solid lines indicate the Fermi levels which correspond to the highest occupied band.

CONCLUSIONS

The effects of hydrogen impurity in the a-IGZO is investigated, based on DFT calculations. For all of the DFT relaxed a-IGZO:H models, the hydrogen forms O-H bonds even if the hydrogen atom is located far from an oxygen ion in the initial structure. The addition of hydrogen raises the Fermi level above the CBM and does not form an in-gap state. This result suggests that the hydrogen bonding to an oxygen ion acts as a donor in a-IGZO and increases the carrier density of a-IGZO layers in the TFTs. These results provide important suggestions that hydrogen can influence the threshold voltage and the on-to-off current ratio, and this would be a reason why electrical characteristics of a-IGZO TFTs are sensitive to their exporsure to hydrogen-related species such as H_2.

ACKNOWLEDGMENTS

The authors are grateful to T. Noda and K. Tanaka for their technical assistance and to M. Shimada, T. Watanabe, M. Watanabe, T. Aiba, N. Itagaki, G. Amita, S. Yaginuma, T. Shoyama, S. Suzuki, K. Takahashi, M. Ofuji, N. Kaji, Y. Tateishi, H. Shimiz, H. Yabuta, A. Sato, M. Sano, K. Abe, R. Hayashi, S. Miyazawa, and M. Okuda for their valuable discussion.

REFERENCES

[1] K. Nomura, H. Ohta, A. Takagi, T. Kamiya, M. Hirano, and H. Hosono, Nature (London) 432, 488 (2004).

[2] A. Takagi, K. Nomura, H. Ohta, H. Yanagi, T. Kamiya, M. Hirano, and H. Hosono, Thin Solid Films 486, 38 (2005).

[3] H. Yabuta, M. Sano, K. Abe, T. Aiba, T. Den, and H. Kumomi, K. Nomura, T. Kamiya, and H. Hosono, Appl. Phys. Lett. 89, 112113(2006).

[4] M. Ito, M. Kon, M. Ishizaki, and N. Sekine, Proc. 12th IDW/AD, 845 (2005); M. Ito, M. Kon, M. Ishizaki, C. Miyazaki, K. Imayoshi, M. Tamakoshi, Y. Ugajin, and N. Sekine, Proc. 13th IDW, 585 (2006).

[5] R. Hayashi, M. Ofuji, N. Kaji, K. Takahashi, K. Abe, H. Yabuta, M. Sano, H. Kumomi, K. Nomura, T. Kamiya, M. Hirano, and H. Hosono, J. SID 15/11, 915 (2007).

[6] C. J. Kim, D. Kang, I. Song, J. C. Park, H. Lim, S. Kim,E. Lee, R. Chung, J. C. Lee and Y. Park, Tech. Dig. –Int. Electron Devices Meet. 2006, 307 (2006).

[7] P. Görrn, M. Sander, J. Meyer, M. Kröger, E. Becker, H. –H. Johannes, W. Kowalsky, and T. Riedl, Adv. Mater. 18, 738 (2006)

[8] H. N. Lee, J. W. Kyung, S. K. Kang, D. Y. Kim, M. C. Sung, S. J. Kim, C. N. Kim, H. G. Kim and S. T. Kim, Proc. 13th IDW, 663 (2006); SID 07 Digest, 1826 (2007); M.-C. Sung, H. -N. Lee, C. N. Kim, S. K. Kang, D. Y. Kim, S. -J. Kim, S. K. Kim, S. -K. Kim, H. -G. Kim and S. -t. Kim, IMID '07 Digest, 133 (2007)

[9] J. Y. Kwon, K. S. Son, J. S. Jung, T. S. Kim, M. K. Ryu, K. B. Park, J. W. Kim, Y. G. Lee, C. J. Kim, S. I. Kim, Y. S. Park, S. Y. Lee and J. M. Kim, IMID '07 Digest, 141 (2007); K. -S. Son, T. -S. Kim, J. -S. Jung, M. -K. Ryu,K. -B. Park, B. -W. Yoo, J. -W. Kim, Y. -G. Lee,J. -Y. Kwon, S. -Y. Lee and J. -M. Kim, SID '08 Digest, 633 (2008).

[10] J. K. Jeong, M. Kim, J. H. Jeong, H. J. Lee, T. K. Ahn, H. S. Shin, K. Y. Kang, H. Seo, J. S. Park, H. Yang, H. J. Chung, Y. G. Mo, and H. D. Kim, IMID '07 Digest, 145 (2007); SID'08 Digest, 1 (2008)

[11] J.-h. Lee, D.-h. Kim, D. -j. Yang, S.-y. Hong, K.-s. Yoon,P.-s. Hong, C. -o. Jeong, H. -s. Park, S. Y. Kim, S. K. Lim, and S. S. Kim, SID 08 Digest, 625 (2008)

[12] M. Ito, M. Kon, C. Miyazaki, N. Ikeda, M. Ishizaki, Y. Ugajin, and S. Sekine, IEICE TRANS. ELECTRON., VOL. E90-C. 2105 (2007).

[13] J. -S. Park, J. K. Jeong, H. -J. Chung, Y. -G. Mo, and H. D. Kim, Appl. Phys. Lett. 92, 072104 (2008)

[14] R. Hayashi, A. Sato, M. Ofuji, K. Abe, H. Yabuta, M. Sano, H. Kumomi, K. Nomura, T. Kamiya, M. Hirano, and H. Hosono, SID 08 Digest, 621 (2008)

[15] S. Kohiki, M. Nishitani, T. Wada, and T. Hirao,Appl. Phys. Lett. 64, 2876 (1994).

[16] S. Narshima, M. Orita, M. Hirano, and H, HosonoPhys. Rev. B 66, 035203 (2002).

[17] J. P. Perdew, K. Burke, and M. Ernzerhof, Phys. Rev. Lett. **77**, 3865 (1996).

[18] G. Kresse and J. Hafner, Phys. Rev. B **47**, 558 (1993); ibid. **49**, 14 251 (1994).

[19] G. Kresse and J. Furthmuller, Comput. Mat. Sci. **6**, 15 (1996).

[20] G. Kresse and J. Furthmuller, Phys. Rev. B **54**, 11169 (1996).

[21] P. E. Blochl, Phys. Rev. B **50**, 17953 (1994).

[22] G. Kresse and D. Joubert, Phys. Rev. B **59**, 1758 (1999).

[23] K. Nomura, T. Kamiya, H. Ohta, T. Uruga, M. Hirano, and H. Hosono, Phys. Rev. B **75**, 035212 (2007)

Mater. Res. Soc. Symp. Proc. Vol. 1109 © 2009 Materials Research Society

Effect of thermal annealing on deep and near-band edge emission from ZnO films grown by plasma-assisted MBE

V. Avrutin,[a] M. A. Reshchikov,[b] N. Izyumskaya,[a] R. Shimada,[a] S.W. Novak,[c] and H. Morkoç[a,b]

[a]Department of Electrical and Computer Engineering, Virginia Commonwealth University, Richmond, Virginia 23284, USA

[b]Physics Department, Virginia Commonwealth University, Richmond, VA 23284, USA

[c]Evans Analytical Group, East Windsor, NJ 08520, USA

ABSTRACT

We found a dramatic enhancement of Cu-related green luminescence (GL) band at 2.45 eV with characteristic fine structure in undoped MBE-grown ZnO layers upon annealing in air at temperatures >600°C. SIMS profiles revealed a significant increase of Cu concentration with annealing temperature. GL intensity shows exponential dependence on the annealing temperature with activation energy of 1.2 eV and linear dependence on the Cu concentration. Temperature-dependent photoluminescence measurements revealed quenching of the GL band at temperatures >250 K due to escape of holes from the excited state of Cu_{Zn} acceptor to the valence band.

Exciton bound to hydrogen-related. donor (the 3.363- eV line) quenched at temperatures above 750°C which was attributed to out-diffusion of hydrogen.

INTRODUCTION

Large exciton binding energy and high quantum efficiency of luminescence make ZnO a very attractive material for light emitting application. However, the major road block for wide commercialization of this material is lack of reliable p-type material. The difficulties with p-type doping can arise from a variety of causes: p-type dopants may be compensated by low-energy native defects and/or background impurities which give rise to propensity to n-type doping. In this view, understanding of defect and defect-impurity interactions in ZnO is of vital importance for controlling the defect composition of ZnO. Photoluminescence (PL) is a powerful tool for studying point defects and impurities in wide band-gap semiconductors. Electrically active defects and impurities manifest themselves as broad emissions in the visible part of the spectrum and as narrow excitonic lines in the near-band-edge emission region.

One of the most studied defects in ZnO is the Cu_{Zn} acceptor, which gives rise to the green luminescence (GL) band peaking at 2.45 eV with a characteristic phonon-related fine structure and zero-phonon line (ZPL) at 2.859 eV at low temperature [1]. This structured luminescence band has been assigned to the internal transition of a hole in Cu_{Zn} center from the excited state at $\sim E_V + 0.4\,eV$ to the ground state at $\sim E_C - 0.2\,eV$ [1-4]. In the ground state the d shell of copper is completely filled ($3d^{10}$). An excited state of the Cu_{Zn} center can be envisaged as a hole bound to the d^{10} shell or as an acceptor-type exciton bound to the neutral d^9 configuration [2]. The GL is considered as a charge transfer of a hole from an orbital encompassing neighboring oxygen atoms to a highly shielded d shell of the copper atom. The fine structure of the GL band is due to strong electron coupling to LO and local or pseudo-local phonons. In fact, many more broad bands can be observed in the visible part of the PL spectrum from undoped n-type ZnO. Most of them are insufficiently studied and remain unidentified [5].

In this contribution we report on the effects of thermal annealing on deep defect-related emission bands as well as on sharp near-band-edge lines in ZnO films grown by plasma-assisted molecular-beam epitaxy (MBE) on a-plane sapphire. The PL data are correlated with the changes in concentration and depth distribution of Al shallow donors and Cu deep acceptors measured by secondary ion mass-spectrometry (SIMS).

EXPERIMENTAL DETAILS

Unintentionally doped ZnO layers, 0.25-0.3 μm thick, were grown on a-plane sapphire substrates by RF plasma-assisted MBE. One sample was cut into several pieces and each piece was annealed in air for one hour at a temperature ranging from 300°C to 900°C. The Hall effect measurements at room temperature indicated n-type conductivity with electron mobility of about 30 cm^2/Vs and a free electron concentration of about 10^{18} cm^{-3} in as-grown sample and up to 2×10^{18} cm^{-3} in the samples annealed at high temperatures. The SIMS data showed that in as-grown ZnO layers the major donor is Al with concentration of ~3×10^{18} cm^{-3} near the ZnO/sapphire interface and about an order of magnitude smaller concentration in the middle of the layer which indicates that the out-diffusion from the substrate during the MBE growth is the major source of Al (see Fig. 1a). The measured concentration of Cu atoms in as-grown samples was ~6×10^{16} cm^{-3}. However, this value should be considered as the upper limit of Cu atoms in as-grown samples since the ratio between Cu isotopes 63 and 65 did not correspond to the natural abundance ratio of Cu, and part of the signal might originate from Al-O molecules. Among the acceptors, Li was detected with concentration of about 10^{15} cm^{-3} (not shown in the Figure). In the ZnO layer annealed at 800°C, the concentration of Al in the middle of the film increased and the Al depth distribution became more uniform, concentration of Cu increased to about 4×10^{17} cm^{-3} (and the two isotopes agreed with the natural abundance) (see Fig. 1b).

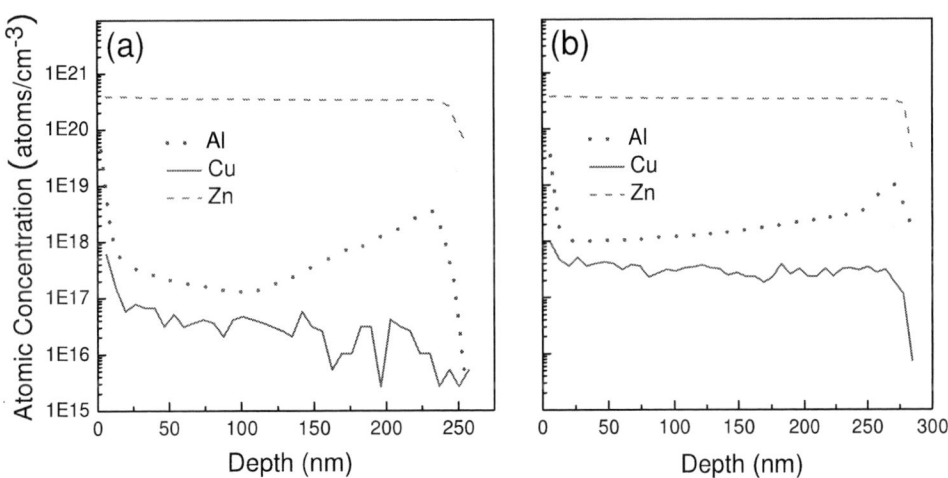

Fig. 1. SIMS depth profiles measured for (a) as-grown sample and (b) sample annealed at 800°C.

Steady-state PL was excited with a *cw* He-Cd laser (50 mW, photon energy 3.81 eV) and the PL signal was dispersed by a 1200 rules/mm grating in a 0.3 m monochromator and detected by a cooled photomultiplier tube. A closed-cycle optical cryostat was used for the measurements at temperatures between 13 and 320 K. The PL spectra were corrected for the response of the optical system. Accuracy and reproducibility of the peak positions is ±1 meV.

RESULTS AND DISCUSSION

A. Effect of annealing on excitonic PL

In as-grown ZnO layers, two peaks, a broad one with a maximum at ~3.363 eV and a sharp one at 3.330 eV, dominated in the excitonic region of PL spectrum at 15 K (Fig. 2). The peak at 3.330 eV (Y line) is attributed to an exciton bound to structural defects [6]. The full width at half maximum (FWHM) of the 3.363 eV line is 9 meV and it has a pronounced low-energy shoulder which is indicative of its composition nature. The peak position of the 3.363 eV line is consistent with the donor bound exciton (DBE) with hydrogen as a donor [6]. Upon thermal annealing at temperatures above 600°C, the low temperature shoulder of the DBE line at 3.363 eV transforms into a sharp line centered at 3.359 eV (FWHM = 2.3 meV), while the higher energy part of this line becomes weaker (Fig. 2). The intensity ratio $I_{3.363}/I_{3.359}$ decreases with increasing annealing temperature (Fig. 3). The line at 3.359 eV dominates in the PL spectra of the samples annealed at 800°C and 900°C, and the 3.363-eV peak reduces to a barely resolved high-energy shoulder. The 3.359-eV peak can be attributed to an exciton bound to *Al* donors [6]. This result is in good agreement with the SIMS data which show the gradual increase in *Al* concentration in the middle of the ZnO film (the region giving the major contribution to PL). On the other hand, quenching of the 3.363 eV line in annealed samples is in line with rapid diffusion of hydrogen at temperatures above 500°C [7]. The Y line at 3.330 eV remained virtually unchanged upon annealing at temperatures from 300°C to 750°C and nearly disappeared in the samples annealed at 800°C and 900°C.

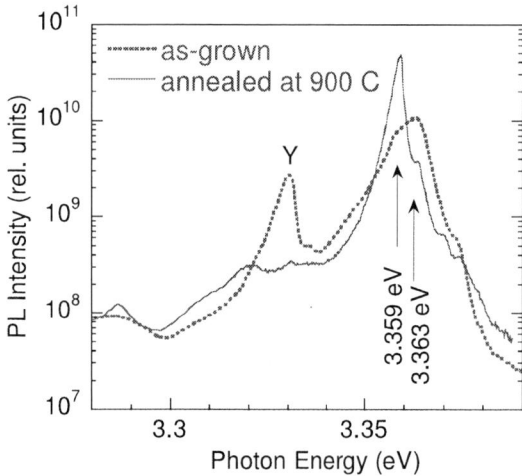

Fig. 2. Excitonic PL spectrum of as-grown and annealed ZnO samples. T = 15 K.

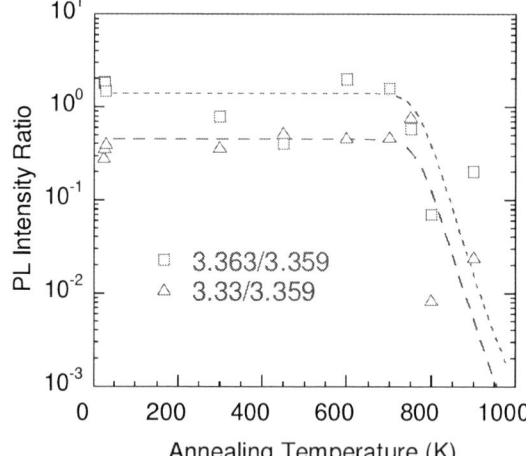

Fig. 3. Intensity ratios of the 3.363 eV peak and 3.330 eV peak to the 3.359 eV peak. T = 15 K. Lines are guides to eye.

This finding rules out the diffusion of point defects towards sinks (such as surfaces and interfaces) as the mechanism responsible for the removal of defect-related centers giving rise to the 3.330-eV emission, because it would result in the Arrhenius dependence of PL intensity on the annealing temperature. The abrupt quenching of the Y line can be explained by annealing of defects or, more likely, defect complexes responsible for the Y line.

B. Effect of annealing on defect-related PL

In the defect-related part of the PL spectrum from as-grown samples the yellow luminescence (YL) band dominates with a maximum at 2.2 eV (Fig. 4). The fine structure of the *Cu*-related GL band can be barely resolved on its high-energy wing. Annealing at temperatures above 600°C resulted in abrupt increase in intensity of the GL band associated with *Cu* which is recognized by a characteristic fine structure caused by electron-phonon coupling [1]. Interestingly, the YL band disappeared and the red luminescence (RL) band peaking at 1.7 eV appeared with annealing (Fig. 4).

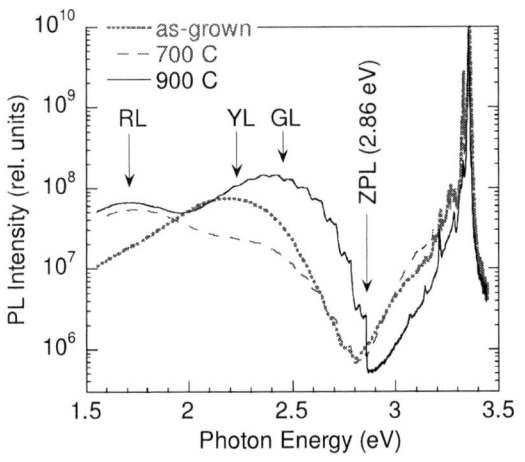 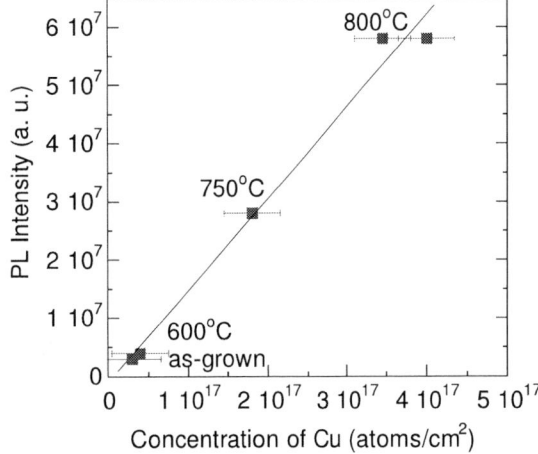

Fig. 4. PL spectra of as-grown and two annealed samples. T = 15 K

Fig. 5. GL band intensity as a function of *Cu* concentration at a depth of about 70 nm derived from the SIMS data. The straight line is to guide eye.

SIMS results indicate that, while concentration of *Cu* in our as-grown ZnO layers is very low, significant amount of copper can be adsorbed from annealing environment. The SIMS measurements revealed the gradual increases in the *Cu* concentration with increasing annealing temperature. Figure 5 shows the intensity of the *Cu*-related GL band in MBE-grown ZnO samples as a function of *Cu* concentration measured by SIMS. We used the values of *Cu* concentration measured at a depth of about 70 nm from the sample surface assuming that the most efficient PL is created at this depth [8]. As seen from the figure, the PL intensity increases nearly linearly with *Cu* concentration. We conclude that the majority of *Cu* atoms in as-grown sample appear as Cu_{Zn} acceptors, and concentration of the *Cu-H* complexes, detected in bulk ZnO by infrared absorption under uniaxial stress [9], is relatively low in our MBE-grown ZnO layers.

Figure 6 presents the dependence of the GL intensity on annealing temperature. For

129

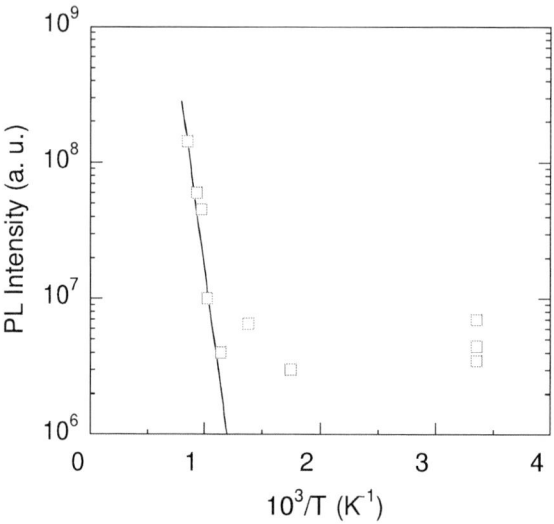

Fig. 6. GL intensity at 15 K as a function of annealing temperature. The line indicates an Arrhenius plot with activation energy of 1.2 eV.

samples annealed at T > 600°C, the dependence shows the characteristic Arrhenius behavior with an activation energy of 1.2 eV that is substantially smaller than the activation energy of 4.2 eV reported for Cu diffusion in bulk ZnO by Müller and Helbig [10]. However, the Cu diffusivity in thin films can be higher than that in ZnO single crystals. To clarify the nature of process responsible for the GL enhancement in our MBE-gown ZnO films, further studies involving the simulations of Cu depth distributions are needed.

We point out that a drastic increase of the Cu-related GL band after annealing in air for one hour at 900°C has been previously reported [4]. To explain this observation, Garces et al. [4] proposed that annealing in air at 900°C lowers the Fermi level below the ground state of Cu_{Zn}, thus converting Cu^+ ions to Cu^{2+} state and activating the GL band. However, in our experiments after annealing under similar conditions an increase of the electron concentration (deduced from both the Hall effect and consistent with SIMS data) and therefore rise of the Fermi level are followed by significant increase in intensity of the Cu-related GL band. Below we suggest the model explaining our results.

When the Fermi level is close to the conduction band, above the ground state of Cu_{Zn}, the $Cu^+(d^{10})$ ion acts as a negatively charged acceptor A^- with the 0/- energy level located ~0.2 eV below the conduction band [1]. Above-bandgap excitation creates free electrons and holes. The holes are efficiently captured by the negatively charged Cu_{Zn} acceptors and stay in the excited state with the level at ~0.4 eV above the valence band (the hole is located at one of the binding orbitals of the surrounding oxygen ligands) until the hole is transferred to the $3d$ shell to form the $Cu^{2+}(d^9)$ state [1,3]. The cycle is closed when an electron from the conduction band is captured by the neutral Cu_{Zn} acceptor. The GL band quenched above 250 K with the activation energy of about 0.4 eV [11]. The quenching is explained by escape of the trapped holes from the excited state to the valence band.

CONCLUSIONS

We studied the *Cu*-related green luminescence band in undoped ZnO layers grown by MBE and annealed in air at temperatures up to 900°C. Annealing at high temperatures resulted in a strong enhancement of the *Cu*-related GL band due to a notable increase in concentration of Cu_{Zn} acceptors. The GL band is attributed to transitions of photo-generated holes trapped at the excited state of Cu_{Zn} acceptor located at ~0.4 eV above the valence band to the ground state of this center located at ~0.2 eV below the conduction band. We argue that observation of the GL band excited with above-bandgap light in *n*-type ZnO samples with the Fermi level located above the ground state of the Cu_{Zn} acceptor is consistent with the widely accepted model of copper in ZnO. Quenching of the donor-bound exciton at 3.363 eV in samples annealed at temperatures exceeding 750°C is attributed to out-diffusion of hydrogen. Enhancement of the donor-bound exciton at 3.359 eV in annealed samples is related to an increase of *Al* concentration.

ACKNOWLEDGMENTS

The research is funded by AFOSR with Dr. Kitt Reinhardt and Dr. Don Silversmith as program monitors.

REFERENCES

1. R. Dingle, Phys. Rev. Lett. **23**, 579 (1969).

2. H.-J. Schulz and M. Triede, Phys. Rev. B **35**, 18 (1987).

3. P. Dahan, V. Fleurov, P. Thurian, R. Heitz, A. Hoffmann, and I. Broser, J. Phys.: Condens. Matter **10**, 2007 (1998).

4. N. Y. Garces, L. Wang, L. Bai, N. C. Giles, E. Halliburton, and G. Cantwell, Appl. Phys. Lett. **81**, 622 (2002).

5. M. A. Reshchikov, H. Morkoç, B. Nemeth, J. Nause, J. Xie, B. Hertog, and A. Osinsky, Physica B **401-402**, 358 (2007).

6. B. K. Meyer, H. Alves, D. M. Hofmann, W. Kriegseis, D. Foster, F. Bertram, J. Christen, A. Hoffmann, M. Strassburg, M. Dworzak, U. Haboeck, and A. V. Rodina, Phys. Stat. Sol. (b) **241**, 231 (2004).

7. K. Ip, M. E. Overberg, Y. W. Heo, D. P. Norton, S. J. Pearton, C. E. Stutz, B. Luo, F. Ren, D. C. Look, and J. M. Zavada, Appl. Phys. Lett. **82**, 385 (2003).

8. Emission from the first 50 nm is suppressed since recombination of carriers in this region is mostly nonradiative due to high density of surface states and defects near the surface, while emission from depth exceeding 100 nm decreases exponentially due to exponentially decreasing penetration of laser light.

9. E. V. Lavrov and J. Weber, Phys. Stat. Sol. (b) **243**, 2657 (2006).

10. G. Müller and R. Helbig, J. Phys. Chem. Solids **32**, 1971 (1971).

11. M. A. Reshchikov, V. Avrutin, N. Izyumskaya, R. Shimada, S.W. Novak, and H. Morkoç, J. Vac. Sci. Technol. (B), **27**, 1749 (2009).

Mater. Res. Soc. Symp. Proc. Vol. 1109 © 2009 Materials Research Society 1109-B06-12

Fabrication and Characterization of Indium Tin Oxide Thin Films on Nanoimprinted Glasses

Yasuyuki Akita[1], Yuki Sugimoto[1], Makoto Hosaka[1], Yushi Kato[1], Yusaburo Ono[1], Osami Sakata[2], Masahiro Mita[3], Hideo Oi[3], and Mamoru Yoshimoto[1]

[1]Innovative & Engineered Materials, Tokyo Institute of Technology,
4259-J2-46, Nagatsuta, Midori, Yokohama, 226-8503, Japan

[2]Japan Synchrotron Radiation Research Institute/ Spring-8, Kouto, Sayo-cho, Sayo-gun, Hyogo 679-5198, Japan

[3]Kyodo International Inc., 8-5-1 Chiyogaoka, Asao, Kawasaki, 215-0005, Japan

ABSTRACT

We fabricated indium tin oxide (ITO) thin films on nanoimprinted glass substrates using pulsed laser deposition (PLD). The nanoimprinted glass had a regular nanostepped pattern (step height of about 2 nm and step separation of about 1 µm). The surface of the ITO thin films well reflected the nanopattern of the glass substrate surface. The degree of crystalline orientation of the ITO thin films fabricated on the nanoimprinted glasses was more intense than that of the ITO thin films on the non-patterned commercial glass substrates. The resistivity of the ITO thin films deposited on the nanoimprinted glasses was lower by about 30% than that on the non-patterned commercial glasses, which was probably due to the higher crystal orientation of the films on the nanopatterned glass surfaces.

INTRODUCTION

Indium tin oxide (ITO) is a highly degenerated n-type semiconductor with a wide bandgap (3.3 eV to 4.3 eV). ITO thin films have been widely used as transparent conductive electrodes for flat panel displays, liquid crystal displays, solar cells, and organic light emitting devices because of their low resistivity ($\sim10^{-4}$ Ω cm) and high transmittance ($\sim90\%$) in the visible region. Currently, researchers are attempting to decrease the temperature at which ITO films are deposited due to the requirements of the device processes, especially for heat-sensitive substrates such as organic polymers. There are several deposition techniques for ITO thin films on glass and plastic substrates, namely sputtering [1, 2], chemical vapor deposition (CVD) [3, 4], and pulsed laser deposition (PLD) [5-8].

The PLD employed in this work has a particular advantage: the composition of the deposited film is quite close to that of the target used in laser ablation. Highly conducting ITO thin films (resistivity of 8.5×10^{-5} Ω cm) fabricated on the glass substrates by PLD was reported [9]. Extremely low resistivity (7.7×10^{-5} Ω cm) was found for the epitaxial ITO film grown on the single crystal yttria-stabilized zirconia substrate by the PLD method [10]. In addition, the PLD process enhances low temperature epitaxial growth because the film precursors laser-ablated from the target impinge on the substrate with high kinetic energies [11]. So far, we have attained room temperature epitaxy of the ITO films on CeO_2-buffered Si (111) substrates using the PLD technique [12].

Recently, we reported the nano-scale surface modifications of borosilicate glass plates by applying a thermal nanoimprint technique in which we used self-organized nanopattern molds of oxides (NiO, α-Al_2O_3) [13, 14]. The use of these nanoimprinted glass substrates for ITO thin film deposition is expected to result in a reduction of the resistivity, probably due to the homogenization of crystal nucleation sites on the regular nanopattern (step height of about 2 nm and step separation of about 1 μm). In this study, we examine the crystal growth and structural and electrical properties of ITO thin films deposited on this nanoimprinted glass by PLD.

EXPERIMENT

We performed thermal nanoimprints using our internally developed nanostepped sapphire (α-Al_2O_3) molds on borosilicate glass plates in air to obtain the nanopatterned glass substrates. The composition (wt. %) of the borosilicate glass was 68.9SiO_2–8.8Na_2O–8.4K_2O–10.1B_2O_3–2.8BaO–1.0MgO and the glass transition temperature (T_g) was 521°C. The stepped sapphire molds for the glass nanoimprints were prepared by annealing a mirror-polished commercial sapphire (0001) substrate at 1400°C for 3 h in air. We then pressed the mold onto the surface of the glass plate at 1–3 kPa and heated it at 600°C for 60 min in air. The sample was cooled to 40°C. We deposited the ITO films on the nanoimprinted glass and the non-patterned commercial glass substrates by the PLD method. A pulsed KrF excimer laser (λ=248 nm, energy density of 3.5 J/cm^2 and repetition of 5 Hz) was focused onto the sintered target of 5 wt. %-Sn doped In_2O_3 (ITO).

Firstly, film deposition was conducted at room temperature (RT) under a 1×10^{-2} Torr O_2 atmosphere. We then annealed as-deposited films for crystallization in vacuum (1×10^{-7} Torr) at temperatures from 100°C to 300°C for 3 h. The crystallinity of the ITO films was examined by *in situ* reflection high-energy electron diffraction (RHEED) and *ex situ* x-ray diffraction (XRD). We observed the surface morphology of the specimens with atomic force microscopy (AFM). The resistivity of the films was determined as a function of temperature (20 K–300 K) by the four probe method.

RESULTS AND DISCUSSION

Figure 1 shows the AFM surface morphology of (a) the stepped sapphire mold and (b) its cross-sectional profile. The regular step morphology (step height of about 2 nm and step separation of about 1 μm) was observed. The large steps of 2 nm high observed in Figure 1 (a) are most likely formed by step-bunching through enhanced atom migration at high temperatures (1400°C). In the case of annealing the sapphire (0001) substrate at 1000°C, we obtained the nanostepped surface with a single step of 0.2 nm high [15], indicating no occurrences of step-bunching.

Figure 2 (a) shows the AFM surface morphology of the non-patterned commercial glass substrate. There are some dents on the glass surface, and the RMS roughness value was about 0.7 nm. Figure 2 (b) shows the surface morphology of the nanoimprinted glass substrate. Here, we observed the regular steps and terraces (step height of about 2 nm and step separation of about 1 μm). The RMS roughness value of the terrace was about 0.4 nm. Figure 2 (c) shows the AFM surface image of the ITO film (500 nm thick) annealed at 300°C after being deposited on the non-patterned commercial glass. The ITO film has a rough surface, and we observed crystal grains of several tens of nanometers in size. On the other hand, Figure 2 (d) shows the surface morphology of the ITO film annealed at 300°C after being deposited on the nanoimprinted glass. The surface image of the ITO film as shown in Figure 2 (d) reflects the step and terrace morphology of the nanoimprinted glass surface. ITO thin films with 500 nm in thickness on the nanoimprinted glasses also reflected the nanopattern of the glass surface. It seems that the crystal nucleation and growth of ITO during annealing occurred, reflecting the step edge of the nanoimprinted glass.

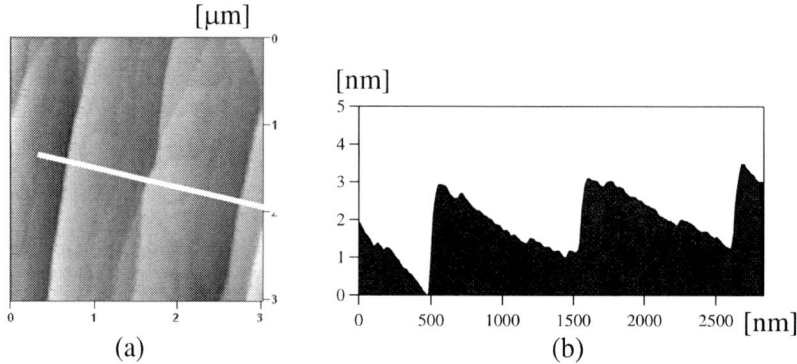

(a)   (b)

Figure 1. (a) AFM surface image (3×3 μm^2) of the stepped sapphire mold and (b) a cross-sectional profile.

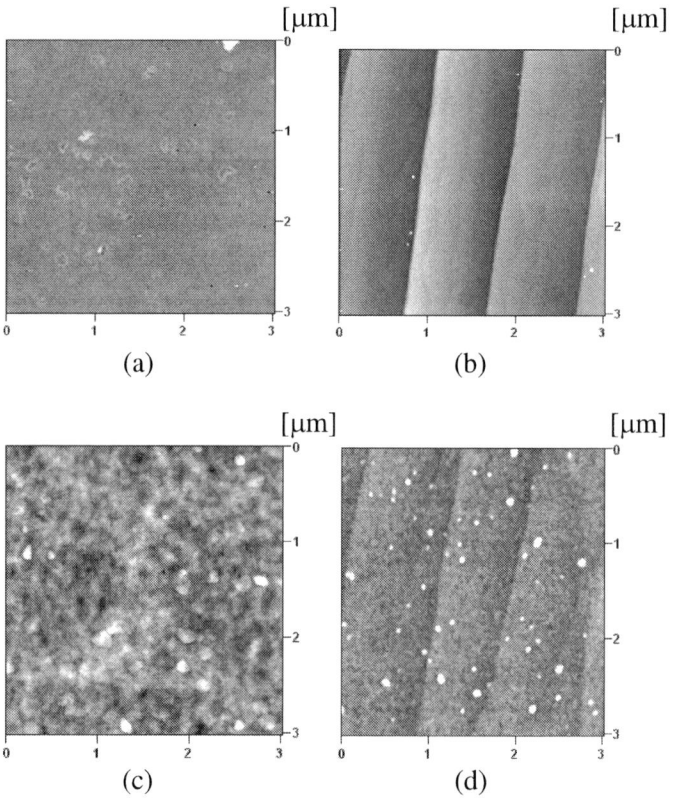

Figure 2. AFM surface images (3×3 μm^2) of (a) the commercial glass, (b) the nanoimprinted glass, (c) ITO film annealed at 300°C after deposition on non-patterned commercial glass, and (d) ITO film annealed at 300°C after deposition on nanoimprinted glass.

Figure 3 shows a RHEED pattern of crystallized ITO film fabricated on the nanoimprinted glass. The pattern seems to have a weak streak. As this streak pattern did not change by rotating the sample azimuth, the obtained ITO film was a verified to be strongly oriented polycrystalline film. On the other hand, we observed a ring pattern in the RHEED image of the crystallized ITO film fabricated on the non-patterned commercial glass.

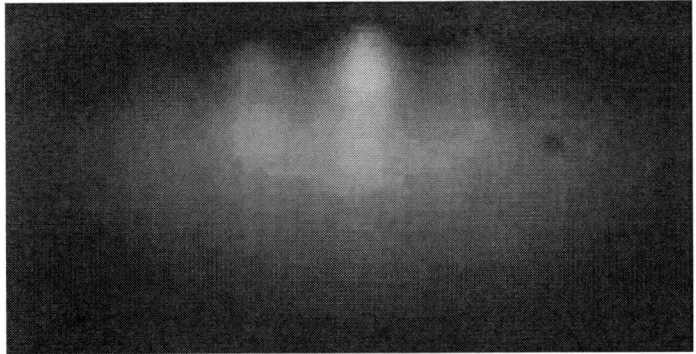

Figure 3. RHEED image taken of the crystallized ITO film on the nanoimprinted glass after annealing at 300°C.

Figure 4 shows the XRD patterns of ITO films fabricated on the imprinted glasses. From XRD patterns, it is verified that we achieved (111)-oriented polycrystalline growth of ITO films for the films annealed at 300°C after deposition on nanoimprinted glass. We also found that the orientation of the ITO film fabricated on the nanoimprinted glass was stronger than that on the non-patterned glass. The preferential orientation of ITO film growth on the nanoimprinted glass resulted from the regular flat terrace of the nanoimprinted glass enhancing the homogeneous crystal growth.

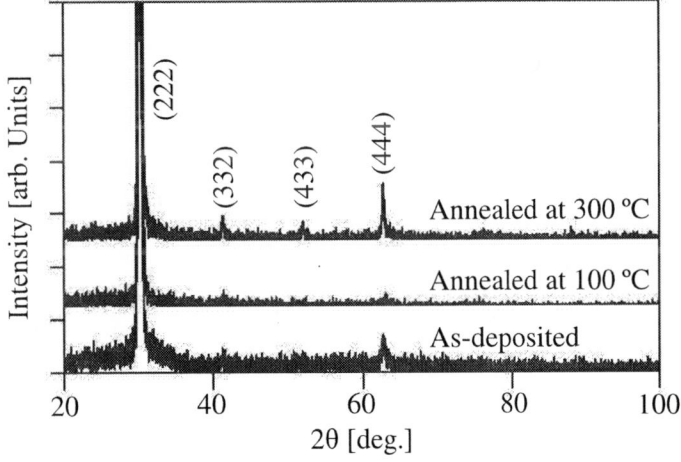

Figure 4. XRD 2θ-θ profiles of the ITO films fabricated on nanoimprinted glass.

Figure 5 reveals the temperature dependences of the resistivity of the ITO films (500 nm thick) on the commercial glass and nanoimprinted glasses after annealing at 300°C. The resistivity of the ITO films on the nanoimprinted glass after annealing at 100°C was 7.1×10^{-4} Ω cm at RT, and the lowest resistivity of 4.6×10^{-4} Ω cm at RT was obtained for the ITO film on the nanoimprinted glass after annealing at 300°C as shown in Figure 5. The high resistivity of the film after being annealed at 100°C was related to the existence of the amorphous and crystalline phases. From Figure 5, we can see that the resistivity of the film on the nanoimprinted glass was lower than that on the non-patterned glass. We believe that the stronger orientation of the films prepared on the nanoimprinted glasses under the appropriate conditions leads to a decrease of carrier scattering and to an increase of carrier mobility.

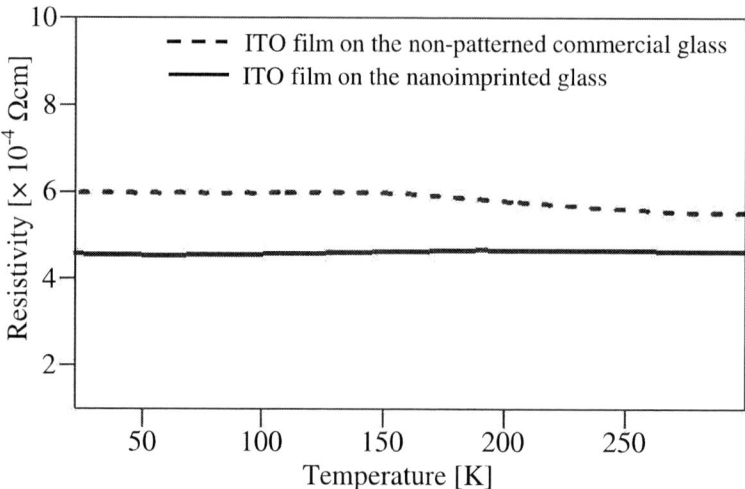

Figure 5. Temperature dependence of the resistivity of the ITO films deposited on the non-patterned commercial glass and that on the imprinted glass substrates after being annealed at 300°C.

CONCLUSIONS

We fabricated ITO thin films on nanoimprinted glass substrates with PLD. The crystal growth of the ITO thin films deposited on the nanoimprinted glasses was different from that on the non-patterned commercial glasses. The crystalline orientation of the ITO films fabricated on the nanoimprinted glasses was stronger than that of the ITO films on the non-patterned commercial glass substrates. The resistivity of the ITO films deposited on the nanoimprinted glasses was lower than that on the non-patterned commercial glass.

ACKNOWLEDGMENTS

This work was supported in part by the Ministry of Education, Culture, Sports, Science and Technology of Japan, the National Institute of Advanced Industrial Science and Technology of Japan, the New Energy and Industrial Technology Development Organization of Japan, and the Regional Innovation Creation R&D Program from the Ministry of Economy, Trade and Industry of Japan.

REFERENCES

1. C. Guillen, J. Herreo, Thin solid films **480,** 129 (2005).
2. E.Nishimura, M. Ando, K. Onisawa, M. Takabatake, and T. Minemura, Jpn. J. Appl. Phys. **35,** 2788 (1996).
3. T. Maruyama, and K. Fukui, Thin solid films **70,** 3848 (1991).
4. M. Girtan, and G. Folcher, Suf. Coat. Technol. **172,** 242 (2003).
5. H. Izumi, T. Ishihara, H. Yoshioka, and M. Motoyama, Thin solid films **411,** 32 (2002).
6. H. Ohta, M. Orita, M. Hirano, and H, Hosono, Mat. Res. Soc. Symp. Proc. **666,** F3.15.1 (2001).
7. S.Y. Kim, N.M. Park, T.Y. Kim, and G.Y. Sung, Thin solid films **475,** 262 (2005).
8. T.Y. Yong, and Y.Y. Tou, B.S. Teo, Appl. Surf. Sci. **248,** 388 (2005).
9. A. Suzuki, T. Matsushita, T. Aoki, A. Mori, and M. Okuda, Thin solid films **411,** 23 (2002).
10. H. Ohta, M. Orita, and M. Hirano, Appl. Phys. Lett. **76,** 2740 (2000).
11. L. Lynds, B.R. Weinberger, G.G. Peterson, and H.A. krasinski, Appl. Phys. Lett. **52,** 320 (1988).
12. J. Tashiro, A. Sasaki, S. Akiba, S. Satoh, T. Watanabe, H. Funakubo, and M.Yoshimoto, Thin Solid Films **415,** 272 (2002).
13. S. Akiba, W. Hara, T. Watanabe, A. Matsuda, M. Kasahara and M. Yoshimoto, Appl. Surf. Sci. **253,** 4512 (2007).
14. Y. Akita, T. Watanabe, W. Hara, A. Matusda, and M. Yoshimoto, Jpn. J. Appl. Phys. **46,** L342 (2007).
15. M. Yoshimoto, T. Maeda, T. Ohnishi, H. Koinuma, O. Ishikawa, M. Shinohara, M. Kubo, R. Miura, and A. Miyamoto, Appl. Phys. Lett. **67,** 2615 (1995).

High Quality ZnO Thin Films for TCOs and Transistors by MOCVD

Bruce I. Willner[1], Shangzhu Sun[1], and Gary S. Tompa[1]
[1]Structured Materials Industries, Inc., 201 Circle Drive North, Unit 102/103, Piscataway, NJ, USA.

ABSTRACT

The wide bandgap semiconductor Zinc Oxide (ZnO) and its alloys are used as transparent contact layers in applications such as solar cells and LEDs. ZnO-based transistors have recently gained significant interest because of their potential for use in displays and as low-noise, high-voltage, high-power devices. Wide bandgap ZnO, and related materials offer the potential for high power and performance transistors at low cost, with the proper deposition techniques. Metal-organic chemical vapor deposition (MOCVD) is an optimal scalable production approach to manufacturing ZnO and its alloys. We have developed high speed rotation susceptor style reactors scaled from deposition planes of 3" to 16" in diameter. As we report herein, we have shown that MOCVD can produce high quality films of desired composition uniformly over large areas on a variety of substrates. In particular, we report on deposition parameters for compositionally uniform pin-hole free smooth morphology insulating and doped films of varying composition as well as multi-layer dielectric-ZnO structures and the characteristics of the films.

INTRODUCTION

The unusual properties of Zinc oxide (ZnO) have been drawing interest to the material for a number of electro-optic and electronic applications[1,2]. A direct bandgap of 3.33eV (similar to GaN) and a large exciton binding energy of 59 meV suggest possibilities for LED[3] and power transistor[4,5,6] applications. High transparency, even when heavily (degeneratively) doped, make it a good candidate for transparent transistors[7] and as a transparent conducting oxides[8]. These applications require high quality, well controlled material deposition. Doping levels, surface morphology, and other characteristics must be well controlled. Metal organic chemical vapor deposition (MOCVD) provides the deposition parameter control required to produce these high quality materials[9]. In this paper, the authors discuss the advantages of ZnO for a variety of semiconductor device applications and the ease with which MOCVD produces ZnO.

A MOCVD tool, designed specifically for ZnO deposition, was used to deposit high quality transparent conducting oxide (TCO) material for use as contact layers on optoelectronic devices. The parameters for these depositions are reported. ZnO thin films were studied with a variety of techniques including photoluminescence, electroluminescence, optical transmission spectroscopy, and sheet resistance measurements. The TCO films were also deposited on LED device structures.

Several properties of ZnO offer great potential for semiconductor devices. ZnO has a wide bandgap and strong exciton bonding energy, a high field effect mobility, low cost deposition and substrates, and compatibility with many other materials, especially oxides. Also, ZnO can be heavily doped (even degenerately so) while retaining high crystal quality and little change in optical performance. The optical bandgap of ZnO is about 3.33eV, and ZnO may be alloyed with MgO, CdO, or MnO to change the bandgap between 3 and 4 eV without a large variation in lattice constant [10].

Transparent conductive layers have become highly desirable for electro-optic applications. Thin metal films, typically gold (or a gold alloy such as nickel/gold), indium tin oxide (ITO), and ZnO [11] are used as electrical contacts for LEDs and for photovoltaic devices. The qualities of these different contacts are compared in Table 1. ZnO has a number of advantages over the alternatives, particularly with respect to bandgap and refractive index control, thermal qualities, and optical transparency. In addition, ZnO is significantly less expensive than ITO due to the limited availability of indium.

Table 1 - Comparison of contact material properties for high-brightness GaN based LED's.

Material	Ni/Au	ITO	ZnO
Bandgap (eV)	Metal	~3.6	3.33
Optical Transparency	Poor	Good	Good
Practical resistivity (ohm-cm)	NA	~1E-4	~1E-4
Lattice Match to GaN	Poor	Moderate	Good
Thermal Stability	Poor	Moderate	Good
Thermal Conductivity	Good	Poor <0.1 W/cmK	Good >0.13 W/cmK
Index of Refraction (@550nm)	metal	2	2.1
Refractive Index and Bandgap Grading (Engineering)	No	Minimal	Excellent (ZnMgO – UV)
Columnar Microstructure	No	Maybe	**Yes**
Achievable Contact Area (current spreading)	Small	Large	Large

THEORY

Structured Materials Industries, Inc. (SMI) has recently developed new ZnO MOCVD systems[12]. The MOCVD system, specifically designed for ZnO deposition, is a vertical reactor with a high speed rotating disc substrate holder. The reactor uses a new multi-zone filament heater technology[13]. Zinc and dopants are deposited from metalorganic sources. Oxygen gas is used as a source as well. In order to improve the oxygen incorporation into films, the oxygen is plasma activated by a plasma source upstream from the deposition chamber. This allows the oxygen plasma to be controlled independently of other source parameters. The MOCVD tool can produce smooth amorphous, polycrystalline, or crystalline ZnO films, free of pinholes.

ZnO can also be deposited to form different surface morphologies (Figure 1) including highly uniform nanotips or nanowires. The material structure is well controlled by modifying the deposition parameters. Accordingly, layers of multiple forms of ZnO may be deposited in single deposition run.

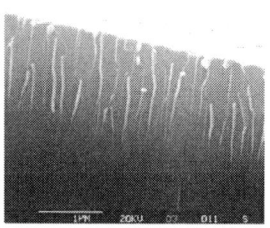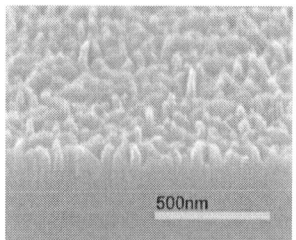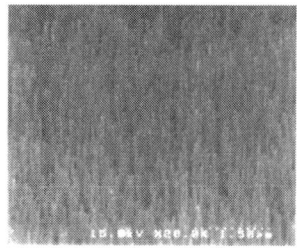

Figure 1 - SEM images of different ZnO morphologies fabricated by MOCVD deposition: smooth polycrystalline (left), surface rough (center), and nanotips (right).

A number of devices would benefit from the incorporation of ZnO TCOs, particularly light emitting diodes (LEDs) and photovoltaic (PV) cells. Currently, a number of manufacturers use ZnO as a contact layer for photovoltaics, particularly with copper indium gallium diselenide structures[14]. There is also growing interest in using transparent contacts for LEDs, reducing the emissions blocked by metals. The ability to extend the transparency window into the UV by alloying with Mg is also an advantage for aluminum gallium nitride (AlGaN) devices. Current GaN structures without a TCO use a thick semiconductor layer for current spreading. This layer is resistive and is a significant source of loss in the device. The use of a TCO and a thinner p-GaN layer will reduce the electrical current losses in the device and reduce the forward operating voltage of the device, especially at higher currents. There is also growing interest in new, more varied, LED device structures and packaging, often involving substrate removal. The use of TCOs offer greater flexibility in device and packaging design.

Metal organic chemical vapor deposition (MOCVD) has several advantages over other deposition techniques. Table 2 shows a comparison of film deposition processes. MOCVD has a higher deposition rate, is scalable, and is capable of composition grading and multi-layer structures, unlike other deposition techniques. Doping densities lower than 10^{15} cm^{-3} and as high as 10^{21} cm^{-3} are possible. This control of the deposition parameters allows precise control of the electrical and optical properties of the material. MOCVD is also the only technique capable of good coverage of step features from device processing on the target.

Table 2 - Comparison of contact deposition processes for high brightness GaN based LED's.

Deposition Technique	Pulsed Laser Deposition	e-Beam Evaporation	Sputtering	MBE	MOCVD
Step Coverage	Poor	Poor	Poor	Poor	**Good**
Scalable to High Volume	Yes	Maybe	Yes	Oxide - No	**Yes**
Deposition Rate	Slow	Moderate	Moderate	Moderate	**High**
Post Deposition Annealing Required	Probable	Yes	Yes	Maybe	**Optional**
Composition Grading	No	No	No	Yes	**Yes**
Structure control	Somewhat	Somewhat	Somewhat	Somewhat	**Yes**
Interface Damage	Small	Some	YES	No	**NO**

EXPERIMENT

To produce highly conductive ZnO for use as a TCO, the material is degenerately doped. A variety of dopants have been used to produce conductive n-type ZnO. The authors most commonly use aluminum or gallium as the dopant, although several other dopants can be used effectively in this material system. Both perform similarly. The deposition is a relatively low temperature process with substrate temperatures of 350° to 650°C. For these samples, the material was deposited wit chamber pressures of 5 – 100 Torr. The layer is grown at 100 to 300 Angstroms/min, depending upon parameter settings. Trimethyl aluminum (TMAl) was used as the dopant. Growth parameters are shown in Table 3.

Table 3 - Summary of Al-doped ZnO film growth parameters.

Substrate temperature:	350 – 650° C
Chamber pressure:	5 – 100 Torr
Oxygen flow rate:	200 – 5000 sccm
Carrier gas (Ar) flow rate:	4000-10000 sccm
Ar flow rate through DEZn:	35-70 sccm @ 17° C and 350 Torr
Ar flow rate through TMAl:	0 – 20 sccm @ 15° C and 350 Torr
Sample rotation speed:	750 rpm
Growth rate:	100 – 300 A/min

The n+ doped films produced in this fashion achieve resistivities of about 5×10^{-4} Ω cm and carrier concentrations of 10^{20} cm^{-3}. The electron mobility in the films is 5 cm^2/Vs. X-ray diffraction was used to assure the quality of the polycrystalline films.

DISCUSSION

ZnO Transparent Conducting Oxide Results

The conductivity and gallium concentration were measured for several films with different deposition parameters to control the gallium dopant incorporated into ZnO thin films, Figure 2. At the optimal gallium concentration, the sheet resistance is below 20 Ω/sq.

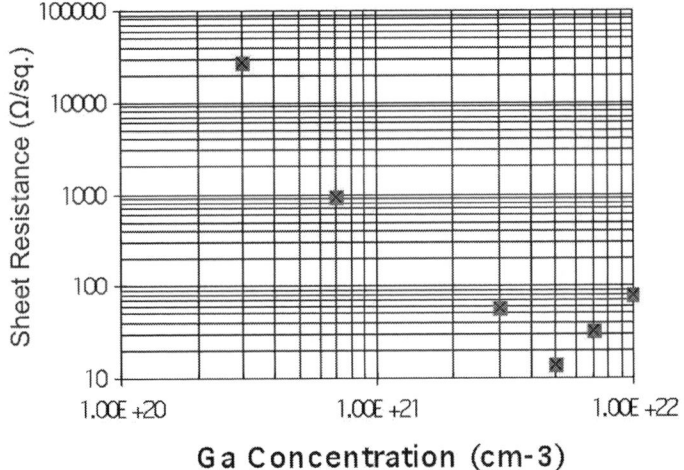

Figure 2 – Sheet resistance versus incorporated gallium concentration for doped ZnO thin films.

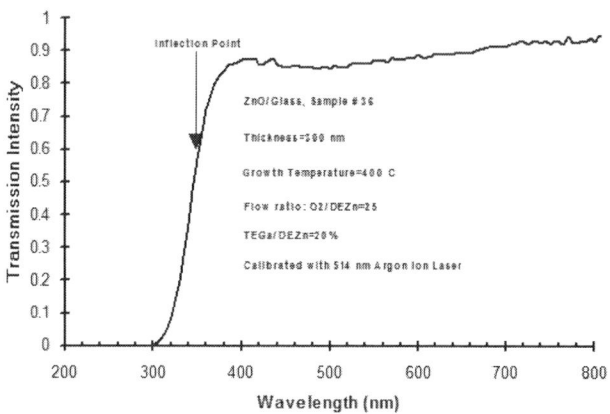

Figure 3 -A typical transmittance measurement on an as grown ZnO films on a glass substrate grown at 420°C with Ga-doping in mid 10^{21} cm^{-3}. The cutoff wavelength is about 375 nm which matches the ZnO bandgap of 3.3 eV.

Optical transmission spectroscopy was performed on degenerately doped ZnO:Ga films in the visible and UV wavelength range as shown in Figure 3. These films were deposited on glass for this purpose. Figure 3 shows the transmission spectrum for one such film. This transmission spectrum is typical for our MOCVD films. The spectrum shows a bandgap of approximately 360 nm, as expected for a ZnO film. The optical transmission is approximately 90% throughout the visible regime. The heavy doping has little effect on the optical transmission of the ZnO. The photoluminescence (PL) spectrum of several samples of degenerately aluminum-doped zinc oxide (AZO) is shown in Figure 4. The samples show a strong, well-defined peak at about 375nm. Figure 5 and Figure 6 show the electroluminescence (EL) spectrum and EL intensity versus bias curve for the ZnO:Al film. The narrow, well-defined EL peak at the ZnO bandgap, the strong EL intensity, and the well-defined PL peaks show the good crystalline quality of this degenerately doped material. The incorporation of large mole fractions of aluminum in place of zinc in the material structure has not had a significant effect on the bandgap or crystalline structure.

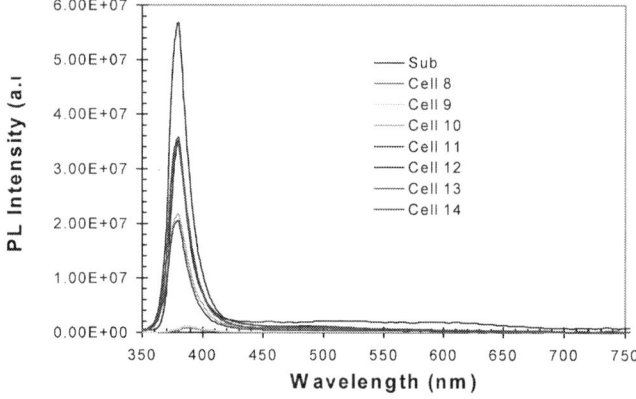

Figure 4 - Photoluminescence spectrum of several degenerately aluminum-doped ZnO thin films.

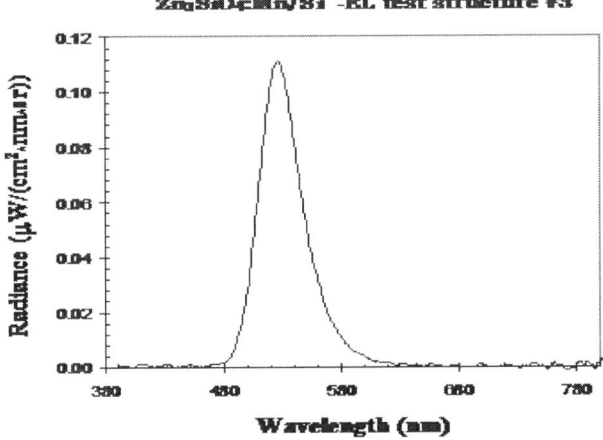

Figure 5 – Electroluminescence spectrum of aluminum-doped ZnO TCO film.

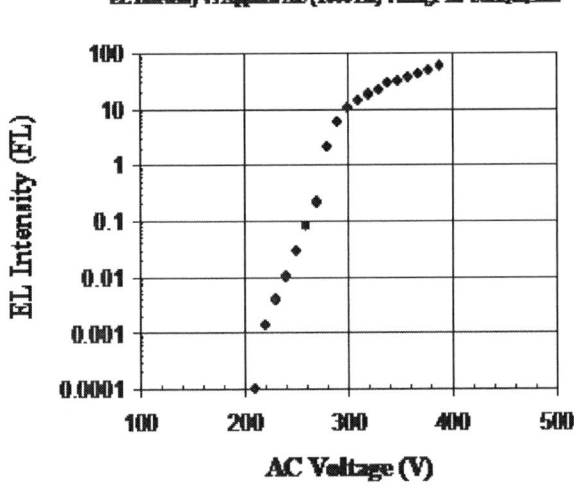

Figure 6 - Electroluminescence intensity versus drive voltage for an aluminum-doped ZnO TCO film.

Preliminary Device Tests

Several LEDs, from the same GaN wafer were tested with aluminum-doped ZnO, ITO, and Ni/Au contact layers to compare performance. The ZnO and ITO contact layers covered the LED surface, while the Ni/Au contact covered a small area, less than 10% of the device surface. Table 4 shows results comparing Ni/Au and ZnO contacts. While there is a significant power increase with a ZnO contact, the most important aspect to note is the drop in the forward voltage when using the ZnO contact. Figure 7 shows a plot of output illumination versus drive current for all three contact materials on the same device structure.

Table 4 - Quick test results for an aluminum doped ZnO contact versus a Ni/Au contact on identical LEDs from the same wafer.

	Ni/Au contact	ZnO contact
Power @ 20 mA	0.5 mW	0.625 mW
Power @ 100 mA	1.82 mW	2.4 mW
Wavelength @ 20mA	464 nm	464.74 nm
Wavelength @ 100mA	463 nm	462.92 nm
Vf @ 20mA	4.73 V	3.4 V
Vf @ 100mA	7.21V	5.68 V

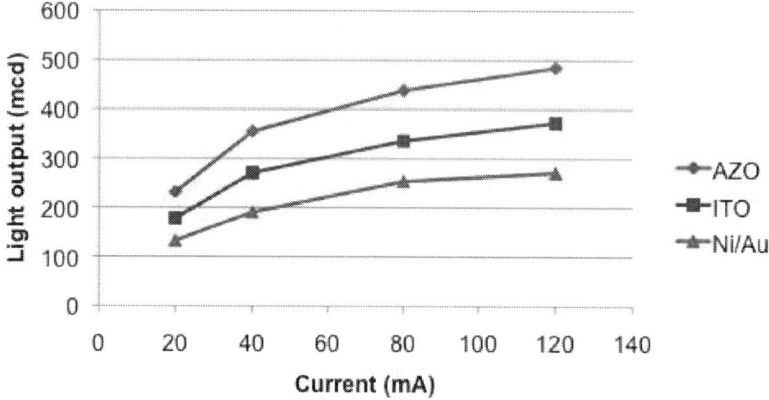

Figure 7 - LED light output versus current for LEDs with aluminum-doped zinc oxide (AZO), indium tin oxide (ITO), and nickel/gold (Ni/Au) surface contacts.

CONCLUSIONS

In summary, high quality ZnO thin films were deposited by MOCVD on a variety of substrates. The MOCVD system was designed specifically for the deposition of ZnO and alloyed films. ZnO film morphology and crystallinity can be controlled through deposition parameters. This MOCVD system can be scaled to large area deposition for production purposes. ZnO, a wide bandgap semiconductor which can be alloyed to tune the bandgap and can be heavily doped without significantly reducing crystalline quality or transparency, has a number of attractive applications. Furthermore, the low cost of deposition on a variety of substrates and ease of processing open possibilities to new devices and applications.

Transparent conductive oxides of degenerately-doped ZnO were fabricated for use as a transparent conducting oxide. These films were found to exhibit excellent polycrystalline quality, exhibit good conductivity (resistivity of 5×10^{-4} Ω cm), and high transparency throughout the visible range. Photoluminescence, electroluminescence, and other measurements demonstrated the high material crystalline quality of the films.

The ZnO TCOs were deposited on LED device structures as contact layers. These devices were compared with identical devices with different contact technologies, ITO and metal. The ZnO contacted devices performed significantly better than the others, demonstrating higher light output and a lower forward voltage. ZnO thin film TCOs will be advantageous in developing new LED device structures and packaging strategies. With further research and refinement, transparent ZnO films will an advantageous element in many devices and applications.

REFERENCES

[1] Zhong Lin Wang, J. Phys.: Condens. Matter **16** R829-R858 (2004)

[2] Jagadish, C., and S. J. Pearton. *Zinc Oxide Bulk, Thin Films and Nanostructures: Processing, Properties and Applications.* (Elsevier, Amsterdam, 2006).

[3] S. J. Jiao, Z. Z. Zhang, Y. M. Lu, D. Z. Shen, B. Yao, J. Y. Zhang, B. H. Li, D. X. Zhao, X. W. Fan, and Z. K. Tang, Appl. Phys. Lett. **88**, 031911 (2006)

[4] B. Bayraktaroglu, IEEE El. Dev. Let. **29** (9), 1024-1026 (2008).

[5] J. D. Albrecht, P. P. Ruden, S. Limpijumnong, W. R. Lambrecht, and K. F. Brennan, *J. Appl. Phys.*, **86** (12), pp. 6864–6867 (1999).

[6] K. Nomura, H. Ohta, K. Ueda, T. Kamiya, M. Hirano, and H. Hosono, *Science*, **300** (5623), pp. 1269–1272 (2003).

[7] E. M. C. Fortunato, P. M. C. Barquinha, A. C. M. Pimentel, A. M. F. Goncalves, A. J. S. Marquues, R. F. P. Martins, and L. M. N. Pereira, *Appl. Phys. Lett,* **85** (13), pp. 2541-2543 (2004).

[8] Y. Li, G. S. Tompa, S. Liang, C. Gorla, Y. Lu, and John Doyle, J. Vac. Sci. Technol. A **15,** p.1063 (1997).

[9] S. Sun, G.S. Tompa, C. Rice, X.W. Sun, Z.S. Lee, S.C. Lien, C.W. Huang, L.C. Cheng, Z.C. Feng, Thin Solid Films, **516**, p. 5571 – 5576 (2008).

[10] T. Makino, Y. Segawa, M. Kawasaki, A. Ohtomo, R. Shiroki, K. Tamura, T. Yasuda, and H. Koinuma, Appl. Phys. Lett. **78**, p.1237 (2001)

[11] E. W. Forsythe, Yongli Gao, L. G. Provost, and G. S. Tompa, J. Vac. Sci. Technol. A **17**, p. 1761 (1999)

[12] Structured Materials Industries, Inc. www.structuredmaterials.com

[13] SMI patent pending.

[14] B. von Roedern and H. S. Ullal, *Solid State Tech.* **51** (2), 52-54 (2008)

Comparison on Optimized Optical Transmission and Electrical Resistivity between Indium Tin Oxide and Gallium Doped Zinc Oxide

Wei-Lun Hsu, Fan-Shuen Meng, Cheng-Tao Lin, Kuang-Chung Liu, Tzu-Huan Cheng, Chee-Wee Liu, JianJang Huang, and Gong-Ru Lin*

Institute of Photonics and Optoelectronics, and Department of Electrical Engineering,
National Taiwan University
No. 1 Roosevelt Road Sec. 4, Taipei 106, Taiwan
Phone: 886-2-33663700 ext. 235, Fax: 886-2-33669598,
*E-Mail: grlin@ntu.edu.tw

ABSTRACT

The comparison on the optical transmission and electrical resistivity of RF-sputtered ITO and GZO thin films is demonstrated. After post-annealing process, the ITO film has better near-ultraviolet transmittance and can be enhanced to over 40%. However, the GZO film has higher optical transparency at visible-light region. At blue, green-light region, the optimized optical transmittance of the GZO films can be observed over 80%, and at red-light region, that of the GZO film is even up to 96%. As for electrical resistivity, the ITO film has much lower sheet resistance than the GZO film by 2 orders. The minimum sheet resistance of the ITO and GZO film is 26 Ω/square and 2900 Ω/square, respectively. The FTIR spectra of the ITO film shows little change of chemical composition after post-annealing process, but there is a growing Ga_2O_3 peak of the GZO sample at the annealing temperature of 450°C or higher, leading to an increase of sheet resistance. After all, ITO film is a referable choice as a transparent contact material.

INTRODUCTION

Transmission and resistivity of performances of the transparent conducting oxide (TCO) are of great interest due to the increasing demand for LEDs or OLEDs applications [1, 2], such as indium tin oxide (ITO) and gallium doped zinc oxide (GZO) films [3, 4]. Both ITO and GZO are wide-bandgap n-type semiconductors. ITO is notable for high visible-light transmittance as well as acceptable sheet resistance. Zinc oxide (ZnO) is famous for even higher visible-light transmittance, and ZnO has advantage of ITO because zinc is abundant while indium is expensive metal. Nevertheless, zinc oxide has a fatal drawback that the electrical conductivity is quite low. In order to enhance the conductivity, III-group elements doped zinc oxide materials have been developed, such as GZO. Several methods for preparing ITO or GZO films were demonstrated to obtain optimized performance of electrical contact, such as RF magnetron sputtering [5, 6], sol-gel process [7, 8], ion beam sputtering [9], pulsed laser deposition [10, 11] and chemical vapor deposition [12]. However, the performance of the as-deposited ITO and GZO films on optical transmittance and electrical resistivity were not acceptable simultaneously for most applications. Therefore, post-annealing process is required to enhance optical and electrical properties. In this work, the ITO and GZO thin films are prepared on quartz substrates by RF sputter where high-quality films can be produced. The comparison on optimized optical transmission and electrical resistivity between the post-annealed ITO and GZO thin films are demonstrated. Fourier transform infrared (FTIR) spectroscopy is performed to show chemical composition of the ITO and GZO thin films after post annealing conditions.

EXPERIMENTAL DETAILS

The ITO and GZO thin films are prepared by RF sputter system. At first, the quartz substrates are cleaned in acetone and methanol solution and then stuck onto the sample holder of the sputter system. During the ITO process, argon and oxygen gas are introduced to the chamber. The flow rate of argon and oxygen are 12 sccm and 1 sccm, respectively, and the pressure of the chamber is set at $\sim 10^{-3}$ Pa. RF power of 75W is supplied, and the deposition rate is about 5 nm per minute. On the other hand, during the GZO process, only argon is introduced. The pressure of the chamber is set at 0.02 torr. RF power of 100W is supplied, and the deposition rate is around 6 nm per minute. The thickness of both ITO and GZO thin films are determined 200 nm by alpha-step system. Then, two different furnace annealing processes with flowing nitrogen gas in atmosphere were conducted, either the annealing time between 5 and 20 minutes with 5-min increment at 450°C or the annealing temperature between 400°C and 500°C with 25°C-increment for 15 minutes were performed. Four different laser sources are used to determine the optical transmittance of the ITO and GZO films, which are 325-nm He-Cd laser, 405-nm laser diode, 532-nm frequency-doubled Nd:YAG laser, and 632.8-nm He-Ne laser. The transmission power is measured by Ophir power meter, and a four-point probe system is used to measure the sheet resistance of the films. MCT/A sensor is used in FTIR spectroscopy system to determine whether the composition of the ITO and GZO films is changed, and the output results is the average of 16-time scans.

DISCUSSION

Optical Transmittance

The optical transmittance of the as-deposited and annealed ITO and GZO thin films are shown in Fig.1. The transmittance of the as-deposited ITO films is quite low at the near-ultraviolet region, which is merely over 10%, and compared to the as-deposited ITO films, the transparency of the as-deposited GZO films is even worse. A transmittance of 4% can be observed due to its above-bandgap absorption. After post-annealing process, the near-ultraviolet transmittance of the ITO film can be enhanced to over 40% while that of the GZO film shows little improvement. Despite the disadvantage of the GZO films at near-ultraviolet region, the GZO film has higher optical transparency at visible-light region. At blue-light region, the transmittance of the as-deposited GZO film is nearly 60%, and that of the as-deposited ITO film is 53%, though. Then, it can be further improved to 79% and 74% for the GZO film and ITO film, respectively. At green-light region, the transmittance of the as-deposited GZO film is almost 80%, but the green-light transmittance of the as-deposited ITO film is relatively low in contrast to that of the GZO film. Finally at red-light region, both GZO and ITO films shows relatively high transparency. The transmittance of as-deposited films both are over 80%, which can be even enhanced to over 90% after post-annealing process. In general, the optical transmittance of both ITO and GZO films grow higher with the red-shift of wavelength from near-ultraviolet to red-light region, and the GZO film has advantage of better transparency throughout visible region, but the ITO film performs better at near-ultraviolet region.

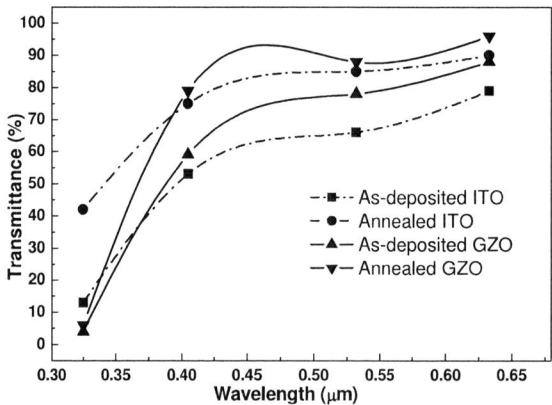

Fig.1. The optical transmittance of both as-deposited
and annealed ITO and GZO films

Electrical Resistivity

The sheet resistances of the ITO and GZO films with different annealing time are displayed in Fig.2. The sheet resistance of the as-deposited ITO film is about 40 Ω/square while that of the as-deposited GZO film is nearly 3500 Ω/square, higher by 2 orders. With longer annealing time at 450°C, the electrical resistivity of both ITO and GZO film first become lower and further increase sharply. The minimum sheet resistance of the ITO film can be observed only 26 Ω/square. Nevertheless, the minimum sheet resistance of the GZO film is still 2900 Ω/square. After annealing of 20 min, the sheet resistance of the ITO film increases to 156 Ω/square. However, the GZO film exhibits a much higher sheet resistance of over 3700 Ω/square. Fig.3. shows the sheet resistance of the ITO and GZO films with different annealing temperature. With annealing temperature from 400°C to 500°C and annealing time of 15 min, the sheet resistances of the ITO and GZO films show the same trend that it decreases and then increase greatly. The sheet resistance of the ITO film can be reduced to 37 Ω/square, while that of GZO film can be observed up to 3000 Ω/square. After annealing at the temperature over 480°C, the sheet resistance of the ITO and GZO film are measured 216 Ω/square and 4450 Ω/square, respectively. Therefore, the electrical conductivity of ITO film is much better than GZO film by 2 orders. In other words, the ITO film has lower contact resistivity.

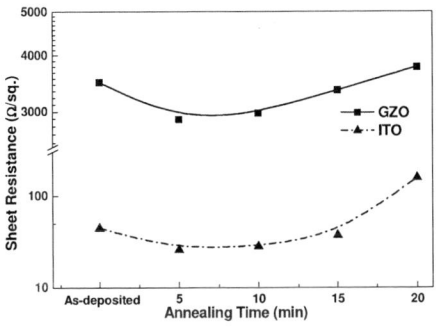

Fig.2. The sheet resistance of the ITO and Fig.3. The sheet of the ITO and GZO film GZO films after annealing at 450°C under same annealing time of 15 min.

FTIR Analysis

Fourier transform infrared (FTIR) spectroscopy is performed to realize the chemical composition of the ITO and GZO thin films. From Fig.4, it can be observed that the absorbance of all the ITO films, inclusive of as-deposited and annealed ITO films, are nearly the same. Except for three major peaks at 786 cm^{-1}, 1128 cm^{-1} and 1272 cm^{-1} representing SiO$_2$ bond, transverse optical mode and longitudinal optical mode of SiO$_2$ due to quartz substrate [13], there is a small disturbance at wavenumbers of 670 cm^{-1}, representing O-Sn-O bond. Compared to the peaks of quartz substrate, the intensity of O-Sn-O peak is extremely low. Despite the small change of O-Sn-O bond, the chemical composition of the ITO film doesn't change, so post-annealing process doesn't affect the chemical composition. Fig. 5 shows the FTIR spectra of the GZO films. The three peaks of quartz substrate at 786 cm^{-1}, 1128 cm^{-1} and 1272 cm^{-1} are still evident. Moreover, there is another minor peak at 2090 cm^{-1} representing Ga$_2$O$_3$ bond. The Ga$_2$O$_3$ peak of the as-deposited film is not clear, as well as that of the 425°C-annealed sample. After annealing at over 450°C, the Ga$_2$O$_3$ peak becomes stronger, leading to an increase of sheet resistance.

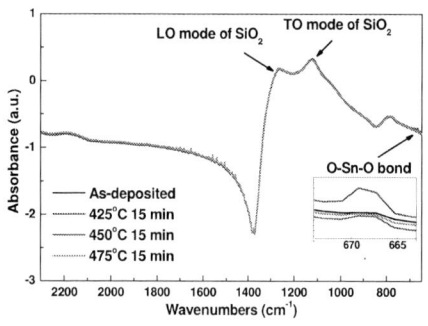

Fig. 4. The FTIR spectra of ITO films   Fig. 5. The FTIR spectra of GZO films

CONCLUSIONS

The comparison on the optical transmission and electrical resistivity of RF-sputtered ITO and GZO thin films is demonstrated. After post-annealing process, the near-ultraviolet transmittance of the ITO film can be enhanced to over 40% while that of the GZO film shows little improvement. However, the GZO film has higher optical transparency at visible-light region. At blue, green-light region, the optimized optical transmittance of the GZO films can be observed over 80%, and at red-light region, that of the GZO film is even up to 96%. On the other hand, the transmittance of the ITO film at visible-light region is a little lower than that of the GZO film. As for electrical resistivity, the ITO film has much lower sheet resistance than the GZO film by 2 orders. The minimum sheet resistance of the ITO and GZO film is 26 Ω/square and 2900 Ω/square, respectively. The FTIR spectra of the ITO film shows little change of chemical composition after post-annealing process, but there is a growing Ga$_2$O$_3$ peak of the GZO sample at the annealing temperature of 450°C or higher, leading to an increase of sheet resistance. After all, the ITO film has an advantage of the GZO film due to much lower sheet resistance. Though, the transmission at visible-light region is a little lower. ITO film is a referable choice as a transparent contact material.

ACKNOWLEDGMENT

The authors thank the National Science Council of Taiwan and the Excellent Research Projects of National Taiwan University for financially supporting this research under grants NSC97-ET-7-002-007-ET, NSC97-2221-E-002-055 and 97R0062-07.

REFERENCES

1. J. F. Wager, "Transparent Electronics", Science, Vol. 300, 1245, 2003.
2. F. Li, H. Tang, J. Shinar, O. Resto, S.Z. Weisz, "effects of aquaregia treatment of indium-tin-oxide substrates on the behavior of double layered organic light-emitting diodes", Appl. Phys. Lett., Vol. 70, 2741-2743, 1997.
3. I. Hamberg, C.G. Granqvist, K.F. Berggren, B.E. Sernelius and L. Engström, "Optical properties of transparent and infra-red-reflecting ITO films in the 0.2–50 µm range", Vacuum, Vol. 35, 207-209, 1985.
4. V. Assuncao, E. Fortunato, A. Marques, H. Aguas, I. Ferreira, M.E.V. Costa, and R. Martins, "Influence of the deposition pressure on the properties of transparent and conductive ZnO:Ga thin-film produced by r.f. sputtering at room temperature", Thin Solid Films, Vol. 427, 401–405, 2003.
5. D.H. Kim , M.R. Park, G.H. Lee, "Preparation of high quality ITO films on a plastic substrate using RF magnetron sputtering", Surface & Coatings Technology, Vol. 201, 927–931, 2006.
6. B.H. Choi, H.B. Im, J.S. Song, K.H. Yoon, "Optical and electrical properties of Ga_2O_3-doped ZnO films prepared by r.f. sputtering", Thin Solid Films, Vol. 193, 712-720, 1990.
7. A. De, P.K. Biswas, J. Manara, "Study of annealing time on sol–gel indium tin oxide films on glass", Materials Characterization, Vol. 58, 629–636, 2007.
8. K.Y. Cheong, N. Muti, S.R. Ramanan," Electrical and optical studies of ZnO : Ga thin films fabricated via the sol-gel technique", Thin Solid Films, Vol. 410, 142-146, 2002.
9. D. Kim, Y. Han, J. Cho, S. Koh, "Low temperature deposition of ITO thin films by ion beam sputtering", Thin Solid Films, Vol. 377&378, 81-86, 2000.
10. G.A. Hirata, J. McKittrick, T. Cheeks, J.M. Siqueiros, J.A. Diaz, O. Contreras,

O.A. Lopez, "Synthesis and optoelectronic characterization of gallium doped zinc oxide transparent electrodes", Thin Solid Films, Vol. 288, 29, 1996.

11. A. Khodorov, M. Piechowiak, M.J.M. Gomes, "Structural, electrical and optical properties of indium–tin–oxide thin films prepared by pulsed laser deposition", Thin Solid Films, Vol. 515, 7829–7833, 2007.

12. J. Hu, R.G. Gordon, "Atmosphereric-pressure chemical vapor-deposition of gallium doped zinc oxide thin films from diethyl zinc, water, and triethyl gallium", J. Appl. Phys. Vol. 72, 5381, 1992.

13. E. Aperathitis, "Properties of rf-sputtered indium–tin-oxynitride thin films", Journal of Applied Physics, Vol. 94, 2, 2003.

Mater. Res. Soc. Symp. Proc. Vol. 1109 © 2009 Materials Research Society

Electrical and optical properties of GaN and ZnO studied by surface photovoltage

M. Foussekis, A. A. Baski, and M. A. Reshchikov
Department of Physics, Virginia Commonwealth University, Richmond, VA, U.S.A.

ABSTRACT

We have studied the effect of ambient on the electrical and optical properties of GaN and ZnO in an original set-up based on a high-vacuum Kelvin probe system combined with an optical cryostat. The surface photovoltage (SPV) signal reached its highest value of 0.64 eV in GaN and 0.4 eV in ZnO, indicating that the initial (dark) upward band bending decreased by this amount under UV light. In air ambient, the SPV signal for GaN increased quickly to a maximum and then gradually decayed under illumination of the sample, whereas for ZnO it gradually increased and saturated. In vacuum, the SPV signal slowly increased under UV illumination for both GaN and ZnO. Relaxation of the SPV after switching off the light followed a logarithmic behavior in GaN and exponential behavior in ZnO samples.

INTRODUCTION

In spite of significant progress in the development of wide-bandgap semiconductors such as GaN and ZnO, the detrimental effects of surfaces and interfaces on the electrical and optical properties of materials and devices based on these semiconductors is often underestimated. It is known that undoped GaN and ZnO, similar to other n-type semiconductors, demonstrate an upward band bending due to negative charge at the surface [1]. Such charge may be localized at intrinsic surface states or at adsorbed surface species from the ambient. The absolute value of band bending is commonly determined from ultraviolet and X-ray photoemission spectroscopy (UPS, XPS) or Kelvin probe measurements. However, in the case of photoemission spectroscopy, band bending may be underestimated due to the photovoltage effect [2], whereas for Kelvin probe the work function of the metal tip and electron affinity of the semiconductor may be influenced by the ambient. The surface photovoltage (SPV), which is the change in band bending caused by illumination, can be measured by Kelvin probe and provides the *lowest* estimate of the absolute band bending in dark [1]. Illumination of GaN with UV light typically reduces the band bending by approximately $0.3 - 0.9$ eV due to the accumulation of photo-generated holes at the surface [2-5]. These values roughly agree with band bending values in n-type GaN which are estimated as 0.4 ± 0.2 eV from UPS and XPS studies [6] and 0.9 ± 0.3 eV from Kelvin probe measurements [4]. With regard to ZnO, there appears to be no band bending for a clean surface cleaved in ultrahigh vacuum [7]. In air ambient, however, chemisorption of oxygen results in upward band bending and electron transfer from the semiconductor to oxygen molecules or atoms [8]. Under illumination in vacuum, desorption of oxygen takes place and results in a surface photovoltage of up to 0.3 eV for single crystal ZnO [8]. Note that GaN and ZnO are polar semiconductors and due to spontaneous polarization may accumulate different charges on opposite sides of the sample grown along the [0001] direction. On the Zn-polar face of ZnO and Ga-polar face of GaN, excess negative charge should cause upward band bending, while on the opposite O- or N-polar faces excess positive charge and downward band bending are expected [9]. In this work, we have investigated the SPV in GaN layers and bulk ZnO in a combined Kelvin probe and optical cryostat with environmental control.

EXPERIMENTAL DETAILS

An undoped GaN layer with a thickness of 2.5 μm and a concentration of free electrons of about 10^{17} cm^{-3} was grown on c-plane sapphire by molecular beam epitaxy (MBE). An undoped n-type bulk ZnO sample with a thickness of 550 μm was grown at Cermet, Inc. and annealed at 1100 °C for 10 hours. The SPV was measured with a Kelvin probe (model KP-6500 from McAllister Technical Services) attached to an optical cryostat (VPF-700 from Janis Research Company, Inc.) Prior to illumination, the sample was maintained in dark for an extended period to minimize any residual SPV from previous light exposure. As shown in Fig. 1(a), the sample was illuminated from the backside through a sapphire window (and sapphire substrate in the case of GaN). A xenon lamp (75 W) was used to illuminate the sample after passing through a 0.25-m grating monochromator and long-pass filters. Neutral-density filters were used to attenuate the standard light power density of 0.03 W/cm^2 by up to nine orders of magnitude. To perform SPV spectra measurements, a constant photon flux was achieved by varying the slit width, and data were acquired for steady-state or nearly saturated conditions.

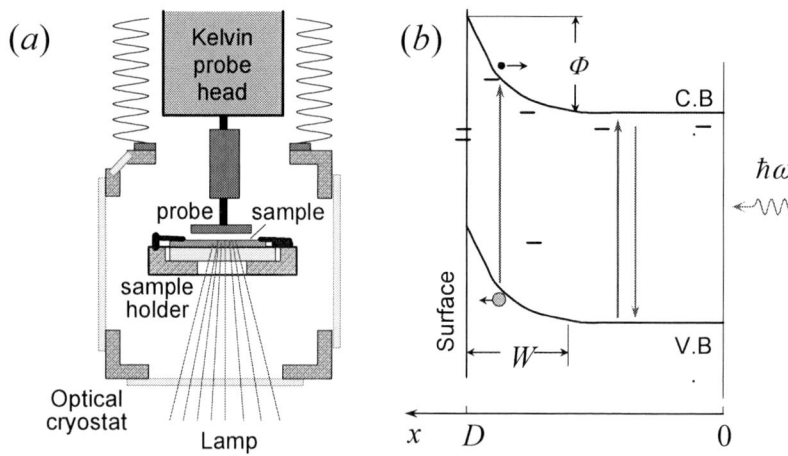

Figure 1. (a) Schematic cross-section of the Kelvin-probe apparatus with illumination from the backside of the sample. (b) Band diagram for an n-type semiconductor with upward band bending and a depletion region near the surface.

Figure 1(b) shows the band diagram for a sample with upward band bending at the surface, where Φ is the barrier height and W is the depletion region width. Illumination through the sample from the backside with photons having energy close to the bandgap creates electron-hole pairs in the depletion region which are quickly separated by the strong electric field. Holes accumulate at the surface and due to their positive charge reduce the band banding. The light flux absorbed in the depletion layer is equal to the rate of holes flowing to the surface R_h given by:

$$R_h = \int_{D-W}^{D} \alpha P_0 e^{-\alpha x} dx = P_0 e^{-\alpha D} \left(e^{\alpha W} - 1 \right) \tag{1}$$

where P_0 is the incident light intensity, α is absorption coefficient and D is the sample thickness. For a 2.5 μm-thick GaN layer with a 0.1 μm-wide depletion region, the maximum amount of light absorbed in the depletion region is 1.5% for $\alpha \approx 4 \times 10^3$ cm^{-1}, which corresponds to photon energies close to the GaN bandgap. For a 550 μm-thick ZnO sample with $W = 0.1$ μm, the maximum amount of light absorbed is 0.007% for $\alpha \approx 20$ cm^{-1}, corresponding to photon energies close to the bandgap of high-quality bulk ZnO [10]. Note that due to the strong electric field in the depletion region, photons with energies slightly less than the bandgap will be effectively absorbed in the depletion region via photon-assisted tunneling (Franz-Keldysh effect), which may significantly increase the fraction of photons absorbed in the depletion region. On the other hand, photons with energy lower than the bandgap for which $\alpha D << 1$ can be captured by surface states, provided that the photon energy is larger than the energy required for excitation of bound electrons to the conduction band. Light absorbed in the bulk region outside the depletion region does not affect the band bending, if the light intensity is not very high [1].

RESULTS

The SPV spectra for GaN in vacuum and air ambient and for ZnO in air ambient are shown in **Fig. 2.** In the case of GaN, the SPV signal exhibited a threshold at ~1.3 eV photon energy, gradually increased with increasing photon energy from 1.3 to 3.2 eV, and then had a relatively sharp maximum at 3.4 eV, which is close to the GaN bandgap at room temperature (3.43 eV). At higher photon energies, the SPV dropped due to a significant rise of the absorption coefficient and the inability of photons to reach the depletion region. The SPV signal in the region above 3.4 eV is due to near-band-edge photoluminescence (PL) penetrating through the sample and being partially absorbed in the depletion region. In vacuum, the SPV slightly decreased at photon energies below 2.2 eV and increased at photon energies above 2.4 eV.

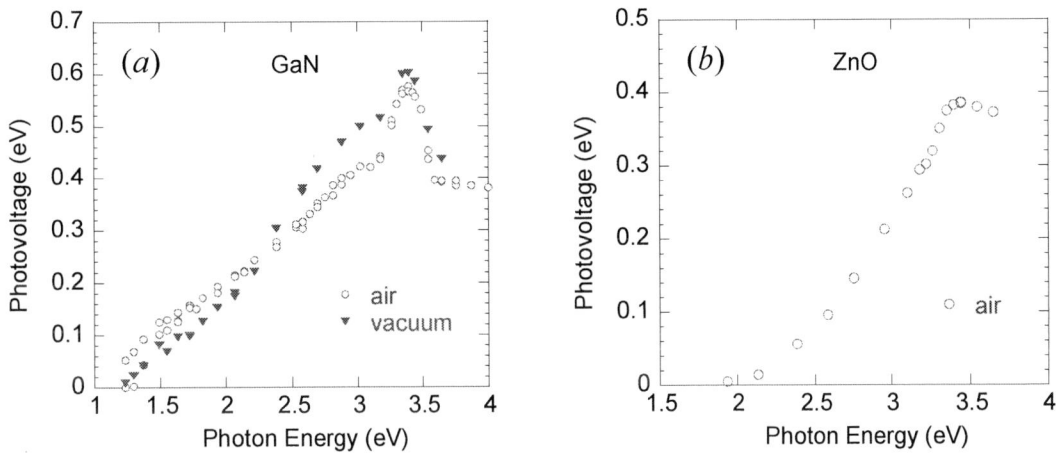

Figure 2. Steady-state SPV spectrum for GaN (*a*) and ZnO (*b*) measured at room temperature in vacuum and air. $P_0 = 0.03$ W/cm^2.

In the case of ZnO in air ambient, the SPV signal exhibited a threshold at the photon energy of ~2.0 eV, gradually increased with increasing photon energy from 2.0 to 3.3 V, and then slightly decreased above the bandgap of ZnO (~3.35 eV).

155

The behavior of the SPV as a function of light intensity for GaN in ambient is shown in **Fig. 3**, where the SPV increases linearly for $P_0 < 10^{-9}$ W/cm^2 and as a logarithm for higher light intensities. This dependence can be empirically fitted with Eq. (2) which uses thermionic emission for current transport in a metal-semiconductor contact [11] and is commonly used in the analysis of photovoltage as a function of light intensity [12,13].

$$SPV = \eta kT \ln\left(\frac{P_0}{J_0} + 1\right) \tag{2}$$

Here, J_0 corresponds to the metal-to-semiconductor saturation current in a Schottky diode and η is the ideality factor [11]. For light intensities $P_0 > 10^{-3}$ W/cm^2, there is a gradual saturation of the SPV signal and subsequent deviation from logarithmic behavior. Such photo-saturation can be explained by flattening of the surface band bending, which indicates that the dark band bending for GaN is not much larger than 0.6 eV.

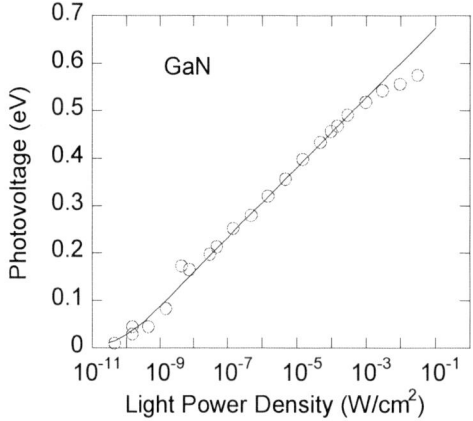

Figure 3. Dependence of the SPV for GaN on band-to-band (3.4 eV) light intensity in air ambient. Solid curve is fit with Eq. (2) with $\eta = 1.26$ and $J_0 = 7 \times 10^{-11}$ W/cm^2.

The time evolution of the SPV signal when the illumination was switched on and off for the GaN and ZnO samples is shown in **Fig. 4**. For the GaN sample, we observed an interesting effect upon illumination with band-to-band light. In air ambient and at the highest light intensity, the SPV reached a maximum value of 0.62 eV in a few seconds and then significantly decreased under UV exposure over the next few hours. This quenching effect decreased with decreasing light intensity and could not be observed for $P_0 < 10^{-3}$ W/cm^2. In contrast, the SPV signal in vacuum initially jumped to 0.53 eV in a few seconds, gradually increased, and then saturated at ~0.64 eV after a 3 h exposure. This behavior of the SPV signal can be explained by the photo-adsorption of oxygen in ambient and photo-desorption of such species in vacuum under UV illumination [14]. A slow decrease of the SPV under UV illumination in ambient corresponds to a gradual increase of the near-surface barrier due to the adsorption of surface species that can be negatively charged. Note that the resulting increase of the depletion region width should also result in a decrease of photoluminescence (PL) intensity, because PL from the depletion region is negligible. Our preliminary studies indicate that PL intensities at 3.35 eV (near-band-edge emission) and 2.2 eV (yellow luminescence band) gradually decrease under continuous UV illumination in ambient, but increase or remain unchanged in vacuum.

In contrast to GaN, the SPV signal for ZnO increased and saturated under UV illumination for both vacuum and ambient conditions. After switching off the illumination, the SPV decreased slowly in air and extremely slowly in vacuum [see **Fig. 4(b)**]. The very slow restoration of the contact potential in vacuum to its dark value hindered most routine measurements for ZnO. We have fitted the decay behavior after switching off the light and found that the SPV decays nearly exponentially with time for ZnO, but decays logarithmically for GaN.

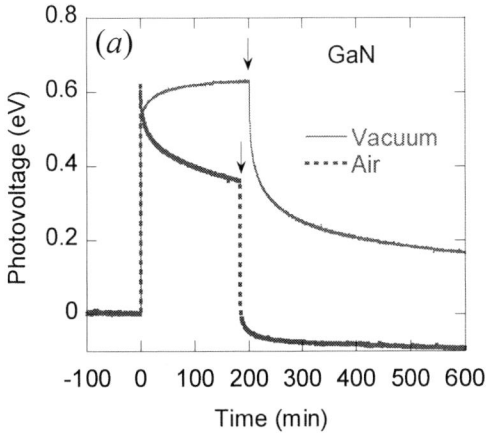 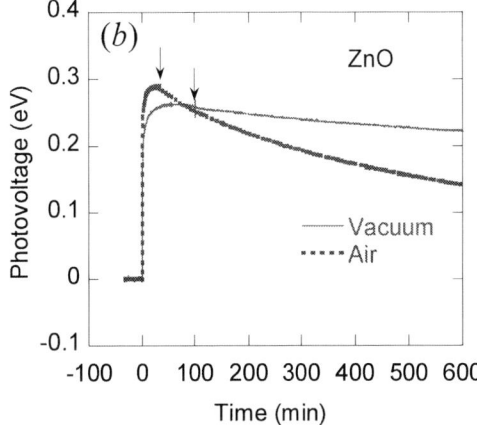

Figure 4. Evolution of the SPV signal in ambient and vacuum under illumination with 365 nm light ($P_0 = 0.03$ W/cm^2) at 295 K for (a) GaN and (b) ZnO. The light is turned on at zero time and turned off at the times indicated by the arrows.

DISCUSSION

In the early studies of SPV in semiconductors, two types of surface states known as "slow" and "fast" were distinguished [15]. The fast states, located at the semiconductor-oxide interface, are believed to be independent of ambient changes, whereas the slow states, located at the surface and associated with adsorbed species, can be influenced by the ambient. In our studies, the initial rise of the SPV under UV illumination is fast for both GaN and ZnO, and is related to the accumulation of photo-generated holes near the surface. After this initial rise, the SPV for GaN is seen to increase or decrease more slowly under continuous UV illumination, depending on the environment. In air, the SPV slowly decreases and is likely due to the transfer of electrons from surface states to adsorbed surface species. The surface states are occupied again by electrons from the bulk and cause the observed gradual increase of band bending (or decrease in SPV). In vacuum, it appears that the UV light causes negatively charged species to desorb from the surface and reduce band bending. Due to the vacuum environment, these desorbed species are not re-adsorbed as would be the case in air ambient. With regard to ZnO, the SPV slightly increases during illumination under both vacuum and ambient conditions. Since the light intensity is significantly lower for the bulk ZnO sample as compared to the GaN film, this behavior is not straightforward to interpret and requires further study. When the illumination is switched off, we observe both fast and slow components in the SPV decay for GaN and only a slow component for ZnO. The initial fast SPV decay in GaN is related to the transfer of free electrons from the bulk to the surface over the barrier. The logarithmic time dependence of the SPV decay in GaN can be explained by the gradual restoration of the near-surface barrier with time. In contrast, the SPV decay for ZnO is slow and demonstrates an exponential dependence on time. Such slow behavior indicates that physisorption and chemisorption processes likely govern the behavior of SPV transients in ZnO.

SUMMARY

We have investigated the transient and steady state behavior of the SPV signal in undoped *n*-type GaN and bulk ZnO with a Kelvin probe in vacuum and air ambient. Under UV illumination, the SPV reaches a maximum value of 0.64 eV in GaN and 0.4 eV in ZnO, and is attributed to the band-to-band excitation of electrons in the depletion region. The SPV spectrum has a threshold energy of 1.3 eV for GaN and ~2 eV for ZnO, and then increases up to a relatively sharp maximum near the bandgap energy (3.4 eV for GaN and 3.3 eV for ZnO). The intensity behavior of the SPV signal in GaN increases with a logarithmic dependence for a wide range of light intensities and starts to saturate at a power density exceeding 1 mW/cm^2. The slow and logarithmic decay of the SPV after ceasing band-to-band illumination in GaN is explained qualitatively by the transfer of electrons from the bulk to surface states. The SPV decay in ZnO is much slower and is attributed to physisorption and chemisorption processes.

ACKNOWLEDGMENTS

This work was supported by the NSF. The authors are grateful to Dr. J. Nause from Cermet, Inc. and Dr. H. Morkoç from VCU for providing ZnO and GaN samples, respectively.

REFERENCES

[1] L. Kronik and Y. Shapira, Surf. Sci. Rep. **37**, 1 (1999).

[2] J. P. Long and V. M. Bermudez, Phys. Rev. B **66**, 121308 (2002).

[3] I. Shalish, Y. Shapira, L. Burstein , and J. Salzman, J. Appl. Phys. **89**, 390 (2001).

[4] S. Sabuktagin, M. A. Reshchikov, D. K. Johnstone, and H. Morkoç, Mat. Res. Soc. Symp. Proc. **798**, Y5.39 (2004).

[5] I. Shalish, L. Kronik, G. Segal, Y. Rosenwaks, Y. Shapira, U. Tisch, and J. Salzman, Phys. Rev. B **59**, 9748 (1999).

[6] V. M. Bermudez, J. Appl. Phys. **80**, 1190 (1996).

[7] R. K. Swank, Phys. Rev. **153**, 844 (1967).

[8] J. Lagowski, E. S. Sproles, Jr., and H. C. Gatos, J. Appl. Phys. **48**, 3566 (1977).

[9] M. W. Allen, P. Miller, R. J. Reeves, and S. M. Durbin, Appl. Phys. Lett. **90**, 062104 (2007).

[10] R. E. Dietz, J. J. Hoppfield, and D. G. Thomas, J. Appl. Phys. **32**, 2282 (1961).

[11] S. M. Sze, *Physics of Semiconductor Devices*, 2nd ed., Wiley, New York, 1981.

[12] S. C. Dahlberg, J. R. Chelikowsky, and W. A. Orr, Phys. Rev. B **15**, 3163 (1977).

[13] A. L. Musatov and S. Yu. Smirnov, Surf. Sci. **269-270**, 1048 (1992).

[14] M. A. Reshchikov, M. Foussekis, and A. A. Baski, unpublished.

[15] W. H. Brattain and J. Bardeen, Bell System Tech. J. **32**, 1 (1953).

Transparent and conductive ZnO:Al powder prepared by Soft Chemical route Process and Design of Experiment Technique

Kuo-Chuang, Chiu, Yi-Wen, Kao, Ren-Der, Jean

Materials Research Laboratories, Industrial Technology Research Institute, Chutung, Hsinchu 31015, Taiwan, R.O.C.

ABSTRACT

Aluminum doped zinc oxide polycrystalline powder (AZO) were prepared by soft-chemical route process. The quantity of aluminum in the sol was varied from 1 to 5 mol%. The structural characteristics studied by X-ray diffractometry were complemented resistivity measurement by AC impedance spectroscopy. Prepared under tartaric acid as chelating agents and sintered at 1400 °C and design of experiment (D.O.E.) method was employed to elucidate the AZO formulation for the powder process. Following the design of experiment method, dominant factors were found for [Al/Zn] mol%, sintering temperature, sintering time, annealing environment was optimized for production. The best conductors were obtained for the AZO powder containing 3 mol% of aluminum.

Keywords: ZnO:Al, Soft Chemical route Process, Design of Experiment, Impedance spectroscopy

INTRODUCTION

Zinc oxide exhibit a combination of interesting piezoelectric, electrical, optical and thermal properties, which are already applied in the fabrication of a number of devices, such as gas sensors, ultrasonic oscillators and transparent electrodes in solar cells, etc.

In the present work we have investigated use Soft Chemical route to prepare powder and measurement by electrochemical impedance spectroscopy to study AZO(Al:ZnO), The Soft Chemical route Process is a particularly attractive synthetic route for the preparation of mulicomponent inorganic oxides, since a homogeneous mixture of the several components at a molecular level can be easily reached in solution. In this paper contain typically AZO solid solution to improve the solid oxide's conductivity and lower the electrical resistance and design of experiment (D.O.E.) method was employed to elucidate the AZO formulation for the powder process.

EXPERIMENTAL PROCEDURE

The AZO oxides powders were prepared by the low temperature Soft Chemical route Process. Zinc nitrate ($Zn(NO_3)_2 \cdot 6H_2O$) and aluminium nitrate ($Al(NO_3)_3 \cdot 9H_2O$) were dissolved in a solution of D.I. water and Nitric acid. The molar ratios of dopant in the solution, [Al/Zn], was varied between 1 and 5%. The solution were heated until gel state and then dried. The powder were calcined at 600 $^\circ$C and sintered at 1000-1400 $^\circ$C.

Experimental procedure of this study was based on design of experiment (D.O.E.) method. The fish-bone causal analysis of materials recipe and process flow on the resistivity is given in Fig.1. The parameters those affects the resistivity includes [Al/Zn] mol%, sintering temperature, sintering time, and annealing environment.

Variables and constants were chosen from Fig.1 as follows

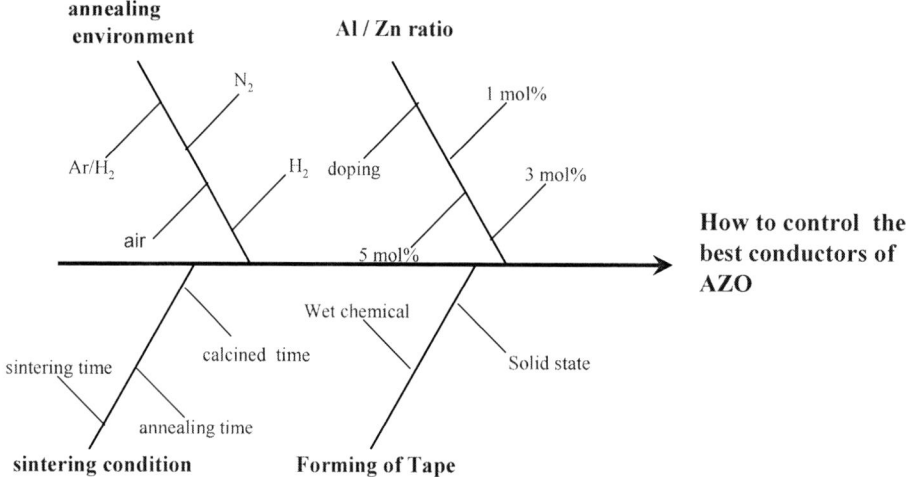

Fig.1 Fish-bone causal analysis of materials recipe and process flow on the Aluminum doped zinc oxide polycrystalline powder.

Variables:
(1) [Al/Zn] mol%
(2) sintering temperature
(3) sintering time
(4) annealing environment
    Constants:
(1) calcined temperature
(2) calcined time

160

(3) annealing temperature

(4) annealing time

The selection guideline was the maximum and minimum amounts allowable by the experimental apparatus. This was a three-level experiment. The variables and chosen levels are listed in Table 1. [Al/Zn] mol% (A), sintering temperature (B), sintering time (C), annealing environment (D) were the experiment variables ; and a suitable point-line graph is shown in Fig2. $L_9 3^4$ orthogonal array and results given in Table 2 translated this chart. This was a 9-test runs experiment.

Table 1. Selected factors and levels for the experiment

Factor		Level 1	Level 2	Level 3	Interaction
Al/Zn ratio(mol%)	A	A1	A2	A3	
Sintering Temperature	B	B1	B2	B3	
sintering time	C	C1	C2	C3	
annealing environment	D	D1	D2	D3	

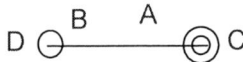

Fig. 2. Point-line graph of various factors in the experiment.

Table 2. Orthogonal array of the first-stage experiment

	1	2	3	4
1	1	1	1	1
2	1	2	2	2
3	1	3	3	3
4	2	1	2	3
5	2	2	3	1
6	2	3	1	2
7	3	1	3	2
8	3	2	1	3
9	3	3	2	1
	a	b	a	a
			b	b2

161

DISCUSSION

In this paper, the numbers following the abbreviation refer to the relative proportions of Al in the material. For example, $Al_{0.01}Zn_{0.99}O$ is designated AZO-1.

The DTA/TGA data of Fig.3 indicate that decomposition of the gel process in several steps. Weight loss calculations from the TGA curve indicate that the pronounced endothermic peak near 150 °C in the DTA curve can be attributed primarily to the dehydration of the gel. The first endothermic peak is followed by an endothermic peak having a minimum near 400 °C. The TGA curve indicates additional weight loss occurring at the temperature. This second endothermic peak signals two superimposed thermal decompositions, the acetate salt into its oxycarbonate, followed by the oxycarbonate into its oxide.

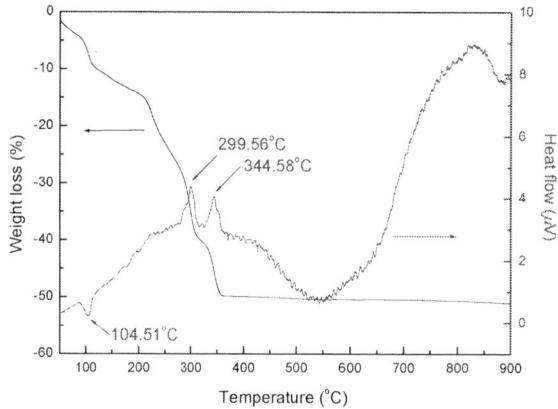

Fig.3 DTA and TGA curves for an AZO powder taken at 1 °C /min in air flowing at 100mL/min

Figs 4 shows XRD patterns of the AZO oxides calcined at 600 °C and sintered at 1000-1400 °C. From the evolution of the diffraction patterns with the doping concentration in the precursor solution it can be observed that the quantity of Al doping determines the orientation of the nanocrystals. When the Al^{3+} doping concentration and sintering temperature exceeded 3 mol% and 1200 °C, the $ZnAl_2O_4$ phase appeared as shown in Fig. 4.

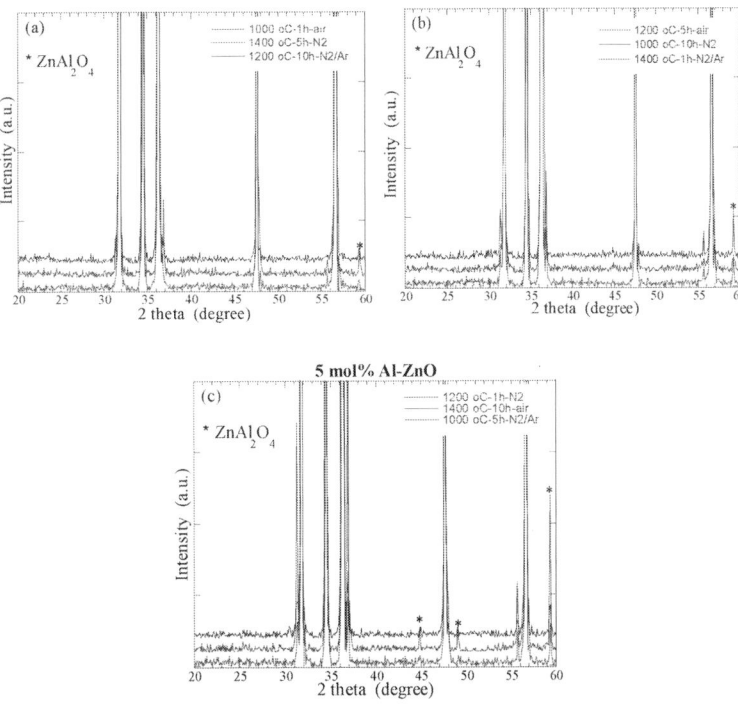

Fig.4 Evolution of the X-ray diffraction patterns of films prepared, at different molar ratios Al/Zn in the oxides (a) 1 mol% Al (b) 3 mol% Al (c) 5 mol% Al.

TEM microphoto for soft-chemical route-prepared powders calcined at 600 °C is shown in Fig. 5. It can be found that the powder is fine and homogeneous, with the mean particle size of around 20-30 nm. The fine and homogeneous powder would have high sinterability. SEM photographs of the surface of the sintered samples are shown in Figs 6. The surface microstructure reveals uniform and fine grain growth about 0.7-2 μm. No pore was observed on the surface of the sample, but there were some pores at 1000 °C and 1200 °C, as shown in Fig 6. (a1), (a2), (b1), (b2), (c1), (c2). According to the measurement, the sintered density is 5.48 g/cm³, which is about 98.7% of the theoretical value.

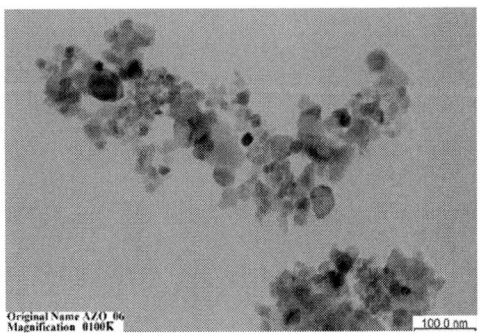

Fig.5 TEM micrograph of AZO powder calcined at 600 °C

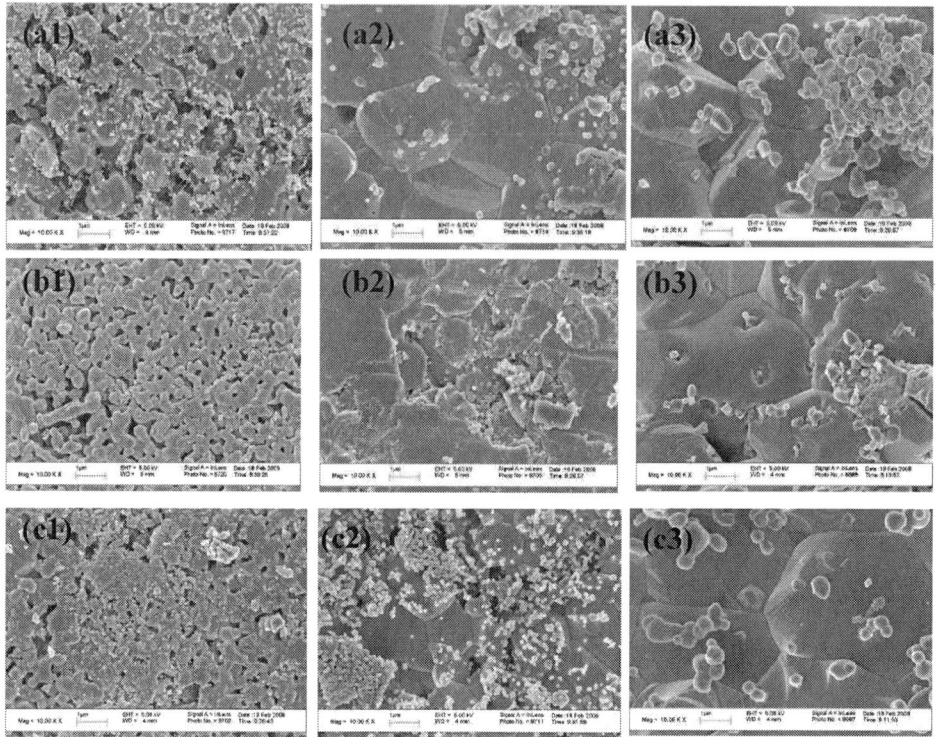

Fig 6. Microstructure of the surfaces of sintered AZO (a) 1 mol% Al (b) 3 mol% Al (c) 5 mol% Al after heat treatment in air, N_2 or N_2/Ar at 1000-1400 °C.

Table 3. show that electrical resistivity as a function of the [Al/Zn] mol%, annealing environments and sintering conditions for AZO. It can be seen that the resistivity for sample with the AZO-3 is lower than others. A further increase in the [Al/Zn] ratio, beyond the optimum value, leads to an increase in the resistivity values of the AZO. The decrease in the resistivity of the AZO has been explained in terms of a gradual replacement of Zn^{2+} ions into Al^{3+}. This process releases a free electron to the conduction band for every Al ions incorporated into the lattice, and hence lead to an increase in the carrier concentration of the AZO.

The goal of controlling the [Al/Zn] = 0.03/0.97 mol% had been successfully achieved and heat treatment in N_2 at 1000 °C was reduced to $2.9582*10^{-1}$ Ωcm

Table 3. Impedance data of AZO (a) 1 mol% Al (b) 3 mol% Al (c) 5 mol% Al after heat treatment in air, N_2 or N_2/Ar at 1000-1400 °C.

1000 °C	1 mol%-1h-air	3 mol%-10h-N_2	5 mol%-5h-N_2/Ar
D(mm)	9.36	9.19	8.9
d(mm)	1.4	1.45	1.38
1 kHz-R	0.59871	1.49E-02	4.55E-02
1 kHz-ρ	12.57518	2.95821E-1	9.21551E-1
1200 °C	1 mol%-10h-N_2/Ar	3 mol%-5h -air	5 mol%-1h-N_2
D(mm)	8.59	8.67	8.56
d(mm)	1.34	1.37	1.5
1 kHz-R	3.76E-01	0.027052	25151
1 kHz-ρ	7.57	5.38E-01	450907.7
1400 °C	1 mol%-5h-N_2	3 mol%-1h-N_2/Ar	5 mol%-10h-air
D(mm)	8.57	8.4	8.11
d(mm)	1.25	1.3	1.4
1 kHz-R	3780.7	0.023619	4298.4
1 kHz-ρ	81431.59	4.79454E-1	78225.71

The main purpose of the present study was to control the best conductor of AZO using Design Experiment method. From the one-stage experiment, following results were concluded:

(1) [Al/Zn] mol%: The variable Al was obviously a critical factor in affecting resistivity. When the Al^{3+} doping concentration exceeded 3 mol%, the $ZnAl_2O_4$ phase appeared.

(2) Sintering Temperature: This variable had more significant effect on resistivity. When the sintering temperature exceeded 1200 °C, the $ZnAl_2O_4$ phase appeared.

(3) Sintering time: This variable had significant effect on resistivity. When sintering time increased, the resistivity decreased.

(4) annealing environment: This variable had no effect upon resistivity.

CONCLUSIONS

We have developed a high-yield method of preparing AZO, a soft-chemical route. AZO single phase could be obtained at 600 °C for 1 h. Particle size of the AZO was found to be in the range 20-30 nm. Minimum resistivity of AZO samples were sintered in N_2 at 1000 °C for 10 h.

REFERENCES

G.G. Valle, P. Hammer, S.H. Pulcinelli, C.V. Santilli, Transparent and conductive ZnO:Al thin films prepared by sol-gel dip-coating, Journal of the European Ceramic Society 24 (2004) 1009–1013

Mater. Res. Soc. Symp. Proc. Vol. 1109 © 2009 Materials Research Society

Enhancement-Mode ZnO Thin-Film Transistor Grown by MOCVD

Jungyol Jo, Junho Yun, and Haemi Kim
Ajou University, Department of Electrical and Computer Engineering, Suwon, 443-749, Korea

ABSTRACT

We studied variation of threshold voltages in zinc oxide (ZnO) thin-film transistors (TFT) grown by metalorganic chemical vapor deposition (MOCVD). We used growth interruptions during MOCVD to encourage complete oxidation of deposited ZnO film. With this method, turn-off characteristics were significantly improved, and threshold voltage was shifted to positive voltages. ZnO TFT's grown at $450^{\circ}C$ showed 10^7 on/off ratio with 18 $cm^2/Vsec$ mobility, and +5 V threshold voltage. We also observed that annealing under different conditions caused significant changes in the threshold voltage.

INTRODUCTION

Zinc oxide (ZnO) has attracted considerable attention due to wide bandgap and high mobility. Various growth methods, such as molecular beam epitaxy [1,2], sputtering [3], pulsed-laser deposition (PLD) [4], and metalorganic chemical vapor deposition (MOCVD) [5,6] have been used. Among these, MOCVD has advantages for industry, since it can be applied to large-size substrates more easily. The problem associated with MOCVD is that turn-off characteristics of ZnO thin-film transistor (TFT) is not good compared to those grown by other methods [5-7]. Sputtering and PLD have produced ZnO TFT's with 20 - 50 $cm^2/Vsec$ mobility and 10^8 on/off ratio, which are superior to those grown by MOCVD. ZnO TFT grown by MOCVD usually shows negative threshold voltage and large off current. The purpose of our study is to realize high quality, enhancement-mode ZnO TFT by using MOCVD.

There are many kinds of defects in ZnO, which behave as n-type dopants [7,8]. In order to have ZnO TFT's with positive threshold voltage, these n-type defects should be removed. ZnO grown by MOCVD usually shows O deficiency, which can form defects of Zn interstitial or O vacancy. It is likely that the O deficiency is one of the reasons for the poor turn-off characteristics. We thought that O deficiency could be improved if we allow sufficient oxidation time by using growth interruptions.

Another problem of ZnO MOCVD is C and H incorporation in ZnO. In our previous work [5] we showed that high concentrations of C and H are found in ZnO grown by MOCVD, where dietyhlzinc (DEZ) and O_2 were used as sources. It is likely that the defects related to the C and H can also contribute to the poor turn-off characteristics. To minimize the effects of C and H, higher growth temperatures should be used.

EXPERIMENTAL DETAILS

Our MOCVD system has a horizontal reactor operating at atmospheric pressure. DEZ and O_2 were used as sources. N_2 was employed as a carrier gas with flow of 4000 sccm. DEZ bubbler was kept at $-10^{\circ}C$, with N_2 flow of 20 sccm. O_2 flow was 1000 sccm. O_2 and DEZ are designed to meet at 2 cm before the substrate. ZnO films were grown on heavily doped n-type Si substrates (10^{19} cm^{-3}) with a thermal oxide of 110 nm thickness. The substrate size is 2 cm x

2cm. The thickness of ZnO film at the center of sample is 30 to 50 nm. The growth temperature was 400°C or 450°C. ZnO TFT's were fabricated by evaporating Al through a shadow mask, and the TFT channel is 15-μm long and 500-μm wide. Current-voltage characteristics were measured by using Keithley 2400 Sourcemeters. Some of the TFT's were annealed in air or in vacuum after device fabrication.

DISCUSSION

Figure 1 is drain current changes measured in two samples, as a function of gate voltages. A schematic diagram of ZnO TFT is also shown in Fig. 1. This result shows the importance of growth interruptions during MOCVD. The two TFT's were grown consecutively at 400°C with different number of growth interruptions. Sample A was grown with 3 interruptions, and sample B with 5 interruptions. All other growth parameters were kept the same. Each growth interruption was 2 min long. After 20-sec ZnO growth, DEZ was closed, and only O_2 was supplied during the 2 min.

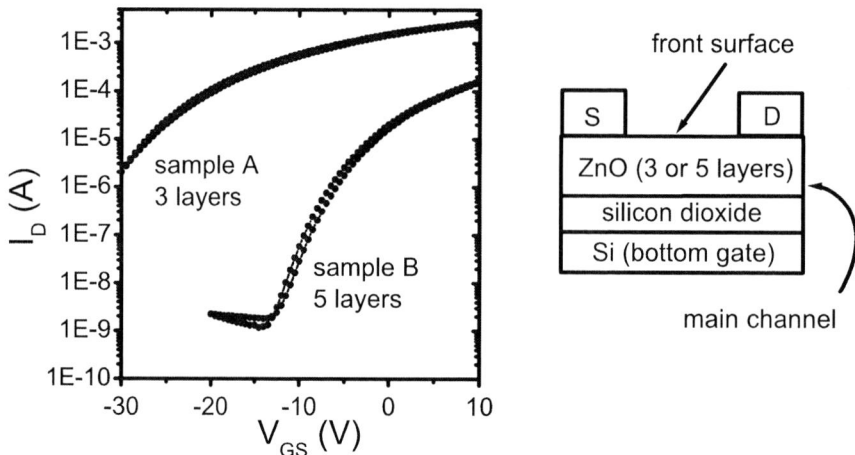

Figure 1. Drain currents as a function of V_{GS}, measured in two ZnO TFT's grown with different number of interruptions during MOCVD. 3 interruptions were used in sample A, and 5 interruptions were used in sample B. $V_{DS} = 10$ V, and channel length = 15 μm, width = 500 μm.

If we assume that ZnO does not grow during the interruptions, then sample A is made of 3 ZnO layers, and sample B is made of 5 ZnO layers. In Fig. 1, sample A shows much larger current than sample B, even though sample A is thinner than sample B. Since sample B is thicker than sample A, sample B should show larger current if each layer carries equal amount of current. However, Fig. 1 shows that this is not the case.

The larger current of sample A indicates that current does not depend on the film thickness. We explain that our results are related to oxygen depletion at the front surface [9], which works as n-type doping for the channel. Since our device is a bottom-gate structure, ZnO/SiO_2 interface is the main channel. In sample A, the front surface is closer to the main channel due to the thinner film thickness, and channel electron density will be higher, resulting in a more negative threshold voltage (V_{th}). When the front surface is more separated from the main channel in sample B, the effect of front-surface doping will be reduced, and V_{th} will shift to a positive direction. We think that the reduced doping effect from the front surface is the reason of positive V_{th} shift and high on/of ratio in sample B.

After optimizing growth parameters, we could obtain high quality ZnO TFT (sample C in Fig. 2). The same 5 periods were used, each made of 20-sec ZnO and 2-min oxidation. Higher growth temperature ($450^{O}C$) showed better on/off ratio and mobility, possibly due to the reduction of C and H incorporation. Figure 2 shows 10^7 on/off ratio and +5V V_{th}, where V_{th} was obtained from the square root of drain current. Mobility was calculated to be 18 cm^2/Vsec.

When we tried similar film thickness without the growth interruptions, TFT always showed strong negative V_{th}, around -30 V. Without the interruptions, ZnO film will have O-deficiency defects due to lack of the oxidation time, and the strongly negative V_{th} can be explained.

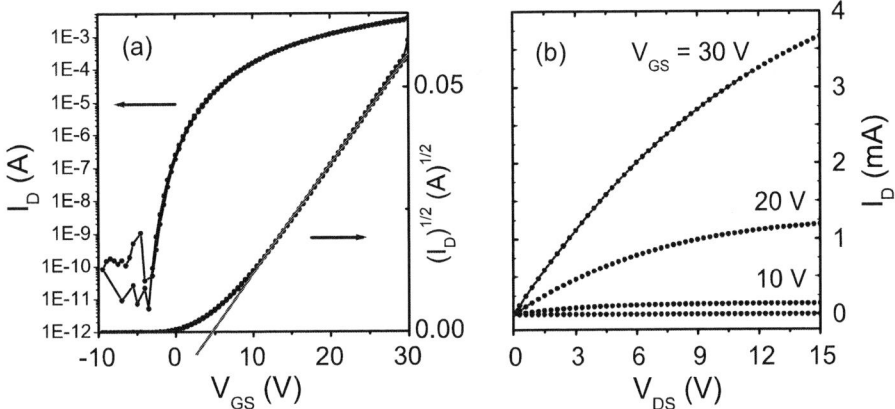

Figure 2. Current-voltage characteristics of sample C. 5 interruptions were used at $450^{O}C$ growth temperature. (a) Drain currents (log scale and square root) measured at V_{DS} = 10 V. (b) Drain currents for gate voltage of 30 to -10 V. The lowest curves are for V_{GS} = 0 and -10 V.

We studied the effect of ambient gas and annealing. The sample used in Fig. 3 was grown using the same growth parameters used in sample C. After initial current measurement, the TFT was stored in a nitrogen chamber. After 4-month storage, the TFT showed negative V_{th} shift (-7 V). When the TFT was subsequently annealed in air at 150 ^{O}C for 30 min, V_{th} showed positive shift (+10 V), and it was more positive than the initial V_{th}. Other TFT's showed similar behavior. We think that this is the result of water or hydrocarbon adsorption/desorption. During storage in nitrogen, residual water or hydrocarbon in the storage chamber could be adsorbed, so that V_{th} would shift to a negative direction. After 150 ^{O}C anneal in air, the adsorbed gases are removed, and V_{th} would shift back to a positive direction. There have been many reports about surface conduction layers formed on ZnO surface [10-12]. They explained that the conduction layer was formed by adsorption of hydrogen or water. We think that similar mechanisms work in Fig. 3.

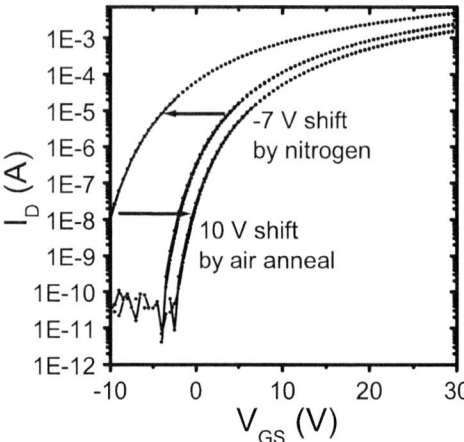

Figure 3. Current changes by storage in nitrogen, and 150^OC annealing in air (V_{DS} = 10 V). Nitrogen storage shifted V_{th} by -7 V, and annealing shifted V_{th} by +10 V.

Annealing at higher temperature was also studied. The samples in Fig. 4 were not grown at optimum conditions, so that they are n^+ type. They were annealed at 450^OC in vacuum for 1 hour. TFT in Fig. 4a was annealed with GaAs wafer on top, and TFT in Fig. 4b was annealed without the GaAs wafer. The purpose of having GaAs wafer is to allow As to move into ZnO TFT. We thought that the high vapor pressure of As at 450^OC can make it a good p-type dopant source. It is well known that column V elements (N, P, As) can work as p-type dopants. We wanted to see if As causes any shift in V_{th}.

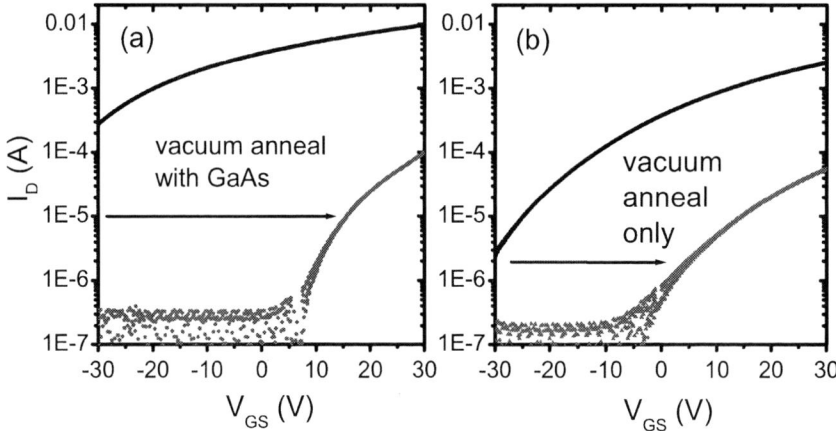

Figure 4. Current-voltage characteristics measured before and after annealing in vacuum. In (a) GaAs wafer was placed on top of TFT during annealing.

In Fig. 4a, current starts to rise at +5 V, while in Fig. 4b, similar behavior is observed at -5 V. We think that this confirms that As can work as a p-type dopant. The As originated from GaAs wafer compensates n-type defects in ZnO, resulting in the positive shift. Since TFT has a thin layer of ZnO, simply placing GaAs wafer on top during annealing can be a good method of p-type doping.

Vacuum annealing always showed better results than air annealing. We think that the vacuum annealing at high temperature removes loosely bound Zn or O atoms in ZnO film. Incomplete oxides can be source of n-type defects. It appears that the high-temperature vacuum annealing removes these n-type defects, and V_{th} shifts to a positive direction as a result.

CONCLUSIONS

We demonstrated that threshold voltage and off-current in ZnO TFT could be controlled by growth interruptions during MOCVD. Turn-off characteristics were significantly improved by the interruptions, and 10^7 on/off ratio with 18 cm^2/Vsec mobility were obtained. Our data show that surface layer is important in determining ZnO-TFT performance, since it works as n-type doping for the conduction channel. We also confirmed that vacuum annealing with GaAs wafer on top of ZnO TFT can cause positive shift of threshold voltage, which is an evidence that As is p-type dopant.

ACKNOWLEDGMENTS

We thank Eun-Ho Lee and Dukwon Lee of the Center for Material Characterization at Ajou University for x-ray diffraction and scanning electron microscope measurements. This work was supported by the Korea Research Foundation Grant funded by the Korean Government (MOEHRD, KRF-2007-313-D00488).

REFERENCES

1. K. Miyamoto, M. Sano, H. Kato, T. Yao, *Jpn. J. Appl. Phys.* **41**, L1203 (2002).
2. A. Tsukazaki, M. Kubota, A. Ohtomo, T. Onuma, K. Ohtani, H. Ohno, S. Chichibu, M. Kawasaki, *Jpn. J. Appl. Phys.* **44**, L643 (2005).
3. H. Q. Chiang, J. F. Wager, R. L. Foffman, J. Jeong, D. A. Keszler, *Appl. Phys. Lett.* **86**, 13503 (2005).
4. K. Nomura, H. Ohta, A.Takagi, T. Kamiya, M. Hirano, H. Hosono, *Nature* **432**, 488 (2004).
5. J. Jo, O. Seo, E. Jeong, H. Seo, B. Lee, Y. I. Choi, *Jpn. J. Appl. Phys.* **46**, 2493 (2007).
6. J. Jo, O. Seo, H. Choi, B. Lee, Appl. Phys. Express **1**, 041202 (2008).
7. X. Li, S. Asher, S. Limpijumnong, S. B. Zhang, S. Wei, T. M. Barnes, T. J. Coutts, R. Noufi, *J. Vac. Sci. Tech.* **A24**, 1213 (2006).
8. S. B. Zhang, S.-H. Wei, Alex Zunger, *Phys. Rev.* **B63**, 075205 (2001).
9. F. Trani, M. Causà, D. Ninno, G. Cantele, V. Barone, *Phys. Rev.* **B77**, 245310 (2008).
10. D. C. Look, H. L. Mosbacker, Y. M. Strzhemechny, L. J. Brilson, *Superlattices and Microstructures* **38**, 406 (2005).
11. O. Schmidt, A. Geis, P. Kiesel, C. G. Van de Walle, N. M. Johnson, A. Bakins, A. Waag, G. H. Dohler, *Superlattices and Microstructures* **39**, 8 (2006).
12. R. Xie, T. Sekiguchi, T. Ishigaki, N. Ohashi, D. Li, D. Yang, B. Liu, Y. Bando, *Appl. Phys. Lett.* **88**, 134103 (2006).

Mater. Res. Soc. Symp. Proc. Vol. 1109 © 2009 Materials Research Society   1109-B08-09

Studies on Electrical Transport in p-ZnO/p-Si Heterojunction

Sayanee Majumdar[1], S. Chattopadhyay[2] and P. Banerji[1*]
[1]Materials Science Centre, Indian Institute of Technology, Kharagpur 721302, India
[2]Department of Physics & Meteorology, Indian Institute of Technology,
Kharagpur 721302, India

ABSTRACT

Nitrogen doped p-ZnO film, with urea as a nitrogen source, is fabricated by pulsed laser deposition (PLD) onto p-type (100) Si substrates. The structural and electrical properties of the p-p heterojunction are investigated by current-voltage (I-V) and capacitance-voltage (C-V) measurements. It shows a diode like behavior with turn-on voltage of 0.5V. C-V results indicate an abrupt interface and a band bending of 0.9 V in the silicon. Heterojunction band diagram for p-ZnO/p-Si is proposed.

INTRODUCTION

ZnO belongs to II-VI group of semiconductors and it is a non-stochiometric compound due to oxygen vacancy and zinc interstitial. It is a typical wide band gap semiconductor with bandgap of 3.37 eV at room temperature (300 K) [1] and exhibits near-ultraviolet (uv) emission [2, 3]. The as-grown ZnO is n-type and p-type doping with good repeatability and consistency is lacking though some reports are available on synthesis of p-type ZnO [4, 5]. Due to difficulties in achieving p-type ZnO, the investigators have been trying to make heterojunctions with ZnO as one of the constituents. Izaki et al. [6] have reported p-Cu_2O/n-ZnO heterojunction diode fabricated by electrodeposition for photovoltaic device application. The pulsed laser deposition of nano-photodiode based on heterojunctions between ZnO nanowires and p-Si substrate has been reported by He et al. [7]. Tansley and Owen [8] deposited n-ZnO both on n-Si and p-Si by rf suputtering and studied in detail the conduction mechanisms. They have obtained ideality factor above 2. Ma et al. [9] have reported fabrication of ZnO based devices on n^+-Si and p^+-Si and observed that such device does not perform rectifying function but emits uv light. Recently Qi et al. [10] reported pulsed laser deposition of n-ZnO on p-Si with varying oxygen pressure. However, we do not find any report on the fabrication of ZnO based p-p heterojunctions such as the formation of p-ZnO on p-Si. This may lead to the integration with the present day Si electronics.

In the present work, we report fabrication of p-ZnO/p-Si heterostructures formed by pulsed laser deposition. The p-type doping of as-grown ZnO has been done by nitrogen with urea as a source of nitrogen. Such type of device can be applied as gas sensors or bio molecule sensors.

EXPERIMENTAL PROCEDURE

The pulsed laser deposition of p-ZnO on p-Si is carried out with 248 nm KrF laser at energy 300 mJ. At first a ZnO buffer layer was deposited at 350 °C substrate temperature in vacuum (10^{-5} mbar) using 300 laser pulses (shots). Then nitrogen (urea) doped ZnO thin film has been deposited at 600 °C substrate temperature at 10^{-1} mbar oxygen pressure using 2000 shots at a frequency of 10 Hz.

The pellet, used as a target in PLD, was prepared chemically from zinc acetate di-hydrate [$Zn(CH_3COOH)_2$], 2-methoxy ethanol ($C_3H_8O_2$) and calculated amount of urea to deposit p-ZnO thin film on the well cleaned Si substrate.

RESULTS AND DISCUSSION

The urea doped ZnO is examined for its conductivity type. The film deposited was found to be p-type with resistivity 5.0 Ω-cm and carrier concentration 1.3×10^{18} cm^{-3}. The resistivity and carrier concentration were measured by van der Pauw technique and Hall measurements, respectively.

The X-ray diffraction clearly showed (002) and (201) diffraction peaks of ZnO and (400) peak of Si.

I-V Characteristics

The current density-voltage (J-V) characteristics of a typical p-ZnO/p-Si heterojunction measured at room temperature is shown in figure 1. It can be seen that the junction exhibits a diode-like behavior with turn-on voltage 0.5 V, a value well below that reported by Osinsky et al. [11] though in a different system. The ideality factor (η) is found to be greater than 2 at 300 K. Such large value of the ideality factor is due to several possible forward current transport mechanisms across the junction barrier, viz. thermionic emission, minority carrier injection, and recombination generation.

The voltage dependence of the junction current can be expressed as [12]

$$J = J_0(T) \exp\left(\frac{eV}{\eta kT}\right) \tag{1}$$

where reverse saturation current (J_0) is a function of temperature (T) and is given by $J_0 = A^* T^2 \exp\left(-\Phi_B / kT\right)$, k is Boltzman constant, A* is Richardson constant (0.248 AK^{-2}cm^{-2}) and Φ_B is the barrier height.

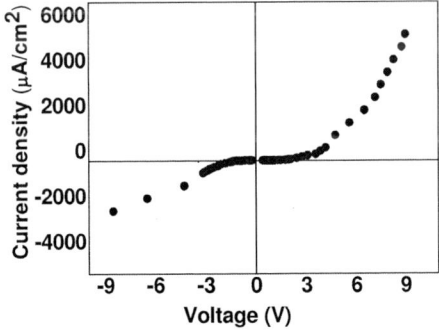

Figure 1. Current-Voltage characteristics of p-ZnO/p-Si at 300 K.

The reverse current at room temperature as a function of $(V_{bi}-V)^{1/2}$ in the range of 0.5 V < V < 1.2 V is found to be almost linear, where V_{bi} is the built-in potential whose value is measured to be 0.9 V by capacitance-voltage (C-V) measurements. This is shown in figure 2. If the generation current in the Si depletion region dominates under reverse bias, the reverse current should be proportional to the width of the depletion region (which varies as $(V_{bi}-V)^{1/2}$ in the case of abrupt junctions [12]). This is observed experimentally and therefore it is reasonable to assume that the reverse current in our p-ZnO/p-Si junctions is mainly due to the generation current in the Si depletion region.

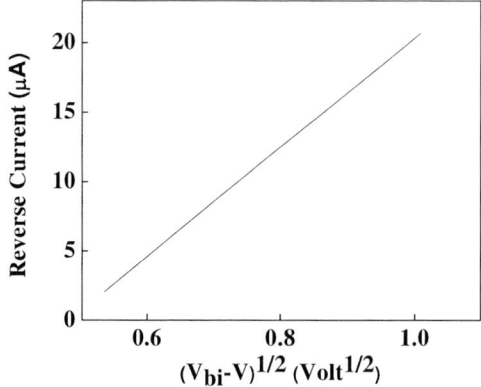

Figure 2. Plot of reverse current as a function of $(V_{bi}-V)^{1/2}$ at 300 K.

C-V Characteristics

C-V analysis is done at 1 MHz as a function of reverse bias at room temperature. The $1/C^2$ versus V plot revealed that the abrupt heterojunction theory was indeed applicable to the p-ZnO/p-Si structure due to good linear nature of the plot. The intercept of $1/C^2$-V plot on V-axis gives a value of 0.9 V which is essentially equal to the diffusion potential within the Si. The $1/C^2$ versus V can be expressed as

$$\frac{A^2}{C^2}=\frac{2\left(V_{bi}-V-\dfrac{kT}{q}\right)}{q\varepsilon_s N_A}, \tag{2}$$

where ε_s is the dielectric constant of ZnO, N_A is the acceptor density and A is the effective junction area.

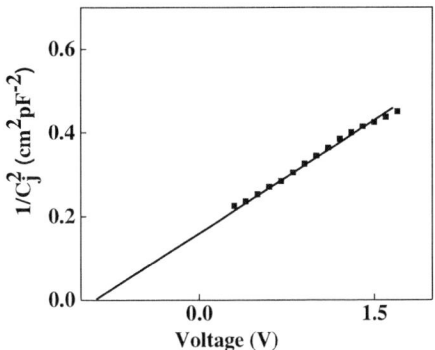

Figure 3. $1/C^2$ *vs* Voltage plot of p-ZnO/p-Si heterojunction at 300K at 1MHz.

The value of N_A is determined from the slope of the plot and it is found to be 9.98×10^{20} cm^{-3}. The depletion width of the junction is determined by the relation

$$w = \sqrt{\frac{\varepsilon_s \varepsilon_0 \left(V_{bi} - V \right)}{q N_A}} \qquad (3)$$

and its value is found to be 0.9 nm. The band diagram of p-ZnO/p-Si is proposed and shown in figure 4. The electron affinity of Si and ZnO are 4.05 eV and 3.7eV respectively which yielded ΔE_c= 0.35 eV. ΔE_v is found to be 2.6 eV. The barrier height calculated from the reverse saturation current is found to be 0.74 eV and matches well with that reported by Gűr et al. [13].

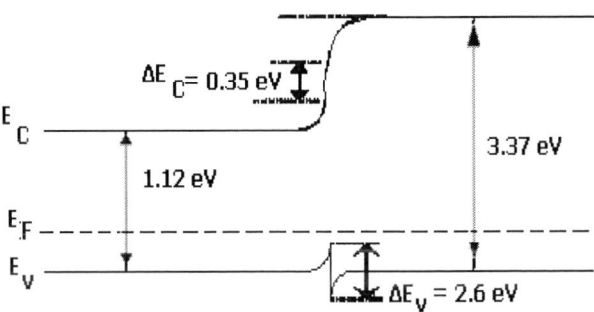

Figure 4. Band diagram of p-ZnO/p-Si.

CONCLUSION

In conclusion, we have investigated the properties of p-ZnO/p-Si heterojunction made by PLD technique. The junctions exhibit diode-like behaviour with turn-on voltage of 0.5 V. A built-in potential of 0.9 V was determined by C-V measurements. The heterojunction band diagram is proposed.

ACKNOWLEDGEMENTS

The authors acknowledge the DST-NSTI project (IIT/SRIC/R/SCM/2006/149) for working with Pulsed Laser Deposition Unit. One of the authors (SM) acknowledges financial support as institute scholar.

REFERENCES

1. Y-J Zeng, Z-Z Ye, W-Z Xu, L-L Chen, D-Y Li, L-P Zhu, B-H Zhao and Y-L Hu, *J. Cryst. Growth.* **283,** 180 (2005).
2. P. T. Hsieh, Y. C. Chen, K. S. Kao and C. M. Wang, *Physica B: Condensed Matter.* **403**, 178 (2008).
3. D. M. Bangall, Y. F. Chen, Z. Zhu, T. Yao, S. Koyama, M. Y. Shen and T. J. Goto. *Appl. Phys.* **70**, 2230 (1997).
4. G. Xiong, J. Wilkinson, B. Mischuch, S. Tuezemen, K. B. Ucer and R. T. Williams *Appl. Phys. Lett.* **80**, 1195 (2002).
5. Y. Ma, G. T. Du, S. R. Yang, Z. T. Li, B. J. Zhao, X. T. Yang, T. P. Yang, Y. T. Zhang and D. L. Lu, *J. Appl. Phys.* **95**, 6268 (1991).
6. M. Izaki, T. Shinagawa, K-T. Mizuno, Y. Ida, M. Inaba and A. Tasaka, *J. Phys. D:Appl. Phys.* **40**, 3326 (2007).
7. J. H. He, S. T. Ho, T. B. Wu, L. J. Chen and Z. L. Wang, *Chem. Phys. Lett.* **435**, 119 (2007)
8. T. L. Tansley and S. J. T. Owen, *J. Appl. Phys.* **55**, 454 (1984).
9. X. Ma, P. Chen, D. Li and D. Young, *Solid State Phenomena.* **131**, 625 (2008).
10. H. Qi, Q. Li, C. Wang, L. Zhang and L. Lv, *Vacuum* **81**, 943 (2007).
11. A. Osinsky, J. W. Dong, M. Z. Kauser, B. Hertog, A. M. Dabiran, P. P. Chow, S. J. Pearton, O. Lopatiuk and L. Chernyak, *Appl. Phys. Lett.* **85**, 4272 (2004).
12. S. M. Sze, *Physics o f Semiconductor Devices* edited by John Wiley & Sons , New Delhi, 1999) p.125-126.
13. E. Gür, S. Tüzemen, B. Kiliç, and C. Coskun, *J. Phys.: Condens. Matter.* **19**, 196206 (2007).

Mater. Res. Soc. Symp. Proc. Vol. 1109

Influence of Single-Wall Carbon Nanotube Length on the Optical and Conductivity Properties of Thin 'Buckypaper' Films

Daneesh Simien, Jeff Fagan, Jack F. Douglas, Kalman Migler and Jan Obrzut

Polymers Division, National Institute of Standards and Technology,
Gaithersburg, Maryland 20899, U.S.A.

ABSTRACT

Thin layers of length-sorted single wall carbon nanotubes (SWNT) were formed in to a "buckypaper" sample through vacuum filtration. These length sorted samples exhibit sharp changes in their optical and conductivity (σ) properties with increasing SWNT surface coverage. At given surface concentrations, longer nanotubes are found to be more transparent and conducting. Changes in σ with SWNT concentration can be quantitatively described by the generalized effective medium (GEM) theory that incorporates both effective medium and percolation theory concepts. The scaling exponents describing the conductivity percolation transition from an insulating to conducting state with increasing concentration are consistent with two-dimensional percolation theory, provided that the SWNTs are sufficiently long. The conductivity percolation threshold, x_c, varies with particle aspect ratio L as, $x_c \sim 1 / L$, which also accords with the expectations of conductivity percolation theory. Our results provide a framework for engineering the properties of thin SWNT layers for the numerous technological applications that are envisioned for buckypaper.

INTRODUCTION

The high conductivity characteristics and large aspect ratio of single wall carbon nanotubes (SWNTs), in conjunction with the transparency of materials containing them at sufficiently low concentrations, has made this type of nanoparticle an attractive material for producing transparent conductive thin films [1]. The development of these materials has previously been limited by issues of impurity, difficulties in dispersing these materials for processing, and by the broad distribution of SWNT lengths that are normally obtained in their synthesis. As the capabilities for purifying, characterizing, and sorting SWNTs continually improve, the ability to study the properties of these unique materials also progresses. In our studies we prepare highly purified and length sorted SWNT samples by using ultracentrifugation techniques [2] to investigate the optical and conductivity properties of thin SWNT layers of length-sorted tubes. This enables us to study the extent to which their properties can be engineered through the control of the SWNT concentration and length for translations into cost effective thin films that are capable of changing the optical and conductivity properties of substrates on which they are placed.

In preparing thin films from length sorted SWNTs, we are able to investigate their transport properties by using the generalized effective medium (GEM) theory. This

---

Official contribution of the National Institute of Standards and Technology; not subject to copyright in the United States.

theory is a generalized version of the effective medium theory [3-4] that formally incorporates analytical results derived from lattice percolation theory. GEM theory has been successfully applied to nanocomposite systems previously studied [5] and therefore was investigated here for its suitability in describing our nearly two dimensional SWNT thin films. The GEM theory can be applied to a broad range of inhomogeneous media involving particles having arbitrary shape and size. Even more importantly the theory exhibits a strong dependence on spatial dimensionality through its connection to percolation theory and the scaling exponents describing the dependence of the conductivity on either side of the percolation threshold of the additive. This theory potentially allows for a description of the crossover between the high and low conductivity states in our films and makes predictions regarding the dependence of the percolation threshold, x_c, where the largest changes in conductivity occur in our length-sorted films, on the aspect ratio of the SWNTs. We find the percolation threshold for these SWNT networks are among the smallest ever reported and x_c is found to scale inversely to the aspect ratio of the nanotubes as expected from percolation theory. In particular, this makes the percolation conditions occur at concentrations well below 1% by volume (or area) in 3-dimensional and 2-dimensional SWNT thin film networks, respectively.

The validated GEM theory can provide a useful framework for engineering the conductivity properties of buckypaper- like layers for numerous applications such as field effect transistors (FETs). In FETs, SWNT thin film networks are potential candidates for the active channel material. The length of the channel is comparable to the length of the nanotubes and only a few nanotubes are required to make the random network conductive. The application of percolation theory can therefore give important information about the network geometrical parameters and using a highly characterized SWNT thin film can result in better predictions of the conductivity of the network. The network conductivity exhibits a power law dependence on the network geometrical parameters, and previously studies were unable to provide values of the critical exponents that were universal. In well characterized and defined systems such as ours, we are able to derive critical exponents which are predicted by existing theories.

Based on these technological and theoretical motivations, we investigated the electrical conductivity and optical transparency percolation transitions of buckypaper layers as a function of the SWNT length where the layers are thin enough to be idealized as nearly two dimensional (2-D), as contrasted to polydispersed SWNT thin films samples which are best described as three-dimensional (3-D) networks [5]. The GEM model, without assuming an infinite contrast between additive and matrix material, is then compared to the measured conductivities of the length sorted SWNT networks. Our findings indicate that GEM theory, and the assumption that the buckypaper layers are two-dimensional structures, provides a good description of the conductive properties of these materials.

EXPERIMENTAL DETAILS

Single-walled carbon nanotubes, SWNTs, grown by the cobalt-molybdenum-catalyst (CoMoCat) process, were purchased from Southwest Nanotechnologies, Inc. (S-P95-02 Grade, Batch NI6-A001). The SWNTs were dispersed at 1 mg soot/mL in a 2 % mass fraction of sodium deoxycholate (DOC) aqueous surfactant solution via sonication (tip sonicator, 0.64 cm, Thomas Scientific) for 1.5 h in an ice water bath at 1 W/mL applied power. Post-sonication purification was performed by centrifugation at an acceleration of 21000 x 9.81 m/s^2 (21 x 10^3 G) for 2 h to pellet the non-SWNT carbonaceous and catalyst impurities. The collected supernatant contained primarily individually dispersed SWNTs. In order to separate the purified SWNTs by length, we used a dense liquid centrifugation method [2, 6]. The difference in scaling with length of the hydrodynamic and buoyancy forces on the individual SWNTs was then used to generate length separation under high acceleration. Separation was performed in a SW-32 rotor using a Beckman Coulter L80 XP ultracentrifuge at 1445 rd/s for 70 h. From the 16 fractions collected, three fractions having mean nanotube lengths of 820 nm, 210 nm and 130 nm respectively were selected. We also used the purified unsorted SWNT suspension, which had a mean tube length ≈ 220 nm, and a standard deviation of ≈ 20 nm. However, the length distribution of this sample is approximately a log-normal, and contains a broader distribution of nanotube of varying length (mixed-length fraction). This was our model polydisperse sample. SWNT length fractions were dialyzed after separation against 0.8 % mass fraction DOC solution to remove the remaining iodixanol and to reduce the total surfactant concentration.

The average length (L_a) of the fractionated SWNTs was characterized by atomic force microscopy (AFM) and transmission electron microscopy (TEM) and was verified by the intensity of the electronic absorption peak at 984 nm on the (near infrared) NIR spectrum [2]. The combined relative uncertainty of L_a mean value was approximately 10%, while each fraction distribution had a standard deviation value of about 20 %. The optical absorbance spectra of the SWNT aqueous suspensions in the (ultraviolet-visible-near infrared) UV-vis-NIR range (300 nm to 1800 nm) were obtained using a Perkin-Elmer Lambda 950 UV-vis-NIR spectrophotometer. Fractionated SWNTs were then deposited on the surface of a porous cellulose ester membrane, diameter of 47 mm, with 0.05 µm diameter pore size (Millipore), via a vacuum filtration process. The active area of the membrane surface was 10.2 cm^2 (diameter 36 mm). A mixture of isopropanol and deionized water (1:4) by volume was used to "condition" the membranes and to rinse away the deoxycholate surfactant. The concentration of nanotubes in fractionated suspensions were typically between 0.05 mg/mL and 0.1 mg/mL, determined by weighing the nanotube deposit on a microbalance after dialyzing and evaporating a fixed volume of fractionated suspension. The networks' surface density was controlled by varying the volume of the fractionated suspension using a micropipette (Oxford Benchmate), then diluting the suspension with DI water to obtain 2 mL of dilute suspension. The deposited SWNTs from the dilute suspension formed a random network on the surface of the membrane. The samples were then dried at 50 °C for 24 h under nitrogen atmosphere. The surface density of these networks was in the range of 0.01 µg/m^2 to 2.0 µg/m^2 with the combined relative uncertainty of 30 %. The optical characteristics of the SWNT networks were determined by measuring the diffuse optical reflectance in the wavelength range of 300 nm to1800 nm, using a diffuse reflectance accessory kit for the Lambda 950. The reflectance spectra were normalized to the reflectivity of the filter membrane without SWNTs, which we used as a non-absorbing reference. Thus, the spectra primarily represent the absorbance of the conducting SWNT film. The scanning electron microscopy images of the deposited SWNT networks were obtained using a Hitachi S4700 field emission scanning electron microscope. To measure the electrical characteristics of the network films, an interdigitated gold electrode pattern was deposited directly on top of the SWNT network through a shadow mask. Electrical measurements were analyzed in terms of complex impedance, yielding the impedance magnitude ($|Z^*|$) and the corresponding phase angle (θ) over the frequency range of 40 Hz to 100 MHz by using a four-terminal fixture attached to an Agilent 4294A precision impedance analyzer [5]. The impedance analyzer was calibrated with a standard extension adapter to short, load, and open standards. The real part (σ') of the complex sheet (surface) conductivity was determined from the measured impedance normalized by the geometry of the electrode pattern $\sigma' = l / [|Z|\cos(\theta) \times d]$ in units of $(\Omega/\text{square})^{-1}$ or Siemens-square (S-square). Where, l is the total length of the finger electrodes, and d is the distance between the fingers. In our measurements, l=9 cm and d=800 μm. The thickness of the deposited gold electrodes was about 0.1 μm. The lowest measurable sheet conductivity in our system was $\sigma' \approx 5 \times 10^{-12}$ S-square at 100 Hz. The sheet conductivity of dried blank membrane substrates, without carbon nanotubes, was typically below 10^{-11} S-square at 100 Hz. The combined relative experimental uncertainty of the measured conductivity magnitude was within 4 %, while the experimental uncertainty of the phase angle measurements was about ± 0.5°. The sheet conductivity σ' of our SWNT films is frequency dependent but at the lowest frequencies exhibits a plateau that persists up to crossover frequency of ω_s. Throughout our paper, we denote this frequency independent real conductivity $\sigma'(\omega \to 0, \theta \approx 0)$ as σ_s. At higher frequencies, $\omega > \omega_s$, we observe that σ' increases with increasing frequency according to power law, $\sigma' \sim \omega^n$, which is commonly described as a "universal" property of disordered solids and applicable to a broad range of semiconducting materials, conductor-dielectric mixtures, and composites [5, 7].

DISCUSSION

Figure 1 shows absorbance spectra of the initial SWNTs in liquid suspensions procured to make our buckypapers. The spectra are normalized to the baseline π-plasmon absorbance at 775 nm.

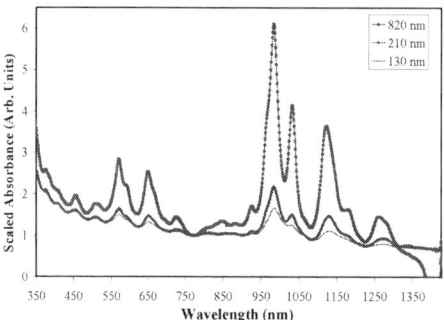

Figure 1. Optical absorbance spectra for length-sorted SWNT fractions scaled to the absorption at 775 nm for the respective SWNT lengths with concentration of 0.0078 mg/mL, 0.036 mg/mL and 0.1125 mg/mL, for 820 nm, 210 nm and 130nm samples [1].

The mean lengths of these initial fractions were 820 nm, 210 nm and 130 nm. In the spectra, the absorption bands at (850 to 1450) nm and at (500 to 850) nm are due to E_{11}^s

and E_{22}^s inter-band electronic transition in semi-conducting tubes [8], the absorption bands between (420 and 520) nm are due to electronic transitions in metallic tubes E_{11}^m, while the 340 nm to 400 nm band correspond to E_{33}^S transition in semiconducting tubes [8-10]. These inter-band transitions as a whole demonstrate that the longest nanotubes (820nm) will display more intense features despite having the lowest concentration of nanotubes in solution which is a recently documented find associated with length sorted SWNTs [2]. These liquid SWNT solutions are then vacuum filtered to produce solid films as shown in Figure 2. The SEM image shown in Fig. 2a is representative of 2D network of nanotubes having a mean length of 820 nm. The network for the 820 nm SWNT thin film samples is near the percolation threshold, when surface density is about 0.02 $\mu g/cm^2$. The nanotubes do not exhibit excessive bundling or re-agglomeration upon deposition and rinse removal of surfactant as deposited on the surface of the membrane. The nanotubes form a continuous network in spite of areas where membrane pores are observed. In Fig 2b the mixed length tubes form a three-dimensional network, with surface density of about 1.4 $\mu g/cm^2$, well above the percolation threshold.

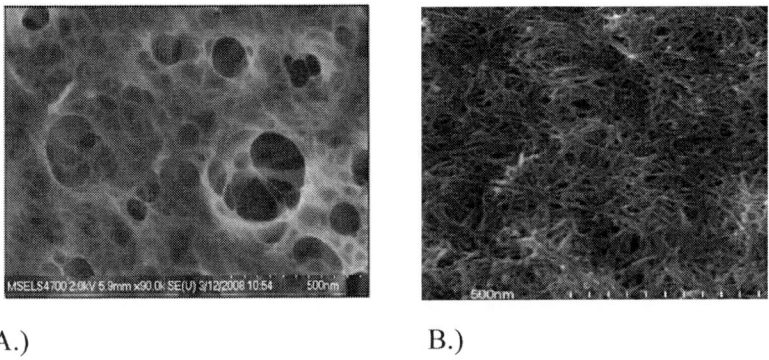

A.)   B.)

Figure 2. Field emission scanning microscope images of SWNTs networks with nanotube lengths and surface densities of: (A) 820 nm, 0.08142 $\mu g/cm^2$ and (B) mixed length, 1.4 $\mu g/cm^2$. The low concentration of SWNTs present allow for some of the pores of the filter membrane to appear more pronounced for figure 2A than in the dense coating displayed in figure 2B, where a higher concentration of SWNTs are present.

The spectra in Figure 3 are shown for selected SWNT solid films for unsorted samples and for fraction lengths of 130 nm, 210 nm and 820 nm. The spectra are referenced to the reflectance of the membrane without the tubes, and therefore, correspond directly to the transmittance, (T) of the SWNT films. These SWNT films are about 20 nm thick, approaching a conductivity plateau above the percolation threshold, while their transparency is about 85 % to 90 % at the wavelength of 775 nm. The absorption peaks that appear in the NIR and in the visible range of the spectrum are characteristic of specific tube chiralities. We note that the normalized intensities of the absorption peaks in the films are lower than those in suspensions shown in Fig. 1 for the corresponding tube length. The effect of decreasing intensity of the characteristic electronic transitions in the network is likely due to interactions between interconnecting tubes. A likely cause of diminished absorbance intensities can be also attributed to doping by oxygen from the ambient atmosphere. These effects should be present in our films where the protective surfactant coating has been rinsed away.

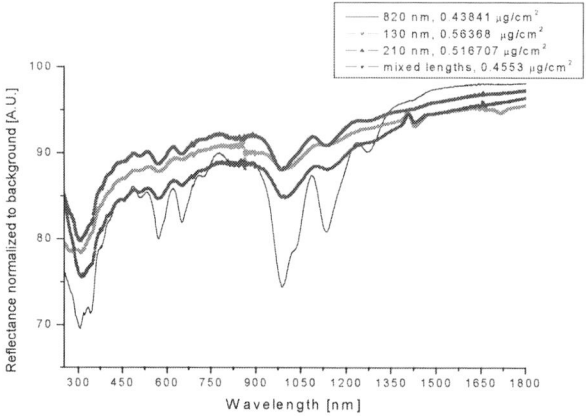

Figure 3. Optical reflectance of SWNTs networks for the following tube lengths with corresponding surface densities: (a) 820 nm, 0.434 $\mu g/cm^2$; (b) 210 nm, 0.517 $\mu g/cm^2$; (c) 130 nm, 0.564 $\mu g/cm^2$; and (d) mixed length tubes, 0.455 $\mu g/cm^2$.

By changing the concentration of SWNTs, the sheet conductivity of our 2D networks changes by six orders of magnitude, from 10^{-11} S to about 10^{-5} S as shown in Fig. 4. The highest σ_s values obtained for thicker SWNT films are in the range of about 0.001 S. The surface conductivity increases dramatically at low concentrations and saturates above a certain critical concentration. The overall character of the conductivity plots is characteristic of a percolation transition. Films made of mixed length nanotubes show the lowest conductivities, with the percolation transition taking place at higher concentration. With increasing nanotube length, the network conductivity increases and the percolation concentration x_c for σ_s shift to lower concentrations. We also see that the crossover transition from the insulating to the conducting state occurs in a narrower concentration range. In the case of 820 nm length tubes, the characteristic step rise in conductivity takes place within a very narrow concentration range of about 0.008 $\mu g/cm^2$.

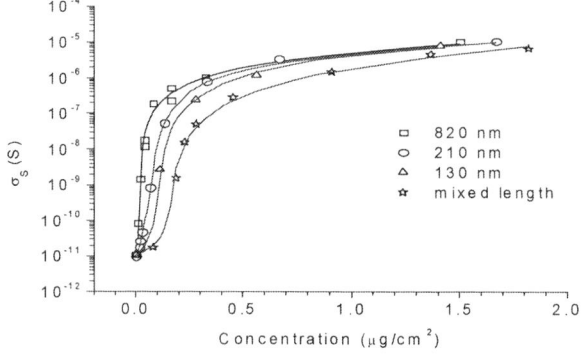

Figure 4. Conductivity as function of tube concentration for length sorted SWNTs. The lines represent the fitted conductivities to the GEM-percolation model (1).

The SWCNT length dependence of conductivity percolation concentration x_c and the dimensionality dependent critical percolation exponents, s and t are determined by analyzing the σ_s data in terms of the generalized effective medium theory (GEM) relation [11, 12]:

$$(1-x)\frac{\sigma_i^{1/s} - \sigma_s^{1/s}}{\sigma_i^{1/s} + A\sigma_s^{1/s}} + x\frac{\sigma_a^{1/t} - \sigma_s^{1/t}}{\sigma_a^{1/s} + A\sigma_s^{1/t}} = 0 \qquad (1)$$

where x is the tube concentration, A is a parameter related to the percolation concentration $A = (1 - x_c)/x_c$, $\sigma_i \approx 10^{-11}$ S is the conductivity of the insulating matrix, and σ_a is the conductivity of the network forming SWNTs. From the asymptotic limit of the experimental data at high SWNT concentrations, we determined σ_a to be in the range of 5×10^{-5} S. Equation (1) fits the data well, and x_c and the percolation exponents t and s can be deduced using a nonlinear least-squares routine. The conductivity of the network in the high concentration limit is not equivalent to that of an isolated nanotube [5]; rather, σ_a depends on the contact resistance between the SWNTs [13-14] and other factors [15] that are not addressed in the simple GEM model. The critical percolation parameters obtained from Eq. 1 are summarized in Table 1 for the three different tube lengths. The universal percolation exponents t and s for the 820 nm tubes and the 210 nm tubes have similar values. These values are close to those theoretically expected for a large conducting 2D network [16-17], where $s = t \approx 1.3$. Kholodenko and Freed, however argue for the exact 2D exponent value, $23/18 \approx 1.28$ [18]. Thus, we find experimental evidence that our SWNT networks can be described as nearly two-dimensional and that GEM-percolation theory provides quantitative description of conductivity changes in these materials through the conductivity percolation transition. We note that the critical percolation concentration that we observe experimentally for the 820 nm tubes ($x_c = 0.018$ µg / cm^2) is one of the smallest reported to date.

Table 1. Percolation parameters of SWNT networks in relation to the tube length

Tube length (nm)	s	t	x_c (µg/cm^2)
820	1.25 ±0.4	1.35 ±0.1	0.018 ±0.01
210	1.25 ±0.5	1.38 ±0.1	0.072 ± 0.07
130	1.2 ±0.8	1.9 ±0.1	0.095 ± 0.07
230 (mixed length)	0.8±0.6	2.1 ±0.1	0.15 ± 0.1

The data in Table 1 indicates that the percolation exponents s and t obtained for networks made of the longer tubes are close to the predicted universal values for 2-dimensional networks, and while the values found for films formed with mixed-length tubes, are in accord with those values predicted for 3D percolation theory where we expect, $s \approx 0.8$ and $t \approx 2.0$. This indicates that an increasing fraction of SWNT connections are being made in the third dimension in these short tube networks [19-20]. The variation of the percolation threshold with the nanotube aspect ratio can be further understood on the basis of percolation and effective medium theory concepts. A polynomial expansion of the conductivity σ_s for a two dimensional layer containing arbitrarily oriented particles of fixed and identical shape relative to an insulating layer without the particles is given by equation (2) [21],

$$(\sigma_s / \sigma_i) = 1 + [\sigma]_\infty \, x + O(x^2) \qquad (2)$$

where $[\sigma]_\infty$ is the leading virial coefficient for the conductivity, which is sometimes referred to as the 'intrinsic conductivity'. $[\sigma]_\infty$ is a function of particle shape and in two dimensions this quantity achieves its minimal value for a circular conducting disc where $[\sigma]_\infty = 2$ [22]. More generally, $[\sigma]_\infty$ for an elliptically-shaped particle of aspect ratio L, is given [23]:

$$[\sigma]_\infty = (1 + L)^2 / 2L \qquad (3)$$

The percolation transition concentration x_c can be estimated based on the conditions that the leading correction to the conductivity becomes on the order, $[\sigma]_\infty x \approx 0(1)$, a Ginzburg criterion [21, 22]. This condition, along with Eq. (3) implies the general scaling relation for the conductivity percolation concentration for slender particles in 2D:

$$x_c \text{ (slender particle)} \sim 1 / L \qquad (4)$$

Equation (4) suggests that in networks exhibiting 2D percolation transition, increasing the CNT length by a factor of 820 nm / 210 nm = 3.9 should decrease the corresponding percolation threshold x_c by a similar factor. Table 1 indicates that x_c is indeed reduced by a factor of 4.0 in our buckypaper films, in quantitative agreement with Eq. (4). In a similar fashion, x_c can be predicted for particles having arbitrary shape [23]. Thus, we find that the optical characteristics of the length fractionated SWNTs are distinct from unsorted polydispersed SWNTs.

CONCLUSIONS

We find that the conductivity properties of thin buckypaper layers formed from fractionated long SWCNT conform well to the predictions of Generalized Effective Medium theory where the derived estimates of the percolation exponents s and t agree rather well with their theoretical counterparts from lattice percolation theory in two dimensions. Moreover, the conductivity percolation threshold x_c itself decreases with SWNT length in accord with percolation theory. Notably, the critical conductivity percolation concentration 0.018 $\mu g/cm^2$ in these layers is the lowest reported to date, an encouraging result pointing to the value of good SWNT dispersion. We also make the novel observation that the IR absorption strength increases linearly with SWNT concentration, regardless of the SWNT length, and that longer fractionated tubes are both more transparent and conducting at the same coverage. Evidently, considerable control of the conductivity of buckypaper layers can be obtained by adjusting SWNT length, concentration and polydispersity and that GEM-percolation theory provides a powerful framework for describing and engineering these changes for nanotechnology applications exploiting these novel materials.

ACKNOWLEDGMENTS

We thank Dr. Thuy Chastek for help in SEM imaging.

REFERENCES

Note: Certain equipment, instruments, or materials are identified in this paper in order to adequately specify the experimental details. Such identification does not imply recommendation by the National Institute of Standards and Technology nor does it imply the materials are necessarily the best available for the purpose.

1. H.E. Unalan, G. Fanchini, A. Kanwal, A. Du Pasquier, M. Chhowalla, *Nano Lett.*, **6**, 677-682 (2006)
2. J. A. Fagan, M. L. Becker, J. Chun, E. K. Hobbie, *Adv. Mat.* **20**, 1609-1613 (2008).
3. D. Stroud, *Phys Rev. B* **12**, 3368 (1975).
4. D.A.G. Bruggeman, *Ann Phys.* **24**, 636 (1935) .
5. J. Obrzut, J. F. Douglas, S.B. Kharchenko, K. B. Migler, *Phys. Rev. B*, **76**, 195420-195420-9, (2007).
6. A. G. Rinzler, J. Liu, P. Nikolaev, C. B. Huffman, F. J. Rodriguez-Macias, P. J. Boul, A. J. Lu, D. Heymann, D. T. Colbert, R. S. Lee, et al. *Appl. Phys. A: Mater. Sci. Process.*, **67**, 29-37, (1998).
7. J. Li, P. C. Ma, W. S. Chow, C. K. To, B. Z. Tang, J. K. Kim, *Adv. Func. Mat.*, **17**, 3207-3215, (2007).
8. R. B.; Weisman, S. M. Bachilo, *Nano Lett.* **3**, 1235-1238, (2003).
9. S. G. Louie, "Electronic Properties, Junctions, and Defects in Carbon Nanotubes", *Carbon Nanotubes: Synthesis, Structure, Properties, and Applications*, ed. M. S. Dresselhaus, G. Dresselhaus, and P. Avouris (Topics in Applied Physics: Springer, NY, 2000) **80**, pp. 113-143.
10. A.P Jorio, H. B. Ribeiro, C. Fantini, M. Souza, J. P. M. Vieira, C. A. Furtado, J. Jiang, R. Saito, L. Balzano, D. E. Resasco, et al., *Phys. Rev. B.*, **72**, 075207-075207-5, (2008)
11. D. Simien, J. Fagan, W. Luo, J. Douglas, K. Migler and J. Obrzut, *ACS Nano*, **2** (9); 1879-1884, (2008).
12. McLachlan, D. S.; Heiss, W. D.; Chiteme, C.; Wu, J. *Phys. Rev. B*, **58**, 13558-13564, (1998).
13. J. Obrzut, K. Migler, L. F. Dong, J. Jiao, APS Conf. Proceedings, **931**, 483-485, (2007).
14. B. Derirda, J. Vannimenus, *J. Phys. A* **15**, L557-L564 (1982).
15. J. L.Blackburn, T. M. Barnes, M. C. Beard, Y. H. Kim, R. C. Tenent, J. McDonald, B. To, T. J. Coutts, M.J. Heben, *ACSNano*, **2**, 1266-1274 (2008).
16. J. P. Clerc, G. Giraud, J. M. Laugier, J. M. Luck, *Advances in Physics*, **39**, 191-309 (1990).
17. D. J. Frank, C. J. Lobb, *Phys. Rev. B*, **37**, 302-307 (1988).
18. A. L. Kholodenko, K. F. Freed, *J. Phys. A: Math. Gen.*, **17**, L55-L59 (1984)
19. J. M. Laugier, J. P. Clerc, G. Giraud, J. M. Luck,. *J. Phys. A*, **19**, 3153-3164 (1986)
20. D. Stauffer, *Introduction to Percolation Theory*, Tayler & Francis Ltd., London and Philadelphia (1985).
21. J.F. Douglas, E.J. Garboczi, *Adv. Chem. Phys.*, **91**, 85-153 (1995).

22. J. Dudowicz, M. Lifschitz, K. F. Freed, J. Douglas, *J. Chem. Phys*. **99**, 4804-4820 (1993),

23. Garboczi, E. J.; Snyder, K. A.; Douglas, J. F.; Thorpe, M. F. *Phys. Rev. E*, **52**, 819-828 (1995).

Mater. Res. Soc. Symp. Proc. Vol. 1109 © 2009 Materials Research Society

Electrical Characteristics and Practical Properties of Amorphous Indium Zinc Oxide Films and Related Materials

M. Kasami[1], K. Yano[1], F. Utsuno[1], T. Shibuya[1], K. Inoue [1],
B. Shinozaki[2], K. Makise[3], M. Funaki[2]

[1]Advanced Technology Research Laboratories, Idemitsu Kosan Co.Ltd, Chiba, 299-0293, Japan
[2]Department of Physics, Kyushu University, Ropponmatsu, Fukuoka 810-8560, Japan
[3]National Institute for Material Science, 3-13 Sakura, Tsukuba 305-3003, Japan

ABSTRACT

Amorphous indium zinc oxide (In_2O_3-ZnO) films have been used for the transparent electrode on TFT-array for the sake of the etching easiness based on its stable amorphous structure. In this study, we have systematically investigated the temperature dependence of resistivity ρ and Hall mobility μ of amorphous In_2O_3-ZnO films (350 nm) in the temperature range 2.0 K to 300 K and studied amorphous structure. The n-μ plots shows the convex characteristic with a broad peak near n $=3$ x10^{20}/cm^3, taking the value $\mu = 50$ cm^2/Vs. In the low n region, it has been found that the μ seems to change as $\mu \propto n^{1/3}$. For films with n > 5x10^{19}/cm^3, the $\rho(T)$ shows the metallic behavior between 20 K and 300 K. On the other hand, for films with n<1 x10^{19} /cm^3, $\rho(T)$ essentially shows the insulating behavior at whole temperatures. We concluded the electron conductive mechanism of the films is mainly due to the electron-phonon and electron-phonon-impurity scattering model. The structural analysis well agreed with the electric measurement. And also, the practical properties of In_2O_3-ZnO films, such as etching rate, contact resistance, internal stress and so on, being required in practical TFT process were evaluated. It was confirmed that the practical properties of In_2O_3-ZnO were the most suitable for the transparent electrode and transistor on TFT-array.

INTRODUCTION

The investigations of amorphous transparent conductive oxide films have received much attention due to the growing demand for the low-cost manufacturing of patterned electrodes, as well as for several new applications such as flexible displays[1-9]. It is well accepted that an amorphous In_2O_3–ZnO film, which is prepared by sputtering method, has an amorphous structure when the substrate remains at room temperature throughout the processing sequence. Recently, it was found that the In_2O_3-ZnO films were used as semiconductor materials[10-13].

In this paper, we will overview the properties of In_2O_3–ZnO system such as electrical properties and structural features. Amorphous In_2O_3–ZnO films have practically attracted considerable interest[1, 2] because these films can give a smooth surface, which enables detailed processing with high accuracy, and have large values of Hall mobility[6-9]. As for electron scattering, it has been assumed that the dominant scattering mechanisms in degenerated TCO films are ionized and neutral scatterings because of the weak temperature dependence of carrier density. However, the mechanism of the temperature dependence of electrical properties has not been investigated extensively. In addition to practical studies on achieving higher quality, understanding of the scattering mechanism of TCO films is important to investigate the fundamental electrical characteristics for degenerated semiconductive materials.

Also, though the electronic and optical properties of amorphous In_2O_3–ZnO films have been well documented, the amorphous structure itself has not been examined in much detail. Transmission electron microscopy and electron diffraction studies have revealed that the In_2O_3–ZnO film has an amorphous structure at an atomic level. Synchrotron radiation is a useful high energy source to analyze amorphous structures by the diffraction and the absorption method. The grazing incident X-ray scattering (GIXS) technique gives structural information from the film without any contribution from the substrate[14]. On the basis of the radial distribution function (RDF) obtained from the GIXS measurement and the nearest neighbor structural information from EXAFS results, we describe here details of the atomic arrangement of the amorphous film by calculating structural models of the amorphous In_2O_3–ZnO film by combination of the molecular dynamics (MD) and reverse Monte-Carlo (RMC) simulations. Then we will discuss the electronic conductive mechanism on the base of the electrical properties and the structural analysis.

RESULTS and DISCUSSIONS

Characteristics of In_2O_3-ZnO film

1) Relationship between composition and electrical properties

Figure 1 shows the relationship between ratio of In_2O_3 to ZnO and the resistivites of the In_2O_3-ZnO films prepared by DC magnetron sputtering with In_2O_3-ZnO targets. The glass substrates were used and cooled by water during sputtering. It is found that the resitivity of films is the lowest in the vicinity of the composition of In_2O_3:ZnO=90:10 (wt.%). The resitivity of the film was 435 $\mu\Omega$cm with the thickness of 100 nm. Though the films with higher content of In_2O_3 and that of ZnO are crystallized easily, the films have amorphous structure stably revealed by XRD and TEM analysis.

The films were heated by 300 °C for an hour in air, the resistivites changed drastically shown as Figure 1. The resistivites did not change in the vicinity of the composition of In_2O_3:ZnO = 90:10 (wt.%), but those of the films with the content of $In_2O_3 > 95$ or $30 < In_2O_3 < 50$ (wt.%) became much higher than the as-deposited ones. The films with the content of $In_2O_3 > 95$ had the crystalline structure and those with $30 < In_2O_3 < 50$ (wt.%) remained amorphous. In recently, both of them with high resistivities were found to have the characters as transistors[8-10]. This will be discussed later in this paper.

2) Practical properties of In_2O_3-ZnO films

Since the In_2O_3-ZnO film has amorphous structure, the film has the several process merits. One is that the films prepared by sputtering have the uniformity of resistivity and thickness in large-scale area (Figure 2). This is why the film was hardly damaged during sputtering in comparison with ITO. The second merit of the film is the etching process, so that the film is easy to be solved by the weak acid solution such as oxalic acid. The etching properties are particularly important in the TFT structure with a top pixel layer. In the case of TFT structure with Al electrode, it is strongly said that the galvanic reaction was generated by the etching process[2]. To protect Al electrode from the reaction, etching process with weak acid

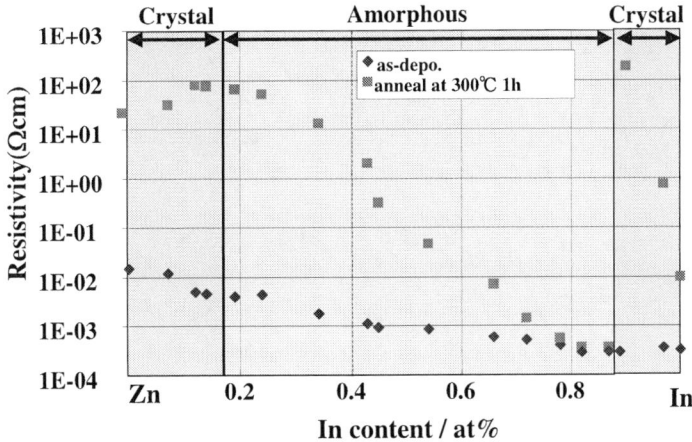

Figure 1. The relationship between In content and resistivities of In_2O_3-ZnO films.

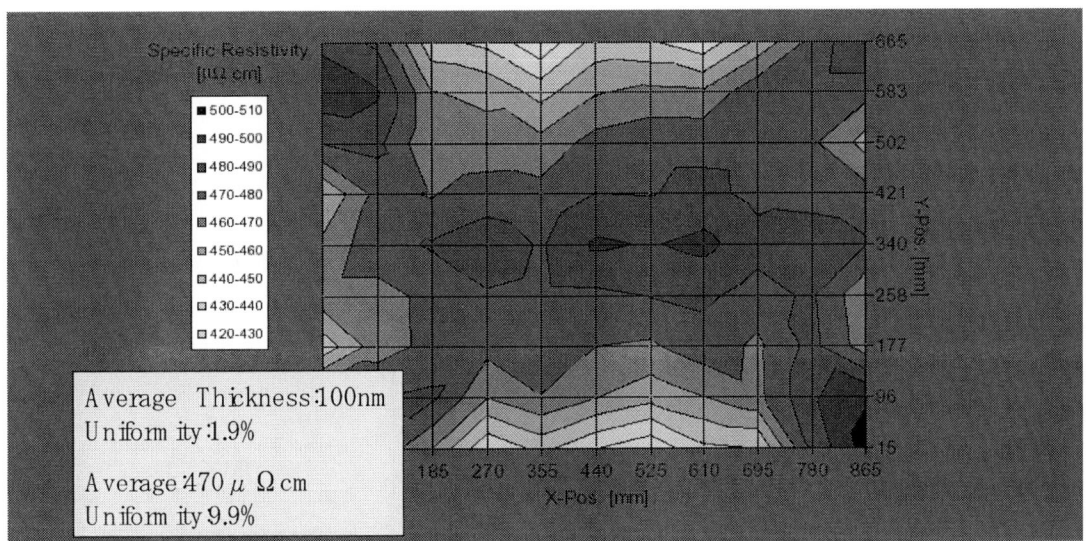

Figure 2. The uniformity of the resistivity of the In_2O_3-ZnO film prepared by sputtering.

is proposed. The taper angle was sharp by wet etching and the under- or over-etching of the film can be controlled easily[2]. Thus, the In_2O_3-ZnO films can be used for the application demanded for finer pitch-pattered electrodes such as TFT-LCD alley. The third merit is that the In_2O_3-ZnO sputtering target at the time of sputtering enables effectively inhibiting particle occurrence, abnormal electrical discharge, and nodule occurrence (Figure 3). It is important to reduce the target maintenance such as cleaning the target surface.

Electric measurement of In_2O_3-ZnO films

In order to investigate the electronic conduction mechanism of the In_2O_3-ZnO films, we carried out the electrical measurements of temperature dependence on the resistivities. The films were deposited on glass substrates by the DC-magnetron sputtering method using the In_2O_3–ZnO

Figure 3. Photographs of the In_2O_3-ZnO (left) and ITO (right) target surface after sputtering discharge test (Power : RF:50W, Ar:0.1Pa)[2].

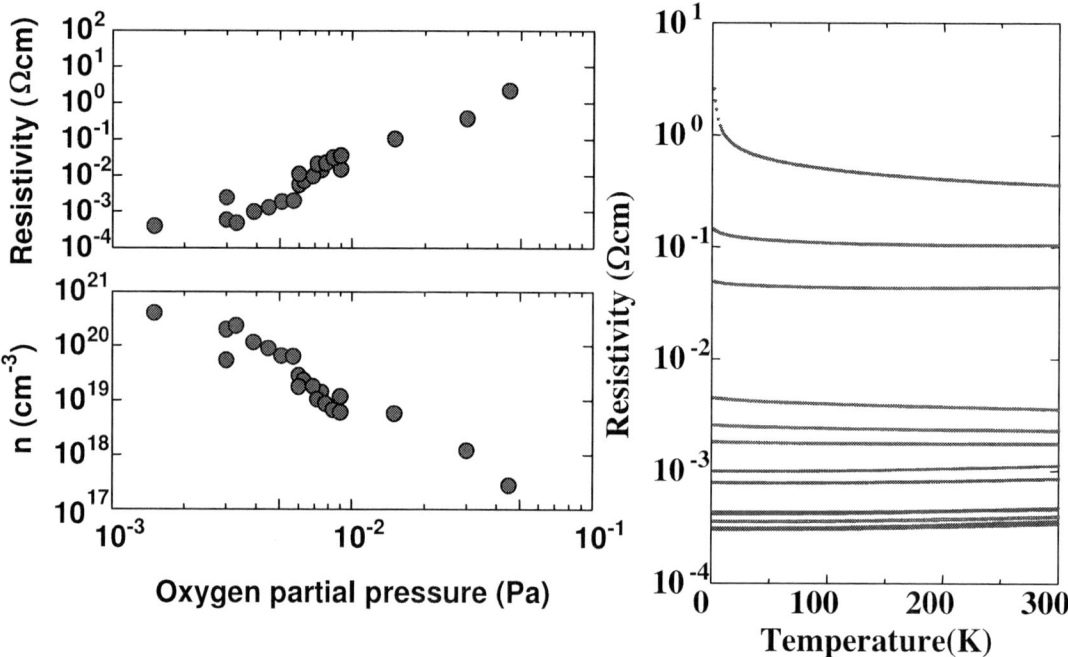

Figure 4. Resistivities (a) and carrier densities (b) of the In_2O_3-ZnO films depended on oxygen partial pressure during sputtering and (c) temperature dependence of their specimens[16].

target (In_2O_3:ZnO=90:10 wt.%). We prepared the films with thickness d = 350 nm, $\rho \approx 10^{-4}$–10^{0} Ω cm (300 K) and carrier density n=10^{17}–10^{20}/cm³ by changing the gas pressure of oxygen during sputtering, as shown in Figures 4(a) and 4(b). The patterned Au electrodes were put on the specimen films for Hall voltage measurement. We measured the resistance using a standard dc four-probe technique and applied magnetic field perpendicular to the film surface up to ± 5 Tesla. Detailed measurement condition was referred in the previous works[15-17]. The temperature dependence on the resistivities were plotted in Figure 4(c). In the case of $d\rho/dT<0$, the scattering mechanism corresponds to the hopping conductive model, and in $d\rho/dT > 0$ it corresponds to the

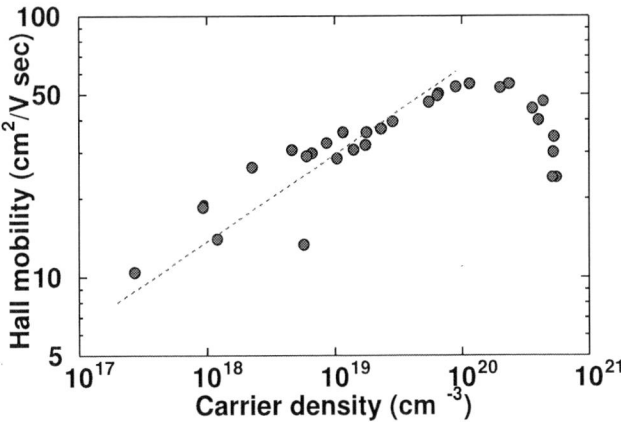

Figure 5. The relationship between carrier densities and Hall mobility of the In_2O_3-ZnO films prepared by sputtering under the oxygen partial pressure.

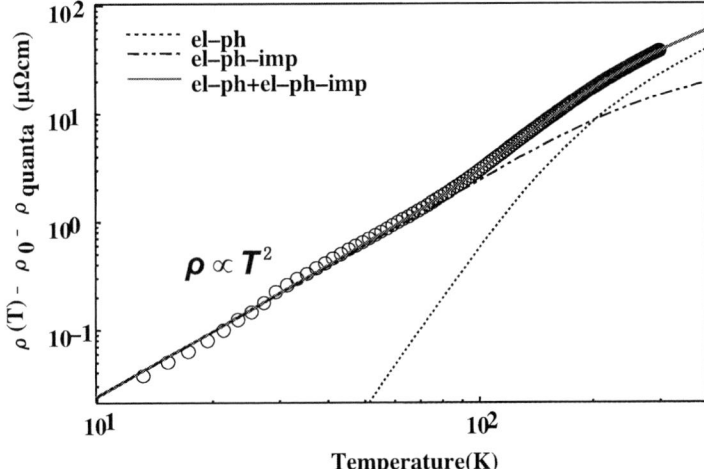

Figure 6. Temperature dependence of the resitivity for the In_2O_3-ZnO film. The solid line is calculated from $\rho(T)-\rho_0-\rho_{quant}(T)=\rho_{el-ph}(T)+\rho_{el-ph-imp}(T)$[16].

metallic conductive model as well known. The n-μ plots (Figure 5) shows the convex characteristic with a broad peak near n =3 x10^{20}/cm^3, taking the value μ = 50 cm^2/Vs. In the low n region, it has been found that the μ seems to change as $\mu \propto n^{1/3}$. We analyzed the scattering mechanism of the metallic conductive films in detail. This ρ-T curve could not to be fitted by the scattering model of amorphous alloy since it is a S-shaped curve (Figure 6). Then we considered the scattering model including with electron-phonon (el–ph) scattering like ITO. For the electron-phonon interaction, contributing to the temperature dependence on resistivity at a whole temperature region, between electron and longitudinal phonon, is well known as Grünissen–Bloch (G-B) formula[18-21]. For the temperature dependence of ρ ∝T observed at high temperatures, the G–B formula can be fitted to the data. In our previous work[15], we have analyze the data on ρ(T) with fitting parameters of residual resistivity ρ_0 and Debye temperature Θ_d. Although the theory seems to agree essentially with data in a wide temperature region, some difference between data and the theory remains at low temperatures below ≈100 K. In order to emphasize the behavior of ρ(T) at low temperatures, we show the relation between log (ρ(T)-ρ_0-

ρ_{quanta}) and log T as shown in Figure 6, where the term of a ρ_{quanta} is due to the quantum effect such as Anderson localization theory. According to the recent theory, the resistivity in disordered metals is given as the theory. The interference (el-ph-imp) between electron–phonon and electron–impurity scattering gives a correction to Matthiessen rule[22]. Interference term is proportional to T^2 at low temperatures and is a constant at sufficiently high temperatures. Figure 7 shows typical data for the film with $\rho_0 \approx 300$ μΩcm. The solid line is calculated from $\rho(T)-\rho_0-\rho_{quanta}(T)=\rho_{el-ph}(T)+\rho_{el-ph-imp}(T)$ with fitting parameters β, $B_{el-ph-imp}$ and Θ_d, where $B_{el-ph-imp}$ is a constant in the interference term, $\rho_{el-ph-imp}(T)$. The theory can well reproduce the experimental data in a large temperature region.

In conclusion of this section, we obtained the following results; the electron scattering of the In_2O_3-ZnO film is mainly due to electron–phonon interaction at high temperature. At temperature below ≈ 100 K, the resistivity shows the relationship of $\rho(T) \propto T^2$ suggesting the electron–phonon–impurity scattering. In the wide temperature region, we analyzed the temperature dependence of ρ by the form of $\rho(T)=\rho_0+\rho_{quanta}(T)+\rho_{el-ph}(T)+\rho_{el-ph-imp}(T)$. We have shown good agreements between the data and the theory, regarding Θ_d and the coefficients β and $B_{el-ph-imp}$ of the third and fourth terms in above equation as fitting parameters and obtained the value of $\Theta_d \approx 1000$ K.

Structural analysis and band structure of In_2O_3-ZnO films

In order to consider the electrical conductive mechanism on the base of the structure of the In_2O_3-ZnO film, we carried out to study the structural analysis using the EXAFS and GIXS method in the BL19B2 and BL47XU facility at SPring-8 (Hyogo, Japan). Amorphous In_2O_3–ZnO films with a thickness of 100 nm were deposited using the sputtering method on a Si wafer for the GIXS measurement. For EXAFS measurements, the films with thickness of 500 nm were deposited on fused quartz glass substrates. The EXAFS spectra near In K-edge and Zn K-edge were measured at the BL01B1 facility in SPring-8 with Si (111) monochromator and a collecting mirror system by means of a fluorescence mode using a Lytle detector. The GIXS measurements were carried out using the BL47XU facility in SPring-8. The detailed measurement condition was shown in the previous works[23-25]. The measured intensity curve was normalized by the Krough–Moe–Norman method, and the radical distribution function (RDF) was obtained from the experimental amplitude function curve $ki(k)$ by the usual method[28]. Then structural models for amorphous In_2O_3–ZnO were prepared using MD and RMC simulations[27, 28]. The distance of M–O, which is determined from the accumulated coordination number of the M–O pairs, and the number of M–O–M' (M, M'=In, Zn) pairs, were calculated with the atomic configuration obtained by the RMC simulations[25].

As for the nearest neighbor In–O and Zn–O pair, the average oxygen coordination number around the In and Zn ions calculated from the EXAFS measurements was about 6.0 and 4.0, respectively (Figure 7). Therefore the framework was built up by InO_6 octahedra and ZnO_4 tetrahedra in amorphous In_2O_3–ZnO. The measured intensities were sufficient to calculate the RDF using the GIXS technique with synchrotron radiation. The observed RDF is shown in Figure 8 with the RDF calculated from the crystalline In_2O_3 with bixbyite type structure. The peak positions of the RDF for crystalline In_2O_3 are close to those of the values observed for the amorphous film. The first peak at 2.12 Å in the observed RDF corresponding to the In–O pair is shorter than that of 2.18 Å in the crystalline In_2O_3. In the In_2O_3 crystal, an In^{3+} ion is coordinated by 6 O^{2-} ions with an average In–O distance of 2.18 Å. This result is well agreed

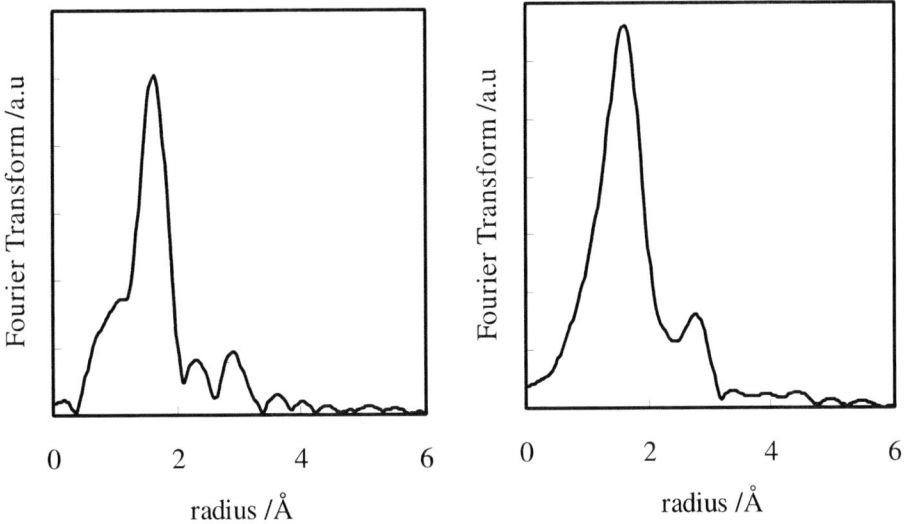

Figure 7. The EXAFS spectra of the In_2O_3-ZnO film as In-K edge and Zn-K edge[25].

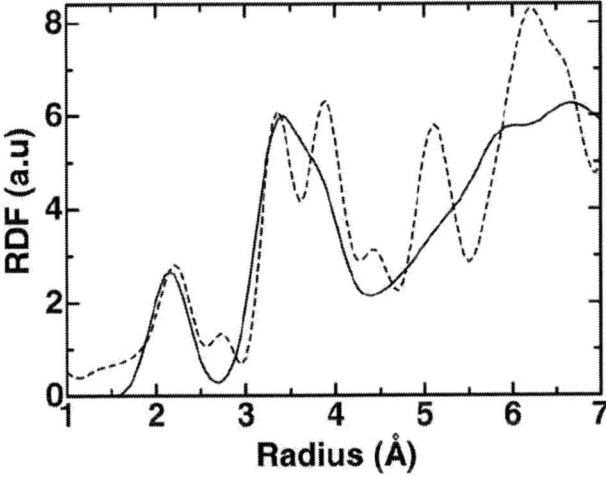

Figure 8. The radial distribution curves of the In_2O_3-ZnO film (solid line) and crystalline In_2O_3 (dotted line)[16, 25].

with Moriga et al[29]. The second peak of the RDF at 3.5 Å was attributed to an In–In pair. There are two kinds of linkage of InO_6 octahedra in bixbyite In_2O_3; one is corner-sharing and another is edge-sharing. Since the peak at 3.5 Å was actually composed of two peaks as shown in Figure 3, it was considered that amorphous In_2O_3–ZnO had both kinds of In–In pairs. To investigate the linkage of the InO_6 octahedra, the numbers of corner- and edge-sharing In–In pairs corresponding to the second peak in the RDF were calculated from the RMC-fitted configuration. In the case of crystalline In_2O_3, the ratio of the corner- and edge-sharing In–In pairs is 0.50 and 0.50, respectively. It was observed that, although the number of edge-sharing pairs decreased slightly, amorphous In_2O_3–ZnO had edge-sharing In–In pairs. Since the bond angle distribution of the corner-sharing pairs is larger than that of the edge-sharing pairs generally, the bond angle distribution of corner-sharing In–O–In and In–O–Zn was much broader than that of the edge sharing pairs. It was considered that the increase in the number of corner-sharing In–In and In–Zn pairs with Zn content, and the broader angle distribution of the corner-sharing In–O–In and In–O–Zn with Zn content caused the decrease of the long range periodicity of amorphous In_2O_3–ZnO films.

The structures of amorphous In_2O_3–ZnO films were investigated using a combination of GIXS and EXAFS with synchrotron radiation and simulations. The average oxygen coordination number around the In and Zn ions was roughly 6 and 4, respectively, as determined from the EXAFS results. For the GIXS and the simulation result, the structural feature of the amorphous In_2O_3–ZnO film substantially agreed with those of crystalline In_2O_3.

This structural analysis was well agreed with the electron scattering model obtained from the above analysis on the basis of the electron-phonon and electron-phonon-impurity interaction. Therefore, we consider that this causes the high mobility of the amorphous In_2O_3–ZnO film. Also, we carried out the HX-PES measurement to investigate the conduction band of the film. The measurement was done at a BL47XU beam line in SPring-8. The result is shown in Figure 9 with that of an ITO film. It was noted that the conduction band of the In_2O_3–ZnO film is quite similar to that of the ITO. This result is well agreed with the above scattering model and structural feature of the In_2O_3–ZnO film.

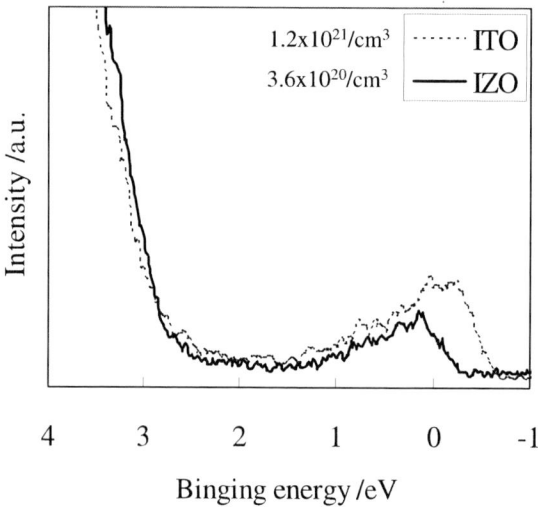

Figure 9. HX-PES spectra of the In_2O_3-ZnO (solid line) and ITO film(dotted line).

In$_2$O$_3$-ZnO films as Semicondoctors

We note that the In_2O_3-ZnO film can be used as the oxide semiconductor. Field-effect transistors (FETs) are an essential component in electronic devices and are included in most consumer electronic products, such as flat panel displays, computers, and cell phones. Until now, various materials have been investigated as potential materials for producing high performance channel semiconductors: poly-Si[30], a-Si:H[31], organic materials[32-35], and oxide materials[10-13, 36, 37]. In particular, recent developments in thin film FETs have focused on the use of oxide materials due to their flexibility, transparency, and low temperature fabrication[36]. Oxide semiconductors have carrier mobility that reach 1–100 cm^2/Vs, which is virtually higher than that of a-Si:H-based FETs. A 300-nm-thick SiO_2 insulating layer on a highly doped n^+-Si wafer was used as a substrate. In the In_2O_3-ZnO formation, the base pressure inside a chamber was reduced to less than 5×10^{-4} Pa. The oxide layer was formed by conventional radio frequency magnetron sputtering using an In_2O_3-ZnO target (In_2O_3:ZnO=95:5 wt.%) with a flow of pure Ar gas keeping the pressure at 1.1×10^{-1} Pa during deposition at room temperature. We controlled the carrier concentration of the In_2O_3-ZnO thin films by annealing the oxide layer at 300 °C in an ambient atmosphere for 1 hour. Although the as-deposited indium zinc oxide thin films have an amorphous morphology, the x-ray diffraction patterns showed sharp crystalline peaks after annealing. This result indicates that the annealing produced the formation of a polycrystalline texture. The carrier concentration of the film was estimated to be 7×10^{14} cm^{-3} in the annealed film based on Hall-effect measurements. Figure 10 shows the IDS-VDS output and transfer characteristics of the FET when a 5-nm-thick indium zinc oxide film was used as the active layer. The FET demonstrated a typical n-type FET operation; IDS increases with an increase in positive VDS and VG. The transfer characteristics showed very good FET characteristics with a saturation mobility of 14 cm^2/Vs, Vth of 3.0 V, and on-to-off current ratio (I_{on} / I_{off}) of 10^4.

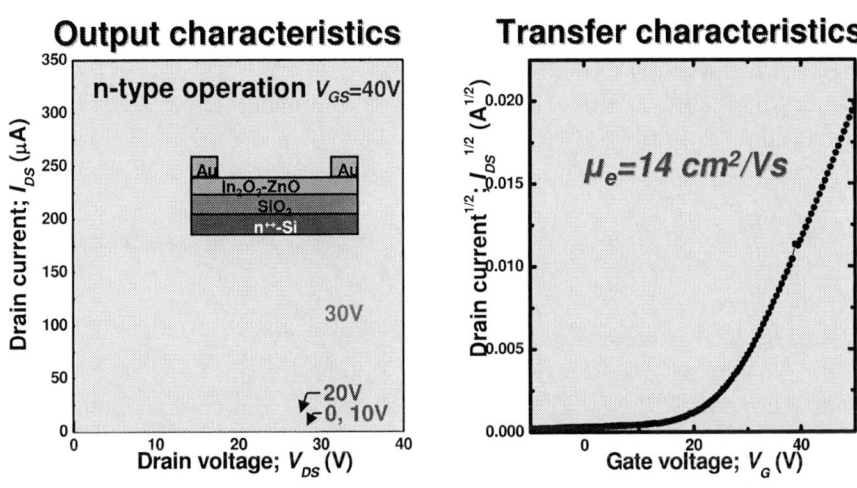

Figure 10. Output and transfer characteristics of the In_2O_3-ZnO TFT[10].

CONCLUSION

The transparent conductive In_2O_3-ZnO films can be used as TFT-LCD electrodes. The practical merits such as resistivity uniformity and the high etching rates are due to amorphous structure. The structural analysis and temperature dependence of the restivities of the films were revealed the electron conductive mechanism for conductivity. Recently, the In_2O_3-ZnO films can be used as semiconductor materials, and then we expect the films will be able to apply new devices.

REFERENCES

1. A. Kaijo, Proc. 3rd Int. Display Workshop, vol. 2, (1996)365.
2. A. Kaijo, K. Inoue, S. Matsuzaki, Y.Shigesato, Proc. 4th Pacific Rim International Conference on Advanced Materials and Processing, (2001)1787.
3. Y.S. Jung, J.Y. Seo, D.W. Lee, D.Y. Jeon, Thin Solid Films 445 (2003) 63.
4. T. Minami, T. Kasumu, S. Takata, J. Vac. Sci. Technol., A 14 (1996) 1704.
5. N. Naghabi, A. Rougier, C. Marcel, C. Guery, J.B. Leriche, J.M. Trascon, Thin Solid Films 360 (2000) 233.
6. 21. N. Ito, Y. Sato, P. K. Song, A. Kaijyo, K. Inoue, and Y. Shigesato, Thin Solid Films 496, (2006)99.
7. Y. Shigesato, D. C. Pines, and T. E. Haynes, J. Appl. Phys. 73, (1993) 3805.
8. A. J. Leenheer, et al., Phys. Rev. B, 77(2008)115215.
9. C. W. Ow-Yang, H. Yeom, D. C. Paine, Thin Solid Films, 516(2008)3105.
10. H. Nakanotani et al., Appl. Phys. Lett. 90(2007)262104.
11. P. Barqyuinha, et al., J. Non.cryst. Sol. 352(2006)1749.
12. H. Kumomi, K. Nomura, T. Kamiya, H. Hosono, Thin Solid Films, 516(7), (2008)1516.
13. D. C. Paine, B. Yaglioglu, Z. Beiley, S. Lee, Thin Solid Films, 516(2008)5894.
14. I. Hirosawa, Y. Uehara, M. Sato, N. Umesaki, J. Ceram. Soc. Jpn. 112 (5) (2004) s1476.
15. B. Shinozaki, K. Makise,Y. Shimane, H. Nakamura, K. Inoue, J. Phys. Soc. Jpn. 76 (2007) 074718.
16. M. Funaki et al. J. Appl. Phys. 103(2008)113701.
17. K. Makise, M. Funaki, B. Shinozaki, K. Yano, Y. Shimane, K. Inoue, and H. Nakamura, Thin Solid Films 515, (2008) 5805.
18. B.L. Altshuler, D. Khmelnitzkii, A.I. Larkin, P.A. Lee, Phys. Rev., B 22 (1980) 5142.
19. B.L. Altshuler, Sov. Phys. JETP 48 (1978) 670.
20. N.G. Ptitsina,G.M. Chulkova,K.S. Il'in, A.V. Sergeev, F.S. Pochinkov, E.M. Gershenson, M.E. Gershenson, Phys. Rev., B 56 (1997) 10089.
21. J.M. Ziman, Electron Phonons, Clarendon Press, Oxford, 1960, p. 364.
22. M. Yu Reizer, A.V. Sergeev, Sov. Phys. JETP 65 (1987) 1291.
23. F. Utsuno, N. Yamada, M. Kamei, I. Yasui, J. Korean Ceram. Soc. 5 (1) (1999) 40.
24. F. Utsuno, H. Inoue, I. Yasui, Y. Shimane, S. Tomai, S. Matuzaki, K.Inoue, I. Hirosawa, M. Sato, and T. Honma, Thin Solid Films 496, (2006)95.
25. F. Utsuno et al. Thin Solid Films 516 (2008) 5818–5821 5821
26. R.L. Mozzi, B.E. Warren, J. Appl. Crystallogr. 3 (1971) 251.
27. R.L. McGreevy, L. Pusztai, Mol. Simul 1 (1988) 359.
28. D.A. Keen, R.L. McGreevy, Nature 344 (1990) 423.
29. T. Moriga, A. Fukushima, Y. Tominari, S. Hosokawa, I. Nakabayashi, K. Tominaga, J. Synchrotron Radiat. 8 (2001) 785.
30. S. C. Wang, C. F. Yeh, C. K. Huang, and Y. T. Dai, Jpn. J. Appl. Phys., Part 1, 42 (2003) 1044.
31. C. S. Yang, L. L. Smith, C. B. Arthur, and G. N. Parsons, J. Vac. Sci. Technol. B, 18 (2000) 683.
32. G. H. Gelinck, T. C. T. Geuns, and D. M. de Leeuw, Appl. Phys. Lett. 77 (2000) 1487.
33. H. Klauk, M. Halik, U. Zschieschang, G. Schmid, W. Radlik, and W. J. Weber, J. Appl. Phys., 92 (2002) 5292.

34. L. A. Majewski, R. Schroeder, and M. Grell, Adv. Funct. Mater. 15 (2005) 1017.
35. S. H. Kim, S. Y. Yang, K. Shin, H. Jeon, J. W. Lee, K. P. Hong, and C. E. Park, Appl. Phys. Lett., 89 (2006) 183516.
36. K. Nomura, H. Ohta, K. Ueda, T. Kamiya, M. Hirano, and H. Hosono, Science, 300 (2003) 1269.
37. K. Nomura, H. Ohta, A. Takagi, T. Kamiya, M. Hirano, and H. Hosono, Nature, 432 (2004) 488.

Wavelength Dependent Contrast Reversal in Reflectivity of Nickel Alloy Nanofilms

Maarij Syed, James Wilkerson, Amanda Barnett, and Azad Siahmakoun
Rose-Hulman Institute of Technology, Terre Haute, IN 47083, U.S.A

ABSTRACT

Nickel alloy films (Nichrome: 80% Nickel, 20% Chrome) were grown by magnetron sputtering on fused silica prisms. In addition, pure metal films of Nickel and Chrome were also grown by magnetron sputtering for comparison. Alloy films of two different nominal thickness values (50 nm and 20 nm) were grown. Reflectivity measurements were carried out in the Kretchmann configuration for a variety of wavelengths that include wavelengths of interest for optoelectronic applications (1320 nm and 1550 nm), and for 633nm. The alloy films showed interesting features in the reflectivity spectra that were not present in the single metal films. Specifically, reflectivity spectra for p-polarized light showed interesting behavior that can be explained by a reflectivity model that incorporates surface plasmon resonance (SPR). Moreover the difference between the reflectivity spectra for the 50 nm film and the 20 nm film for 633 nm wavelength is dramatic. The spectra undergo a contrast reversal wherein the absorption feature present in the 50 nm alloy film is replaced by very high reflectance for the 20 nm alloy film, in the vicinity of the so called plasmon angle.

INTRODUCTION

SPR effect is based on the coupling of a totally internally reflected (TIR) photon's energy and momentum with the free electrons of a metal that are excited as a two dimensional charge density wave on the interface between a metal and a dielectric. SPR effects in thin metal films have led to a renewed and sustained interest in plasmonics applications and devices that range from optoelectronic switches to biochemical sensors [1]. Typically, the materials of choice are gold and silver due to their high figures of merit like long plasmon propagation distances, chemically stable surfaces (critical for repeated use as sensors) and very strong plasmon absorption over an extremely narrow range of angles of incidence and at easily accessible and commercially useful wavelengths. In addition to device applications, SPR has been investigated as a fundamental phenomenon in its own right where dimensionality effects produce interesting departures from bulk optical properties of a given material [2]. The present investigation focuses on alloy films of metals that are not typically used as plasmon supporting systems. Other complimentary techniques like spectroscopic ellipsometry were employed to extract the dielectric function of a layer system from companion slides grown under the same condition as a given film on the prism. This dielectric function was used as an input for the reflectivity model that relied on the dielectric properties of the metal film and the two dielectric media surrounding it (prism and air, in this study). Surface roughness measurements were performed using a white light phase-shifting interferometer. For the samples in this study, the roughness values (RMS) were in the range of 20-30 nm.

THEORY

The model used to account for the behavior of the thin films reflectivity has been discussed by Simon et al [3]. It is a three layer model with the metal layer sandwiched between two dielectrics; these two layers are fused silica prism and air in our case. The main feature of the model is that it allows for a propagating solution at one (or both) of the metal-dielectric interfaces. This solution decays evanescently away from the metal layer in both directions and propagates along the interface with a characteristic decay length in this direction as well. Reflectivity is given by the following expression.

$$R = \left| \frac{r_{12} + r_{23} \exp(-2kd)}{1 + r_{12}r_{23} \exp(-2kd)} \right|^2 \tag{1}$$

Here r_{12} and r_{23} are the Fresnel reflection amplitudes at the dielectric-metal and the metal-air interfaces respectively, d is the film's thickness while k, the exponential decay due to the absorption by the film, is given by the following expression.

$$k = -i\left(\frac{\omega}{c}\right)\left(\varepsilon - n^2 \sin^2 \theta_1\right) \tag{2}$$

Where n is the index of refraction for the dielectric, ε is the complex dielectric function for the metal film, and θ_1 is the angle of incident at the dielectric-metal interface.

EXPERIMENT

Thin films of 80/20 Nichrome were grown using magnetron sputtering. A fused-silica prism was coated with a film thicknesses of 200 ± 20 Å and a BK7 prism is coated with a film of 500 ± 20 Å. For every prism coated with metal films, companion layers of the same nominal thickness and under the same growth conditions were also grown on glass substrate that were later used to carry out spectroscopic ellipsometry measurements. The experimental setup used to observe the SPR effects includes Nd:YAG laser at 1320 nm or 1064 nm wavelengths followed by a polarization rotator and a linear polarizer as shown in figure 1. The coated prism was mounted on a rotational stage with 0.5 arc minute resolution. Optical power meter was mounted on an independent rotational stage with 0.5 arc minute resolution. Each measurement was carried out with the light polarized in TM mode and then repeated in transverse electric (TE) mode. Both the reflected light as well as transmitted light through the film were measured. The ellipsometric measurements were carried out on Woollam spectroscopic ellipsometers for the determination of the ellipsometric parameters Ψ and Δ over a wavelength range from $400 - 2300$ nm.

To model the actual experimental data one has to account for the fact that the metal film was deposited on the hypotenuse of a right angle face of the prism and light was refracted onto the glass side of the prism-metal interface in the TIR geometry and after reflection exited the prism. Fresnel losses at the first and last face of the prism were also accounted for along with the reflection (Eq. 1) that took place at the hypotenuse of the right angled prism.

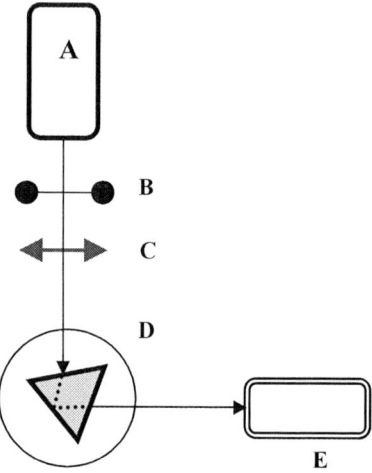

Figure 1. Schematic of experimental setup: (A) laser-diode, (B) polarization rotater, (C) linear polarizer, (D) coated prism on rotation stage and, (E) optical power meter.

DISCUSSION

Sample reflectivity results are shown for various wavelengths as a function of the angle of incidence on the dielectric-metal interface, for the two types of film thicknesses (50 nm and 20 nm). The modeling procedure would involve getting the dielectric function from ellipsometry and getting roughness estimates from ellipsometric model and from measurements using the phase-shifting interferometer setup. Only *p*-polarization data is presented in detail as the *s*-polarization data does not reveal any absorption or reflection features. To illustrate the point, one sample *s*-polarization curve is shown that exhibits a fairly featureless reflectivity curves (nearly monotonic) as predicted by a simple application of Fresnel's equations (see figure 2). In other words, no plasmon solution is expected for the *s*-polarization, and no corresponding feature is ever seen in the *s*-polarization data. As clear from the figures, the theoretical model does a reasonably good job of explaining the data. While figure 2 shows the reflectivity for the 20 nm film with its high reflectance, figure 3 show the results of *p*-polarization reflectivity for the 50 nm Nichrome film with the SPR feature (now absorption) present prominently in at an angle of about 41.5°. Theoretical calculations for the plasmon angle with the appropriate choice of the dielectric functions result in an angle of 41.9 °[3]. While this is close to the observed value, the difference is not insignificant. It is hoped that a more detailed investigation into the nature of the roughness and the dielectric function will result in better agreement. It should also be noted that neither Chromium nor Nickel show any such feature for the *p*-polarization geometry for the corresponding 50 nm single metal films. The theroetical model in general, predicts a slightly higher reflectivity but that is mostly due to losses on the slightly inferior surface of the BK7 prism (as opposed to the fused silica prisms). It is worth noting that the 20 nm film shows fundamentally different behavior. The primary absorption feature is now replaced by a strong reflection peak. In order to model this data, Nichrome dielectric function values for 633 nm are obtained from the ellipsometric measurements. The starting thickness is assumed to be 50 nm but once reasonable agreement is found between experimental data and the corresponding model, thickness is allowed to vary by a small amount and surface roughness is built into the model. The real and imaginary parts of the dielectric function are 2.06 and 25.98, respectively. Typically, for the surface plasmon to be excited the real part of the dielectric function for the metal should be negative. Moreover, for metals like silver and gold that show very large SPR response with reflectivity varying over 98% in the neighborhood of the plasmon angle (compared to ~18% for Nichrome), the imaginary part of the dielectric function is very small. That is also not the case for Nichrome. The relatively large imaginary part of the dielectric function is easier to interpret than the positive real part. The large imaginary part of the dielectric function implies strong damping of the plasmon and that is evident from the relatively shallow absorption feature associated with the plasmon absorption in Nichrome. The positive real part is more difficult to explain. Another complication is the nature of roughness. The interferometer results suggest a roughness that is on the order of the thickness of the thinner of the two sets of films. This really suggests that the surface profile for both types of films may be fairly corrugated and the model should really be refined to make the interface more of a textured surface with some kind of an effective medium approximation. This would also modify our present approach as regards to the dielectric function of the two nichrome layers (50 nm and 20 nm). Presently, the companion layer analysis suggests that the same dielectric function is used for both layers. However, if an effective medium approximation is to be used then clearly, the 20 nm layer incorporates greater void fraction than the 50 nm layer. Indeed, it is surprising that the present model explains the results, as well as it does for all the wavelengths examined.

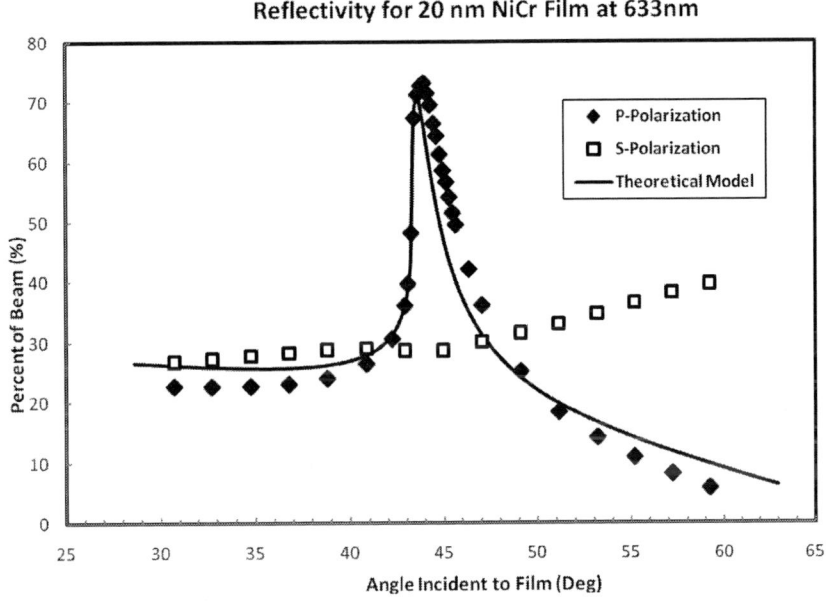

Figure 2. Reflectivity data for 20 nm Nichrome film at 633 nm wavelength.

The next two figures show the results for the reflectivity at 1320 nm wavelengths for the two thickness values. In both cases, the data looks very similar for the 1550 nm wavelengths. Dielectric functions for the nichrome film at 1320 nm and 1550 nm wavelengths are **3.2 + 41i** and **3.5 + 45i**, respectively. While the 50 nm film shows an absorption feature for 633 nm, it also displays a reflection maximum at longer wavelengths, albeit less dramatic than the reflection peak for the 20 nm film.

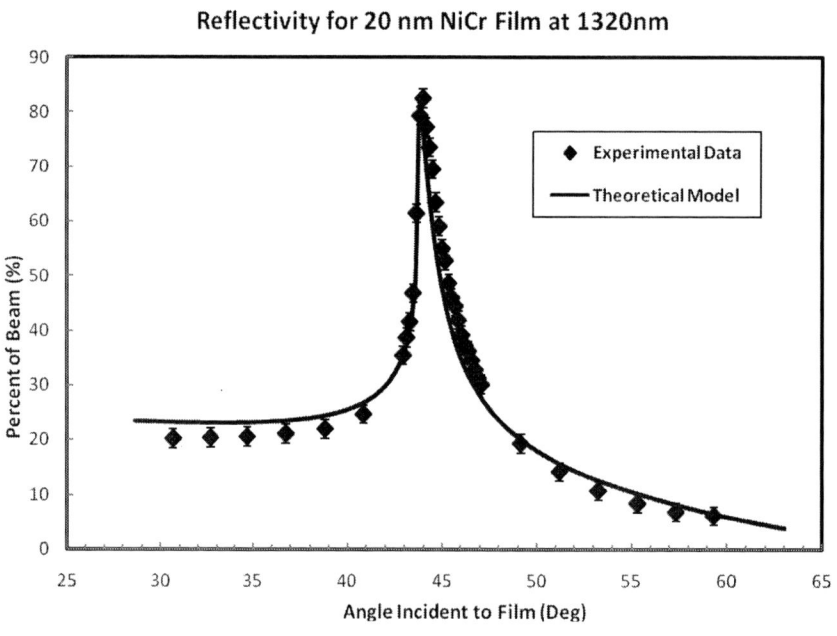

Figure 3. Reflectivity data for 50 nm Nichrome film at 633 nm wavelength.

Figure 4. Reflectivity data for 20 nm Nichrome film at 1320 nm wavelength.

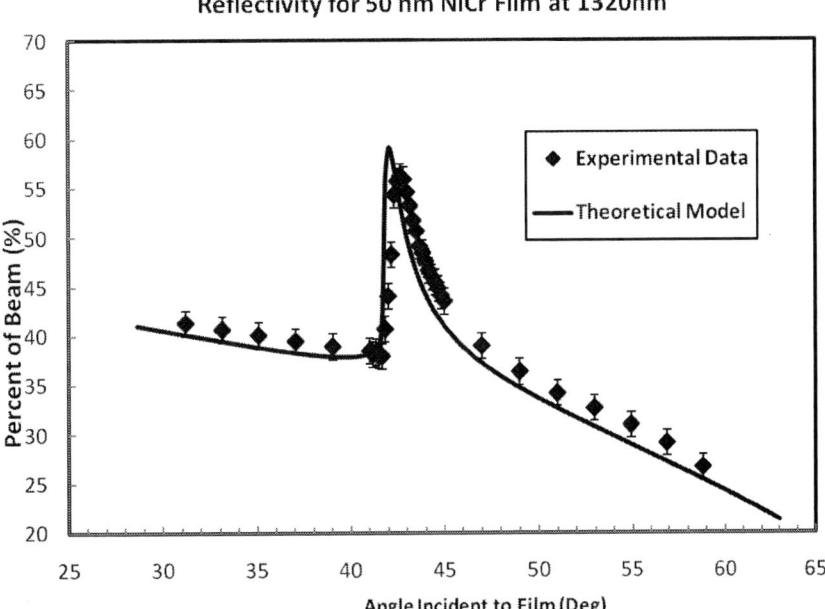

Figure 5. Reflectivity data for 50 nm Nichrome film at 1320 nm wavelength.

CONCLUSIONS

Interesting reflectivity results are obtained for 20 nm and 50 nm nichrome alloy films which show a contrast reversal at 633 nm wavelength. The differences persist at longer wavelengths, although both films show reflection peaks at these wavelengths. While the results are modeled as SPR, interesting questions remain regarding the exact nature of the mode responsible for these peaks. Future work will focus on refining the model and investigating the transmittance data to gain insight regarding the nature of the mode in these films.

REFERENCES

1. K. Kurihara and K. Suzuki, "Theoretical Understanding of an absorption-based surface plasmon sensor based on Kretchmann's theory," *Anal Chem.*, 74, 696 -701, 2002.
2. E. Kretschmann & H. Raether, "Radiative Decay of nonradiative surface plasmons excited by light," *Z. Naturforsch*, A23, 2135-2136, 1968.
3. H. J. Simon, D. E. Mitchell, J. G. Watson, "Surface Plasmons in silver films," *Am. J. Phys.*, vol. 43, (7), 630-636, 1975.

Mater. Res. Soc. Symp. Proc. Vol. 1109 © 2009 Materials Research Society   1109-B06-26

Study of current stability and fluctuations of field emitted electrons from ZnO nanostructure

Kishore Uppireddi, Boqian Yang, Peter Xian Feng, and Gerardo Morell

Department of Physics, University of Puerto Rico, San Juan, Puerto Rico, PR 00931, USA

ABSTRACT

Stable field emission currents with lower fluctuations are important feasibility requirements for the application of materials in field emission devices and displays (FED), than generally considered lower turn on voltages. The current stability and current fluctuations of field emitted electrons from ZnO nanostructure were investigated over the period of 2, 12 and 24 hours. The ZnO nanostructure films were synthesized by pulsed laser deposition (PLD) on silicon substrates by varying the deposition time. The film with nanoneedle structure having a density of ~10 per μm^2 showed better temporal stability over a period of 24 hours. The short- and long-term stability, high frequency current fluctuations are discussed in detail and results were compared with nanostructure density and aspect ratio of the ZnO films.

INTRODUCTION

The field electron emission current stability and life time stability are significant feasibility requirements for the next generation field emission displays (FEDs). The unique advantages of FEDs like fast response times, lower power consumption, wide viewing angle, higher contrast and better image stability made them as constructive display over cathode-ray-tubes (CRT) [1]. ZnO is an important II-VI compound semiconductor whose one-dimensional nanostructure has attracted considerable research interests for its unique electronic [2] and photonic properties [3]. The emission stability of FEDs mainly depends on cathode material and the material should satisfy the following requirements such as being resistant to oxidation, ion bombardment by residual gases, withstanding the pressure differences and also environmental stability (in different gas ambient's), for the realization of emission sources in flat panel displays [4]. Stable emission currents with lower fluctuations and long-term stability are more important than low onset fields for using cold cathode materials in the applications as electron source like in FEDs [5]. Much of the effort has been dedicated to achieve lower operating voltages and the correlation between the aspect ratio and field enhancement factor [6,7,8,9,10,11,12,13,14,15] and little had been reported on the emission stability of ZnO nanostructured films [16]. The ZnO nanostructures prepared by various deposition techniques resulted in nanowires [6,10], nanoneedles [7], nanofibers [9], nanotowers [17], nanopins [9], nanoflowers [15], tetrapods [8], multipods [12] had shown one common characteristic of relatively lower turn on fields comparable to other cold cathode materials [18]. Here we report the temporal stability of ZnO nanostructure films, which is one of key issue for the application of the emitter in FEDs.

EXPERIMENT

The films were prepared by Pulse Laser Deposition (PLD) using an ArF Lambda Physik 1000 excimer laser (193 nm, ~20-30ns, 10 Hz repetition rate, and 200 mJ pulse energy) by irradiating the commercial zinc oxide target (purity up to 99.99%) at a pressure of 2×10^{-5} Torr in the chamber. The laser beam focused with a 30 cm focal length fused silica lens, was incident at 45 degrees relative to the target surface. The diameter of the focused spot of the laser beam on the ZnO target was about 3mm. The power density of the laser on the target was 1.1×10^{8} W/cm² per pulse. The ZnO target was rotated circularly at 200 rpm. The silicon (100) substrates were used, which are placed 4 cm away from the target. A heater and thermocouple were used to obtain and monitor the desired substrate temperature. Prior to laser irradiation, the Si (100) and Cu, and W wafers were rinsed in acetone and methanol. The films were deposited for a period of 10, 20 and 30 minutes and the substrate temperature was kept at 600°C. The morphology of ZnO nanostructures was characterized using Scanning Electron Microscopy (SEM) and the phase structure by X-ray diffraction (XRD) measurements (Siemens D5000 diffractometer using the Cu K_{α} line source (λ=1.5405 Å) in θ-2θ configuration). The field emission (FE) measurements were taken in a custom-made system [19] in which a Ta cylindrical rod of 1.6 mm diameter (area: 0.0201 cm²) attached to a micro-positioner serves as the anode. By using this field emission configuration, we can accurately approximate the macroscopic electric field on the surface of the sample (i.e. cathode) as $E=V/d_{C-A}$, where V is the voltage applied to the anode and d_{C-A} is the distance between anode and cathode. The voltage is applied using a Stanford Research Systems PS350 power supply and the emitted current is measured with a Keithley 6517A electrometer. All the measurements were done at d_{C-A} varied in between 100-130 μm with ±2 μm and at a high vacuum of 1×10^{-7} Torr (1.33×10^{-5} Pa). Currents lower than 1×10^{-12} A were considered as the background noise level. The turn-on field (E_{to}) was defined as the field required to produce a current density of 0.1 μA/cm². For data acquisition, a custom LabView (National Instruments) program was developed. As a measure of precaution to avoid the influence of displacement or charging current in the field-emitted current, a delay of 4 seconds between the change in voltage and the data acquisition was employed.

RESULTS AND DISCUSSION

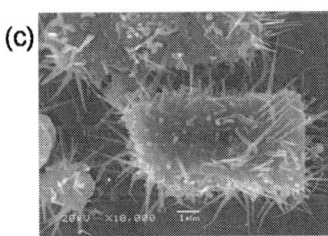

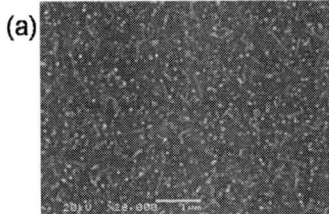

Figure 1 shows the SEM micrograph of self-assembled ZnO nanostructures on Si substrate for a deposition time of 10, 20 and 30 minutes. The films here after identified as Zn_10, Zn_20 and Zn_30 for 10, 20 and 30 minutes deposition time respectively. The uniform distribution of rod-like ZnO structures in large quantities with some presence of nanoneedles can be seen in Figure 1(a) of the film Zn_10. The density of the structures was decreased for a deposition time of 20 minutes as evident from Figure 1(b). The structures are mostly nanoneedles with some presence of nanorods. As the deposition time was increased to 30 minutes there is change in the density and improvement in the needle-like structure with an average length of needles around 2 μm with a tip diameter of ~20 nm while the roots are ~100 nm wide, whereas the rods are ~100 nm wide

Figure 1. The SEM micrographs of ZnO nanoneedles on Si substrate for a deposition time of (a) 10 minutes, (b) 20 minutes and (c) 30 minutes respectively.

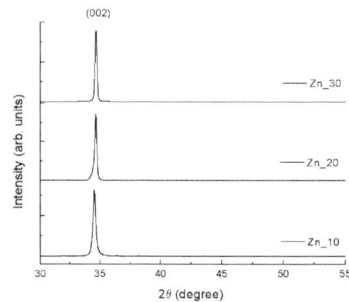

Figure 2. The XRD pattern of ZnO nanostructure films Zn_10, Zn_20 and Zn_30 respectively.

throughout. These ZnO nanoneedles are arranged like many flowers with a uniform diameter of 4 – 8 μm as shown in Figure 1(c). The crystallographic structure of the films was examined by XRD as shown in Figure 2. The peak related to wurtzite ZnO structure and exhibited a (002) preferred orientation. All films showed single-crystal nature with highly c-axis orientation.

The study of the emission current stability has two distinct and individually important aspects. One is the short-term stability associated with high frequency current fluctuations and the other one is the long-term stability linked with the temporal dependence of the average emission current at a fixed electric field value [19]. The temporal field emission behavior of ZnO films under study is shown in Figure 3. Since the amount of electric field required to induce a current of ~ 0.1- 0.33 mA/cm^2 was different for each film, percentage change from the initial current density at start (J_0) was taken to elucidate the degradation of the emission current. Each

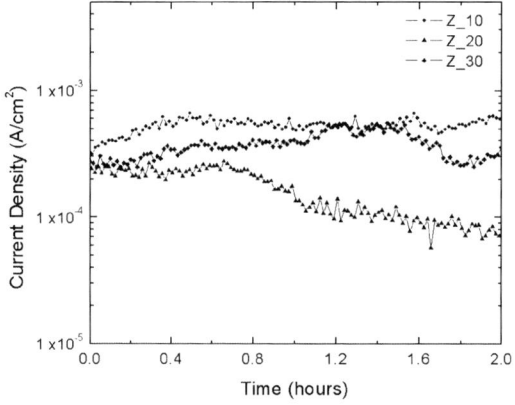

Figure 3. Temporal dependence of emission current vs time.

emission data point in Figure 3 is the average of 10 measurements with time gap of 250 ms. The Zn_10 and Zn_30 films showed an 91% and 25% increase in the current density after two hours (J_{2hr}), whereas Zn_20 films J_0 had fallen by 66% respectively. The field electron emission current density of the films after 12 hours (J_{12hr}) and 24 hours (J_{24hr}) and the corresponding percentage change were tabulated in Table I respectively. After 24 hours of stability test, the Zn_10, Zn_20 and Zn_30 films underwent a 85%, 95%, and 25% decrease respectively as indicated in Table I. The film Zn_30 showed better temporal stability compared to Zn_20 and Zn_10 films.

Table I Stability of field electron emission current from ZnO nanostructures over a period of 2, 12 and 24 hours

Sample Name	J_0 (A/cm^2)	$J_{2\,hr}$ (A/cm^2)	% Change from J_0	$J_{12\,hr}$ (A/cm^2)	% Change from J_0	$J_{24\,hr}$ (A/cm^2)	% Change from J_0
Zn_10	3.229×10^{-4}	6.188×10^{-4}	+91%	1.697×10^{-4}	-47%	4.683×10^{-5}	-85%
Zn_20	2.401×10^{-4}	8.045×10^{-5}	-66%	2.455×10^{-5}	-89%	1.13×10^{-5}	-95%
Zn_30	2.733×10^{-4}	3.423×10^{-4}	+25%	6.932×10^{-6}	-97%	9.427×10^{-5}	-65%

The high-frequency fluctuation in the emission current density of each data point from the average value (10 measurements with 250 ms time gap), which is standard deviation (S.D.) were also calculated to provide accurate analysis on stability. The films showed an S.D. values close to 1×10^{-5} A/cm^2, which accounted to high-frequency fluctuation of about 6%. As seen from Figure 4, Zn_30 sample showed stable value over a period of 2 hours compared to Zn_20 and Zn_10 films. The current fluctuations J/J_{avg} of the ZnO films over a period of 2 hours was shown in Figure 5, which is one of the key parameter for field emission devices. The current fluctuations from Zn_10 film are relatively stable as seen from Figure 5, deviating no more than 33% from J_{avg}, while that of Zn_20 and Zn_30 films deviated up to 65% and 41% from average, respectively. Hence, the Zn_10 and Zn_30 films showed lower emission fluctuations.

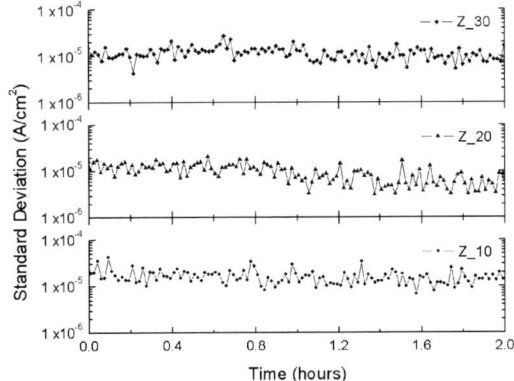

Figure 4. Standard deviation of emission current vs time.

Reliability issues in FEDs are associated to the short- and long-term current fluctuations, and to random current surges. The latter can be prevented by improving the FED device design, whereas the intrinsic fluctuations have to be managed by improvements in the cold cathode material employed. The dominant factors affecting the emission current stability are: (i) adsorbed

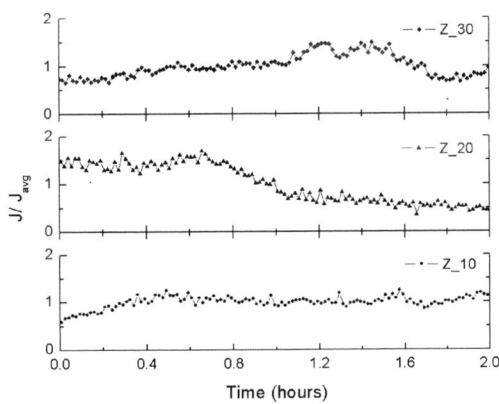

Figure 5. Emission current fluctuations (J/J_{avg}) vs time.

layers of the cathode surface by residual gas molecules, which affect the surface work function (φ) and in turn current by $\partial I/\partial \varphi$ [20] and (ii) ion bombardment of residual gases at high emission current densities, causing spatial changes in the surface topography $\partial I/\partial r$ that causes changes in the local field enhancement factor (β) [20]. Other important factors impacting the field emission stability and reproducibility are mechanical failure of nanostructures under applied field and failure by resistive heating at the contact [21,22]. The effect of adsorbates in our ZnO nanostructure films could be minimum due to their loss by local heating at an emission current values ~6 μA but could not be ruled out completely. Whereas the residual gas analysis during emission stability measurements indicated lower possibility of ion bombardment at a vacuum of ~10^{-7} Torr. Thus the effect of adsorbates and ion bombardment by residual gas can be safely discarded.

Recent study on emission site distribution before and after emission stability test revealed breaking of ZnO nanostructures by resistive heating [23] leading to poor temporal stability. This suggests that uniform emission site density with reasonable geometrical shape (aspect ratio) may be a key parameter for obtaining stable FE properties [24]. The Zn_30 film showed better short-term (high-frequency fluctuations) and long-term (temporal) stability compared to Zn_20 and Zn_10 films. Also Zn_30 and Zn_10 films demonstrated lower current fluctuations compared to Zn_20 film. Contrary, the Zn_30 film with lower nanostructure density and higher aspect ratio performed better than film with higher nanostructure density and lower aspect ratio of Zn_20 and Zn_10. This may be due to the presence of more active emission sites over entire nanoneedle structures in Zn_30 film with relatively high aspect ratio, leading to uniform emission sites with less damage by resistive heating as indicated by emission stability test for a period of 24 hours with a decrease of 65% from J_0. Whereas Zn_20 and Zn_10 film not be able produce more active sites even with higher nanostructure density and the corresponding damage by resistive heating. Resistive heating typically caused due to poor contact between the substrate and the film. Short deposition times during the growth can create poor contact between substrate and nanostructures. Previous work [22] by one of the authors in this regard on carbon fibers showed improved FE behavior by providing a better contact, and efforts are being made in this aspect to improve emission stability from ZnO nanostructures. By comparing above results and analysis with their respective nanostructures density on the film 40, 20 and 10 per μm^2 and aspect ratio of 5, ~10 and 20 of the Zn_10, Zn_20 and Zn_30 films, suggests that the instability in the field electron emission current from ZnO nanostructures was most probably due to the variation in number of active emission sites (emission density) and their degradation over long period of time by resistive heating.

SUMMARY

The temporal current stability and current fluctuations of field emitted electrons from ZnO nanostructure films deposited by PLD have been investigated. Zn_30 film showed the best short- and long-term emission current stability over a time of 2, 12 and 24 hours, compared to the Zn_20 film. The Zn_30 film showed 65% decrease from J_0, where as it is 95% for Zn_20 film over 24 hour period. The films exhibited relatively large current fluctuation from than the acceptable range for FED applications. The emission stability results were analyzed by comparing with nanostructure density, aspect ratio, percentage change in J_0 and current fluctuation over time. The results suggests that the instability in the field electron emission current from ZnO nanostructures was most probably due to the variation in number of active emission sites (emission density) and their degradation over long period of time by resistive heating.

REFERENCES

[1] Fursey George, in *Field emission in vacuum microelectronics*, ed by I. Brodie and P. Schwoebel (Kulwar, New York, 2005) pp. 144-150

[2] M. H. Huang, S. Mao, H. Feick, H. Yan, Y. Wu, H. Kind, E. Weber, R. Russo, and P. Yang, *Science* **292**, 1897 (2001).

[3] F. A. Ponce and D. P. Bour, *Nature* (London) **386**, 351 (1997).

[4] K. A. Dean and B. R. Chalamala, *Appl. Phys. Lett.* **75**, 3017 (1999).

[5] B. Günther, F. Kaldasch, G. Müller, S. Schmitt, T. Henning, R. Huber, and M. Lacher, *J. Vac. Sci. Technol. B* **21**, 427 (2003).

[6] C. J. Lee, T. J. Lee, S. C. Lyu, Y. Zhang, H. Ruh, and H. Lee, *Appl. Phys. Lett.* **81**, 3648 (2002).

[7] Y. W. Zhu, H. Z. Zhang, X. C. Sun, S. Q. Feng, J. Xu, Q. Zhao, B. Xiang, R. M. Wang, and D. P. Yu, *Appl. Phys. Lett.* **83**, 144 (2003).

[8] Q. Wan, K. Yu, T. H. Wang, and C. L. Lin, *Appl. Phys. Lett.* **83**, 2253 (2003).

[9] C. X. Xu and X. W. Sun, *Appl. Phys. Lett.* **83**, 3806 (2003).

[10] S. H. Jo, J. Y. Lao, Z. F. Ren, R. A. Farrer, T. Baldacchini, and J. T. Fourkas, *Appl. Phys. Lett.* **83**, 4821 (2003); S. H. Jo, D. Banerjee, and Z. F. Ren, *ibid.*, **85**, 1407 (2004).

[11] C. X. Xu, X. W. Sun, and B. J. Chen, *Appl. Phys. Lett.* **84**, 1540 (2004).

[12] K. Yu, Y. S. Zhang, F. Xu, Q. Li, Z. Q. Zhu, and Q. Wan, *Appl. Phys. Lett.* **88**, 153123 (2002).

[13] Y. H. Yang, B. Wang, N. S. Xu, and G. W. Yang, *Appl. Phys. Lett.* **89**, 043108 (2006).

[14] C. J. Park, D. K. Choi, J. Yoo, G. C. Yi, C. J. Lee, *Appl. Phys. Lett.* **90**, 083107 (2007).

[15] C. Li, W. Lei, X. Zhang, J. X. Wang, X. W. Sun, and S. T. Tan, *J. Vac. Sci. Technol. B* **25**, 590 (2007).

[16] Q. H. Li, Q. Wan, Y. J. Chen, T. H. Wang, H. B. Jia, D. P. Yu, *Appl. Phys. Lett.* **85**, 636 (2004).

[17] J. Xiao, X. Zhang, and G. Zhang, *Nanotechnology* **19**, 295706 (2008).

[18] M. H. Huang, Y. Y. Wu, H. Feick, N. Tran, E. Weber, and P. D. Yang, *Adv. Mater.* **13**, 113 (2001).

[19] K. Uppireddi, B. R. Weiner, and G. Morell, *J. Appl. Phys.* **103**, 104315 (2008).

[20] Ching-Yin Hong, and Akintunde Ibitayo Akinwande, *IEEE Trans. Electron. Devices* **52**, 2323 (2005).

[21] J. M. Bonard, and C. Kinke, *Phys. Rev. B* **67**, 115406 (2003).

[22] K. Uppireddi, A. G. Berríos, F. Piazza, B. R. Weiner, and G. Morell, *J. Vac. Sci. Technol. B* **24**, 639 (2006).

[23] J. Xiao, G. Zhang, X. Bai, L. Yu, X. Zhao, and D. Guo, *Vacuum* **83**, 265 (2009).

[24] X. Qian, H. Liu, Y. Guo, Y. Song, and Y. Li, *Nanoscale Res. Lett.* **3**, 303 (2008).

Mater. Res. Soc. Symp. Proc. Vol. 1109 © 2009 Materials Research Society

Transparent Non-volatile Memory using Pt Nano-particles embedded in an Amorphous Indium Gallium Zinc Oxide Thin Film Transistor

Arun Suresh, Steven Novak, Patrick Wellenius, Veena Misra, and John F. Muth
Department of Electrical and Computer Engineering, North Carolina State University, Raleigh, NC 27695

ABSTRACT

Amorphous oxide semiconductors (AOS) have attracted considerable attention because of their unique material properties. AOS based high performance thin film transistors (TFT) have been demonstrated for flexible and transparent electronics. Here we report a transparent non-volatile memory device based on a indium gallium zinc oxide (IGZO) TFT. Atomic layer deposited platinum nano-particles were incorporated in the transistor dielectric which acts as a floating gate and a charge trapping medium. The memory effect was determined by a positive shift in the turn-on voltage of the TFT when a positive gate bias was applied and is caused by electrons being trapped in the platinum nano-particles. Large memory operation windows with good charge-retention times have been observed. An asymmetric program/erase characteristic was observed for the IGZO memory TFT.

INTRODUCTION

Recently, "Transparent Electronics" is becoming one of the heavily researched areas due to the development of a relatively new class of materials called amorphous oxide semiconductors (AOS). These wide band gap materials based on post-transition metal oxides with the electronic configuration $(n-1)d^{10}ns^0$, where $4 \leq n \leq 6$ are not only transparent in the visible regime but exhibit relatively high carrier mobilities even in the amorphous state.[1] The origin of the high mobility in oxide semiconductors is attributed to the large, overlapping, isotropic s orbitals of the heavy metal cations that form the conduction band. Several binary and ternary oxide systems[2,3] have been extensively studied for their application as thin film transistor (TFT) channel materials including indium gallium zinc oxide (IGZO).[4,5]

As evidence from recent publications, the majority of research to date in the development of AOS based devices has dealt with channel material exploration and device optimization for TFTs. But to develop fully functional transparent systems it would be advantageous to study other circuit components such as diodes and memory elements. The floating gate type non-volatile memory is a common device in which a charge-trapping layer is sandwiched in the gate insulator, which either stores or releases charge to bring about the memory behavior. Though several materials systems have been considered, metal nano-particles as the charge-trapping layer provides several advantages.[6] Here we report the development and demonstration of a transparent non-volatile memory device consisting of a transparent floating gate transistor that incorporates platinum nano-particles (Pt-NPs), which act as the charge-trapping medium in the gate dielectric. This integration of a transparent TFT with memory functionality can help develop new pixel circuit or memory cell designs.[7]

210

EXPERIMENT

The schematic cross-section of the IGZO memory device is shown in Figure 1(a), which is similar to a bottom, gated TFT structure. Commercial glass substrates coated with 200 nm sputtered indium tin oxide (ITO) and 220 nm atomic layer deposition (ALD) AlO_x-TiO_x superlattice (ATO), which is capped on both ends by AlO_x, were used.[5] ITO acts as the gate electrode and the ATO as the dielectric (blocking oxide) of the device. Pt-NPs were formed on the ATO dielectric at 270 °C using the precursor $MeCpPtMe_3$ and high purity oxygen.[8] The initial nucleation of the Pt film produces small, disbursed crystals, which can be grown further with subsequent ALD cycles. Next a thin layer, ~7 nm, of AlO_x, which acts as the tunneling oxide, was deposited at 200 °C by ALD. This was followed by the deposition and patternings of IGZO (channel) and ITO (source/drain) by pulsed laser deposition (PLD) at room temperature.[5] Control samples without the Pt-NPs (ITO/ATO/AlO_x/IGZO/ITO) were also fabricated for comparison purposes. The fabricated devices were post-annealed at 250 °C in atmosphere for 1 hr. The TFT dimensions are L = 100 μm and W = 400 μm. Transmission spectra were obtained using a Perkin Elmer Lambda 9 UV/Vis/NIR dual beam spectrophotometer. Plan view images of the Pt-NPs grown on ALD AlO_x layer were obtained using an analytical SU-70 FESEM at 1 kV. X-ray photoelectron spectroscopy (XPS) was carried out using a Riber XPS with dual Mg/Al anode source. The IGZO memory device characteristics were measured with an HP 4155B semiconductor parameter analyzer.

DISCUSSION

One of the advantages of amorphous oxide semiconductors over conventional amorphous semiconductors is their transparency in the visible regime. To verify that the proposed device is transparent, optical transmission experiments using a spectrophotometer were carried out. The raw optical transmission versus the wavelength of the substrate and the entire IGZO memory device stack is shown in Figure 1(b). (A) represents the substrate (glass/ITO/ATO) and (B) represents the entire stack of the memory device (glass/ITO/ATO/Pt-NPs/AlO_x/IGZO/ITO). The addition of subsequent layers to the substrate did not cause an appreciable reduction in the transparency of the device and at visible wavelengths the entire stack was still highly transparent.

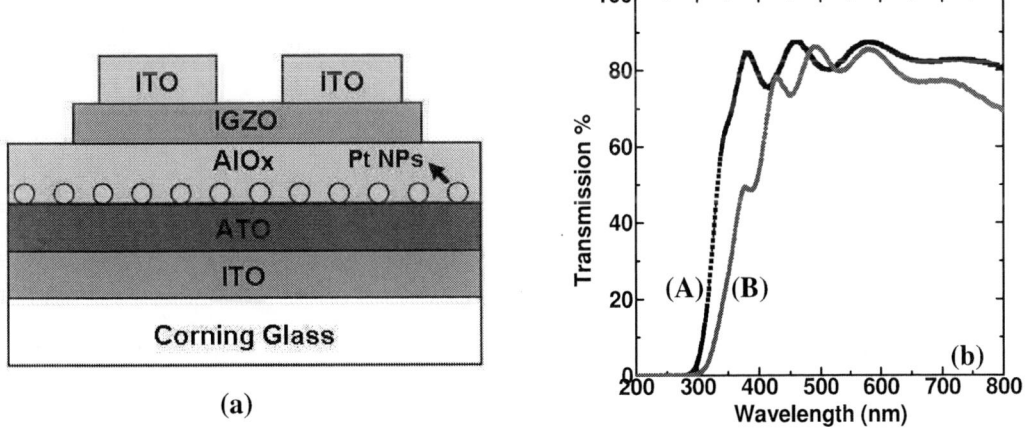

Figure 1. (a) A schematic representation of the IGZO memory device, (b) Transmission spectrum of the substrate (A) and device stack (B).

To verify the presence of Pt-NPs on the substrate several analytical techniques were utilized. Figure 2 shows the XPS data for the platinum 4f and 4d signals after 25 and 35 cycles of ALD Pt on ATO substrates. The signal intensity increases as a function of the cycle count. The

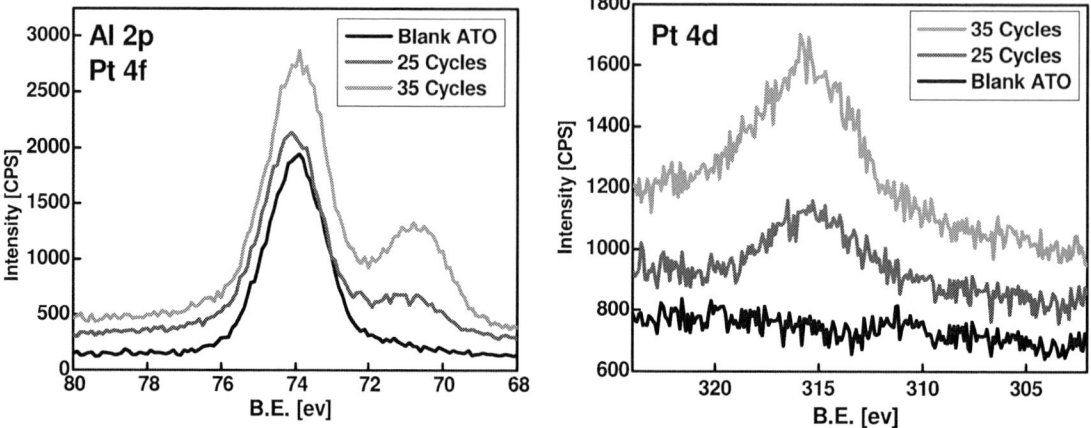

Figure 2. XPS data showing an increase in signal intensity with ALD Pt cycle count corresponding to the increase in Pt on the ATO sample surface.

increase is attributed to the increase in the number of Pt nuclei and the increasing size of each nuclei. The Pt 4f peak overlaps with the Al 2p peak (~ 74 ev) but the Pt 4f doublet (~ 71 ev) becomes more apparent as the cycle count increases. We also use the weaker Pt 4d peak to confirm the presence of Pt after 25 cycles. Similar results were seen for ALD AlO$_x$ samples.

Field emission SEM (FESEM) was used to image the Pt-NPs grown on ALD AlO$_x$. Figure 3(a) shows disbursed Pt-NPs in the 4-5 nm size range. The Pt was grown for 30 cycles and particles have a number density of around 1-1.5 e^{11} cm^{-2}. The cross-section of the IGZO memory device stack has been imaged using TEM and show Pt-NPs in the 2-4 nm range.[9] The memory behavior of the devices were estimated in the following manner. The transfer characteristics [log(I_{DS})–V_{GS}] of the TFT was initially measured. A gate bias, V_G, was applied for a pre-determined time while grounding the source and drain (programming). The transfer characteristics was re-measured and the shift in the turn-on voltage, V_{on}, was estimated to be the memory window. Figure 3(b) shows the V_{on} shifts [$\Delta V_{on} = V_{on}$ (post) – V_{on} (pre)] for IGZO TFTs both with and without Pt-NPs.

In the IGZO memory TFTs (with embedded Pt-NCs, solid markers) a positive gate bias induces a positive V_{on} shift. It is found that the hysteresis memory window increases with an increase in the applied gate bias and the stress duration. The shift in V_{on} tends to saturate for stress times above ~3 s due to the Coulomb blockade effect which raises the nano-particle potential when more electrons are to be stored.[10] On the other hand in the control TFTs (no Pt-NPs, hollow markers) the gate bias does not introduce a shift in V_{on}. The principle by which the IGZO memory device works is that when a positive gate bias is applied to the gate electrode an accumulation layer of electrons (since the IGZO TFT is n-type) is formed at the semiconductor/dielectric interface. These electrons tunnel through the AlO$_x$ (tunneling oxide) and get trapped in the Pt-NPs during programming. The trapped electrons now screen the applied gate bias and a larger V_G is required to turn on the device, which is manifested as a positive V_{on} shift.

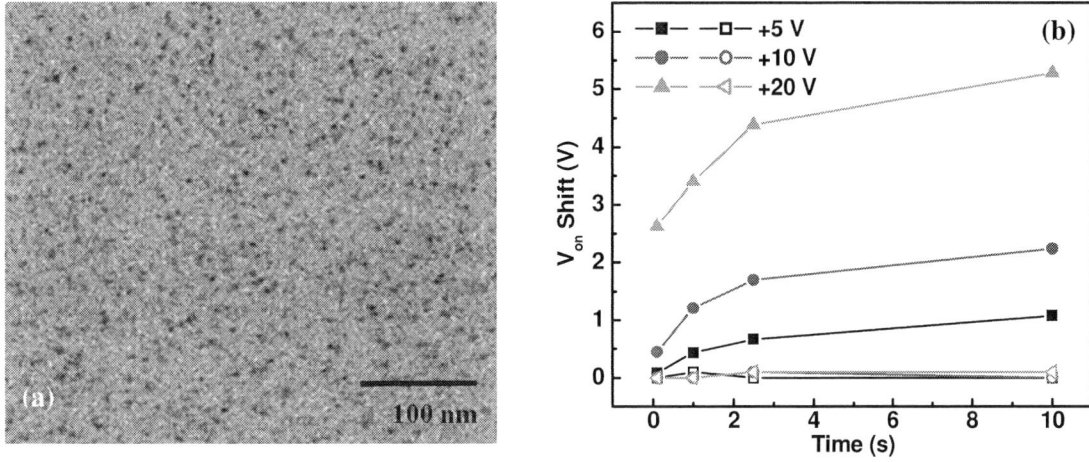

Figure 3. (a) A FESEM image of the Pt-NPs on ALD AlO$_x$ (b) The effect of the presence of Pt-NPs. Pt-NPs grown for 30 cycles. Large V_{on} shifts are observed for the IGZO memory TFTs (solid) . The control TFTs (without Pt-NPs) hardly show any V_{on} shift (hollow).

The control TFTs not showing a V_{on} shift clearly shows that the presence of Pt-NPs enhances the trapping of electrons. Also it was observed that a negative bias on the gate of a IGZO memory TFT does not cause a V_{on} shift[9] which can be explained by the lack of electrons in the channel due to the creation of the depletion layer (when $V_G < 0$) and lack of holes, which are not generated in the semiconductor.[11] This again proves that the positive V_{on} shift seen in the IGZO memory TFTs are due to electrons being trapped in the Pt-NPs.

To study the effect of number of Pt-NP growth cycles on the memory effect, devices with two different cycle counts (25 and 35) were fabricated. Increasing the number of growth cycles has two effects. It increases the size of the existing Pt-NPs and also induces nucleation of new Pt-NPs. Though Novak et al. pointed out that subsequent grown cycles predominantly grows the existing particle size,[8] nevertheless both these effects increase the number of trapping sites for the electrons and which in turn increases the tunneling probability for the electrons. In Figure 4(a) the IGZO memory device with Pt-NPs grown for 35 cycles shows a higher V_{on} shift compared to the one with 25 cycles as expected. Similar to data shown in Figure 3(b) we can see that the V_{on} shift starts to saturate above 3 s.

For the device to function as a memory element, the charge retention characteristic needs to be evaluated. The memory device was programmed at $V_G = 20$ V for 2 s while the source and the drain electrodes were grounded. The gate, drain and source electrodes were left grounded during the charge retention measurements. The stored electron charge loss (measured as shift in V_{on}) was measured periodically making sure to limit the charging effect during the read measurements. The V_{on} shift as a function of time is shown in Figure 4(b). We see an initial fast drop of ~15% in the first 1000 s, this we attribute to the charge loss corresponding to electrons trapped in shallow interface states formed during the Pt-NPs incorporation in the TFT dielectric. Then the charge loss slows down and we get an additional 10% loss over 10^4 s. This shows us that the majority of the electrons are strongly trapped at the Pt-NP sites.

To understand the erase characteristics of the device a $V_G = -10$ V was applied after the device was programmed with $V_G = 20$ V for 2 s. But unlike the charge retention experiment, the gate bias was maintained at -10V when the device was idle and not being read. The charge loss was estimated as before and plotted in Figure 4(b). We can see that the charge loss behavior of the stressed device (V_G = -10 V) is similar to the unstressed device (gate grounded) with a fast component and a slow gradual component and that the negative gate bias is not effective in completely erasing the device showing us the strong trapping of the electrons in the PT-NPs. In fact the charge loss difference between the stressed and unstressed devices (~5% over

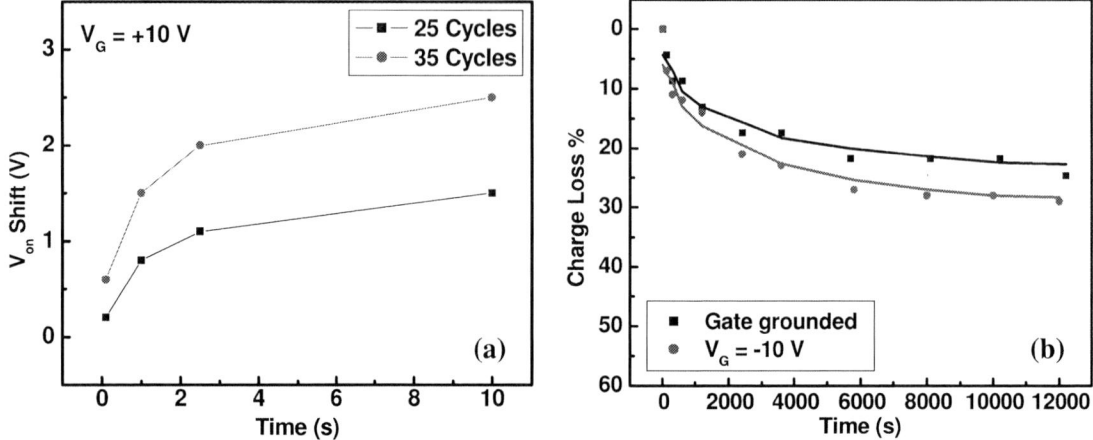

Figure 4. (a) Effect of the Pt growth cycles on the memory window. (b) Charge retention behavior and the erase characteristics of the IGZO memory device.

10^4 s) is the amount of erasing induced by the negative gate bias. The IGZO memory devices are easy to program but difficult to erase and this can be understood by looking at the program/erase mechanisms using simple energy band diagrams shown in Figure 5.

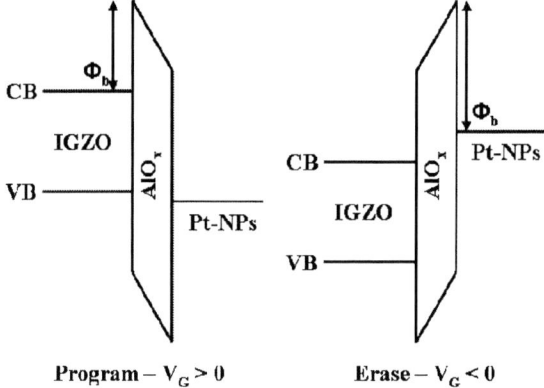

Figure 5. F-N tunneling program/erase mechanisms for IGZO memory devices.

In all the program/erase experiments, a Fowler-Nordheim (F-N) tunneling method was used, in which a positive (program) or a negative (erase) gate bias was applied while the source and drain electrodes were grounded. The number of charges tunneling through the thin AlO_x dielectric, as expressed as a tunneling current density (J), which is also a measure of the V_{on} shift is given by,[12]

$$J = \frac{A}{4\Phi_b} E_{inj}^2 \exp\left[-\frac{2B\Phi_b^{3/2}}{3E_{inj}}\right] \qquad (1)$$

where A and B are constants, E_{inj} and Φ_b are the electric field and the energy barrier at the charge injection interface respectively. We can see from Figure 5 that Φ_b for electron injection during programming from the IGZO channel is about 3.1 ev. Owing to the large work function of the Pt (~5.6 ev), the Φ_b for electron injection during erasing from the Pt-NPs is about 4.6 ev. This large discrepancy between the electron barriers leads to the asymmetric in the program/erase characteristics of the IGZO memory devices.

CONCLUSIONS

In summary a transparent memory device utilizing Pt nano-particles and an amorphous oxide semiconductor TFT based on IGZO has been demonstrated. The charge trapping can be attributed to the presence of Pt-NPs and verified from the turn-on voltage shifts seen during positive gate voltage stress. Charge retention measurements show that the charges are strongly trapped in the Pt nano-particles and the non-volatility of the device was ascertained with only a 25% loss of the stored charge even after 10^4 s. An asymmetric response was observed between the programming and erasing nature of the IGZO memory TFT and was explained using the difference between the energy barriers that the electron encounters during program/erase. The memory characteristics of this device could simplify design of truly transparent circuits with various functions, which will be beneficial for applications in display technology.

REFERENCES

1. M. Orita, H. Ohta, M. Hirano, S. Narushima and H. Hosono, *Philos. Mag. B* **81,** 501 (2001)
2. H. Q. Chiang, D. Hong, C. M. Hung, R. E. Presley, J. F. Wager, C.-H. Park, D. A. Keszler, and G. S. Herman, *J. Vac. Sci. Technol. B* **24**, 2702 (2006)
3. P. Barquinha, A. Pimentel, A. Marques, L. Pereira, R. Martins, E. Fortunato, *J. Non-Cryst. Solids* **352**, 1749 (2006)
4. K. Nomura, H. Ohta, A. Takagi, T. Kamiya, M. Hirano and H. Hosono, *Nature* **432,** 488 (2004)
5. A. Suresh, P. Wellenius, A. Dhawan and J. F. Muth, *Appl. Phys. Lett.* **90,** 123512 (2007)
6. F. M. Yang, T. C. Chang, P. T. Liu, P. H. Yeh, Y. C. Yu, J. Y. Lin, S. M. Sze, and J. C. Luo, *Appl. Phys. Lett.* **90**, 132102 (2007)
7. H. Kimura, T.Maeda, T. Tsunashima, T. Morita, H. Murata, S. Hirota, and H. Sato, *SID Int. Symp. Digest Tech. Papers*, 268 (2001)
8. S. Novak, B. Lee, and V. Misra, *submitted to Appl. Phys. Lett.*
9. A. Suresh, S. Novak, P. Wellenius, V. Misra, and J. F. Muth, *submitted to Appl. Phys. Lett.*
10. J. H. Kim, K. H. Baek, C. K. Kim, Y. B. Kim, and C. S.Yoon, *Appl. Phys. Lett.* **90**, 123118 (2007)
11. H. Hosono, *J. Non-Cryst. Solids*, **352**, 851 (2006)
12. M. Lenzlinger and E. H. Snow, *J. Appl. Phys.* **40**, 278 (1969)

Metal Oxide-based (IZO and ZnO) TFTs for Flexible Electronics

Shahrukh A. Khan and Miltiadis Hatalis
Display Research Laboratory, Lehigh University, Bethlehem, PA 18015

ABSTRACT

This work emphasizes room temperature deposition and fabrication of staggered bottom-gate ZnO and IZO TFTs. We synthesized these oxide thin films by RF sputtering in an Ar/Oxygen ambience with no intentional heating of the substrates. Bottom gate staggered structure ZnO TFTs were fabricated (Ti/Au/Ti gate and Au/Ti source/drain) and characterized. ZnO TFTs retained well-behaved transfer characteristics down to a channel length of 4 μm with field effect mobility of 5 cm^2/V.s, on/off current ratio exceeding 10^6 and threshold voltage around -5V. The IZO TFTs, with ITO as gate metal layer and highly conducting amorphous IZO forming the source/drain material had reasonably high field effect mobility of 20 cm^2/V.s and on/off current ratio exceeding 10^6, which are well suited for active matrix display applications. Finally, to demonstrate the viability of oxide-based device integration, simple circuits such as inverters and pseudo-logic circuits are designed, fabricated and tested.

INTRODUCTION

Oxide-semiconductor based thin-film transistors (TFTs) have advanced tremendously off-late and provides an attractive alternative to silicon-based TFTs. Current industrial approaches to produce thin film transistors for display devices include hydrogenated amorphous silicon (a-Si:H) and low-temperature poly silicon (LTPS). The mainstay of today's display devices based on a-Si:H although mature, is typically incompatible with plastic substrates and poses other limitations such as light-induced degradations and inherent low-mobility. On the other hand, LTPS offers high mobility but suffers from poor threshold voltage variation across display backplanes [1].

Oxide semiconductors composed of heavy-metal cations with $(n-1)d^{10}ns^0$ $(n \geq 4)$ electronic configurations [2, 3] have been widely investigated as they offer several key advantages. They are usually wide-gap materials, transparent in the visible spectrum and thus render possible ubiquitous transparent electronics. Furthermore, large carrier mobilities are achievable due to their unique electronic configuration. The preservation of relatively high mobility in the amorphous phase can be attributed to a high degree of localization [4, 5] and suggests that if the carrier concentration can be controlled, the properties of amorphous oxides are quite suitable for TFT applications. It is desirable to use very thin films of these metal-oxide materials to produce high electronic activity based either on amorphous or nanostructured layers inorder to reduce the role of interface and surface states on carrier transport as required in TFTs. In this study, we present room temperature fabrication and characterization of amorphous ZnO and IZO based TFTs.

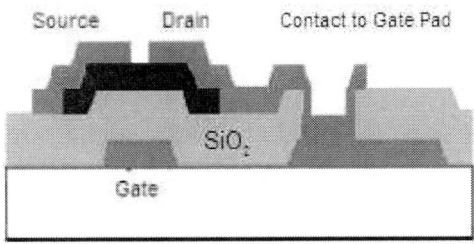

Figure 1: Schematic cross-section of fabricated amorphous oxide-TFT

EXPERIMENTAL DETAILS

In order to optimize the electrical, structural and optical properties of the ZnO and IZO films by RF sputtering, we varied oxygen partial pressure and RF power density. The targets were commercially available 6 inch diameter, dense ZnO ceramic target and In_2O_3:ZnO (90%:10%) with the substrate to target distance set at 75 mm. Prior to deposition, the chamber was cryopumped to ~ 10^{-7} Torr. Substrates were synthetic quartz, oxide coated Si for measurement purposes and patterned Si for device fabrication. Sputtering was carried out at room temperature at a RF power density of 3 W/cm^2 for ZnO deposition and between at 1-2 W/cm^2 for IZO deposition respectively. Generally we sputtered at low RF power to mitigate potentially damaging bombardment of the growing films by energetic negative oxygen ions, which can cause substantial stress in the film [6]. Figure 1 shows the schematic cross-section of the completed bottom-gate TFTs.

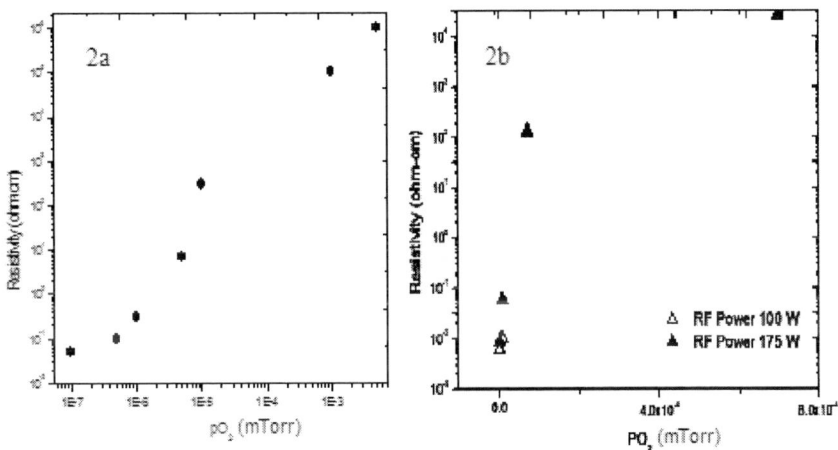

Figure 2: Dependence of resistivity of ZnO and IZO films on PO_2

Ti/Au/Ti were lithographically patterned and deposited by e-beam evaporation to form the gate of amorphous ZnO-TFT. 100 nm of SiO_2 was then deposited by PECVD at 300 °C .Then 40 nm of optimized ZnO thin film was deposited by RF sputtering at a chamber pressure of 10-15 mTorr. Contact openings to the gate pads were accomplished by lithography and selective etching of the oxide layer. Finally, Au/Ti source and drain metallization was done by RF sputtering. In the IZO TFTs, the active layer (40 nm) and source drain layers (150 nm) were sputter deposited at RF power densities of 1 and 2 W/cm^2 and with relative Ar/O$_2$ flow ratios of 10 and 50 respectively. ITO gate was also sputter deposited at 1 W/cm^2 with no oxygen flow. The rest of the fabrication sequence was same as for ZnO TFTs. Both ZnO and IZO TFTs were subjected to post-fabrication anneal in N$_2$ ambience at 350 °C for 30 minutes. The IZO TFTs were further treated in a N$_2$O plasma in a PECVD chamber at 125 °C.

RESULTS and DISCUSSION

The structural, electronic and optical properties and growth rates of sputtered films are strongly influenced by various processing conditions, such as gas phase composition, plasma conditions, deposition temperature and geometry. Transmission is quite high (above 80%) in the visible range of the spectrum (400-700 nm) and it falls off sharply in the UV region due to the onset of fundamental absorption. Surface imaging with AFM reveal that the as-deposited films possess very smooth surface with minimal roughness. Figure 2(a) shows the dependence of the electrical resistivity for sputtered ZnO thin films on oxygen partial pressure. The resistivity underwent an abrupt transition from semiconducting ($\rho \sim 0.01$ ohm-cm) at low oxygen partial pressure to semi-insulating ($\rho \sim 10^6$ ohm-cm) at higher pressure. In undoped ZnO, excess, interstitial Zn ions or oxygen vacancies can contribute free electrons for electrical conduction [7, 8]. In the IZO thin films (Figure 2(b)), as the partial pressure of oxygen pressure (PO$_2$) increases, so does the resistivity of the IZO thin film. This behavior is related to the

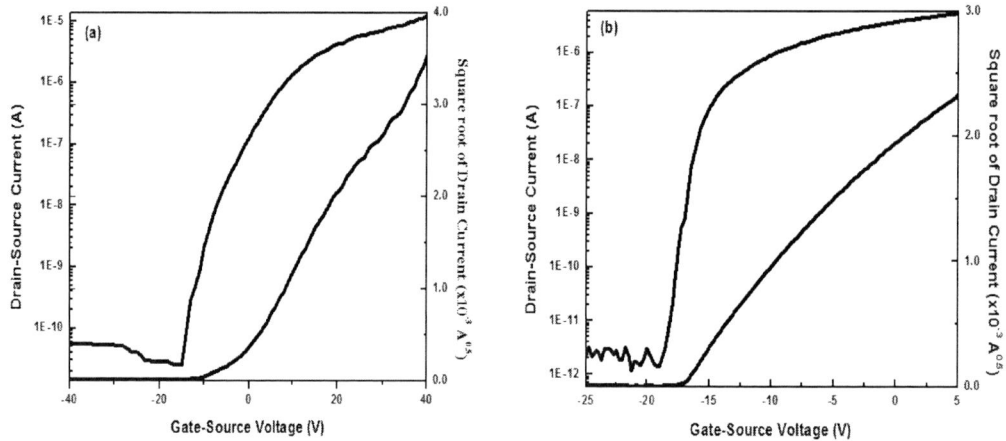

Figure 3: Transfer characteristics of (a) ZnO and (b) IZO TFT

elimination of the oxygen vacancies and consequentially the decrease on the carrier concentration [9]. The highest resistivity (1×10^3 ohm-cm) was obtained for PO$_2$ of 0.8 mTorr. IZO thin films that formed the active channel of our TFT had a resistivity of 40 ohm-cm and films with lower resistivity (1×10^{-4} ohm-cm) formed the source/drain. The Hall mobility measured from Van der Pauw structures were 20cm^2/V.s and 60cm^2/V.s

218

respectively. The specific contact resistance of IZO source/drain layer on IZO active channel layer as measured by Transfer length method was around 1×10^{-5} ohm-cm^2.

Based on the above optimized parameters for film deposition, we fabricated staggered bottom-gate TFTs. Figure 3a and b show the transfer characteristics of amorphous ZnO and IZO TFTs respectively with W/L = 64 μm/16 μm at a drain bias of 1 V. The ZnO transistors operate in n-type depletion mode and exhibit hard saturation. It is noticed that drain currents in the μA range can easily be obtained with voltages below 20 V. From the log(I_D) vs. V_{GS} plot, the subthreshold curve is estimated to be around 0.75 V/decade for amorphous ZnO TFTs and 0.46 V/decade for IZO TFTs. The low values obtained for the gate voltage swing indicates the existence of a low density of surface states at the semiconductor-insulator interface. Both ZnO and IZO TFTs have reasonable on/off current ratio around 10^6. The PECVD SiO$_2$ dielectric layer was effective in suppressing gate-leakage current below 10 pA as is observed from the low drain current for both type of TFTs. In the saturation mode, the field effect mobility calculated was 5 and 20 cm^2/V.s

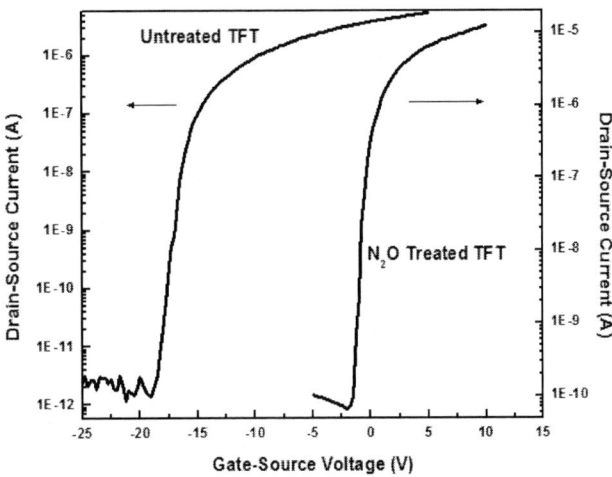

Figure 4: Effect of N$_2$O plasma treatment on IZO TFTs

for ZnO and IZO TFTs. The mobility values are at least an order of magnitude higher than that of conventional a-Si:H TFTs widely employed in display based electronics. As described above, the TFT obtained are depletion-mode devices which means even at zero gate bias there is significant source-drain current. For applications in display devices it is highly desirable to obtain enhancement-mode devices. As such we did a post-fabrication N$_2$O plasma treatment for the IZO TFTs. The idea is that nitrogen species in the plasma can reduce the sensitivity of relative O$_2$/Ar ratio to carrier concentration. Nitrogen is also known to act as an acceptor in ZnO system and can annihilate shallow donor levels. In Figure 4, it is easily seen that we can get enhancement mode TFTs by this simple plasma treatment. The resulting TFTs had steeper gate voltage swing of about 0.3 V/decade with On/Off current ratio going down slightly. Although this may be due to some plasma induced damage but the same treatment also improves device reliability and thus outweighs the drawbacks.

219

Finally, pseudo-NFET type inverters are designed and fabricated as a vehicle to realize circuit implementation utilizing amorphous oxide TFTs. In Figure 5, depletion-depletion mode IZO-NFET inverter operation at operation frequency of 1 kHz is shown. With different W/L ratio of drive TFT and load TFT, beta ratios of 1-16 are obtained. Stable inverter operation up to 50 kHz is possible.

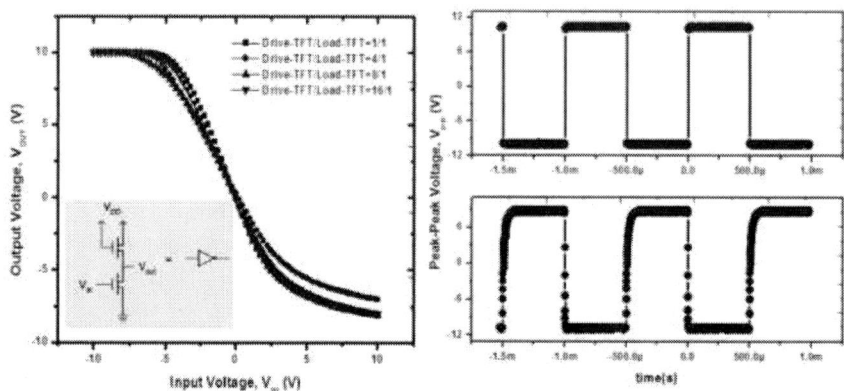

Figure 5: (a) Transfer characteristics of IZO inverter (circuit diagram inset) with different beta ratio (b) Corresponding timing diagram of Inverter operation at 1 KHz.

CONCLUSION

We have demonstrated the fabricated of both depletion mode and enhancement mode room temperature bottom-gate oxide (ZnO, IZO) TFTs by RF magnetron sputtering with good electrical characteristics. Tuning the oxygen partial pressure during deposition provides a simple way to control the carrier density in the oxide thin films. TFTs retained well-behaved transfer characteristics down to a channel length of 4 μm with on/off drain current ratio exceeding 10^5, low subthreshold swing and relatively high field effect mobility in the range of 5-20 cm^2/V.s These results make Oxide TFTs promising candidates for future display devices rendering possible high-resolution applications.

REFERENCES

1. Y. He, R. Hattori, and J. Kanicki, IEEE Electron Devices Letters, 21, 590-592 (2000)
2. K. Nomura et al, Japanese Journal of Applied Physics, 45, 4303-4308 (2006)
3. K. Nomura et al, Nature, 432, 288 (2004).
4. David Paine et al, Thin Solid Films, 516, 5894-5898 (2008)
5. David Hong and John F. Wager, J. Vac. Sci. Tech, B 23(6), L25-L27 (2005)
6. M.E. Thomas, M.P.Holloway and J. E. Mckay, J. Vac. Sci. Tech, A6, 2570 (1988)
7. H. Hosono et al, Journal of Non-Crystalline Solids, 354, 2796-2800 (2008)
8. A. J. Leenheer, Physical Review B **77**, 115215 (2008)
9. K. Uang, Thin Solid Films, 515, 2501–2506 92006)

Mater. Res. Soc. Symp. Proc. Vol. 1109 © 2009 Materials Research Society

Carrier Transport in Homo- and Heteroepitaxial Zinc Oxide Layers

Klaus Ellmer

Helmholtz-Zentrum für Materialien und Energie, dept. Solare Energetik (SE5),
Glienicker Str. 100, 14109 Berlin, Germany

ABSTRACT

Homo- and heteroepitaxial ZnMgO:Al, ZnCaO:Al and ZnO:Ga films have been grown on sapphire and ZnO substrates by RF (13.56 MHz) reactive magnetron sputtering from oxidic targets. The films grow epitaxially, i.e. with a preferred in-plane and out-of plane orientation. However, the heteroepitaxial films on sapphire exhibit a much higher crystallographic defect density, compared to the homepitaxial films. The ZnMeO films (Me – metal)on a-plane sapphire exhibit a lower defect density leading to higher Hall mobilities. Both, homo- and heteroepitaxial ZnO:Ga films with carrier concentrations $N > 10^{20}$ cm^{-3} exhibit the same mobility values, which increase with increasing carrier concentration. This behaviour is typical for electrical grain barrier limited transport, as decribed recently for polycrystalline ZnO:Al(Ga) films on glass. For the ZnCaO:Al films, deposited at similar conditions as the ZnO:Ga films, much lower carrier concentrations were measured, both for sapphire as well as for ZnO substrates. The mobilities of the ZnCaO:Al films on ZnO are much higher than that on the sapphire single crystals. The measured Hall mobilities are compared to single crystalline ZnO transport data.

Additionally, the work functions of the ZnMeO layers have been measured by X-ray and ultra-violet photoelectron spectroscopy. As expected, the work functions are lower compared to unalloyed ZnO, which can be used for ZnO band gap and band alignment engineering.

INTRODUCTION

Since more than a decade zinc oxide (ZnO) and its derivatives see a renaissance (the second one!), due to its prospective use for opto-electronic devices in the blue and ultra-violet spectral range [1]. One of the main research topics is p-type doping of ZnO in order to build pn junctions. ZnO and its derivatives are also of interest for transparent, conductive electrodes, especially for thin film solar cells [2]. In this field high transparency and at the same time low resistivity are the main goals. Also, alloying of ZnO is interesting for improving the band alignement in the heterojunction between the absorber and the window and contact layer in a thin film solar cell. Though many investigations on electrical transport in ZnO can be found in literature (for a review see [3]), a systematic comparison between homo- and heteroepitaxial films has not yet been performed. In this article the transport properties of homo- and heteroepitaxial ZnMe$_x$O$_y$ films are compared.

A series of deposition techniques, including chemical vapour deposition [4], pulsed laser ablation [5], magnetron sputtering [6-8], molecular beam epitaxy [9,10], and ion-beam assisted deposition [11] were used to grow ZnO thin films. In flat panel displays and thin films solar cells the zinc oxide films are degenerately doped with carrier concentrations up to 10^{21} cm^{-3}, which leads to quite low mobilities (due to ionized impurity scattering) limiting the resistivity that can be achieved in such films to about $2 \cdot 10^{-4}$ Ωcm [12]. Further investigations on the influence of crystallographic defects and of ionized impurities on the charge carrier transport are necessary to reduce the resistivity of such films.

The aim of this research project is to prepare highly doped zinc oxide films on sapphire substrates in order to investigate the charge carrier transport in epitaxially grown films and to compare it with that in polycrystalline films. Besides, we wanted to demonstrate that the magnetron sputtering process can be used not only for the preparation of polycrystalline films but also to grow epitaxial films on sapphire at low substrate temperatures. Single crystals of zinc oxide are very expensive. Therefore, heteroepitaxial growth on cheap single crystals, i.e. on sapphire was performed. These data for epitaxial and polycrystalline $ZnMe_xO_y$ films are compared to data from literature and with the reported data of single crystalline ZnO [3]. Furthermore, by photoelectron spectroscopy (PES), the work functions Φ of the different $ZnMe_xO_y$ layers have been measured, which are essential for semiconductor heterojunctions.

EXPERIMENT

The films have been prepared by reactive magnetron sputtering from oxidic targets by RF (13.56 MHz) plasma excitation in a turbo-pumped magnetron sputtering system with a base pressure of better than $1 \cdot 10^{-5}$ Pa, equipped with a load-lock. The following targes were used: $ZnO:Ga_2O_3 2mol\%$, $Zn_{0.88}Ca_{0.1}O:Al_2O_3 2mol\%$ and $Zn_{0.88}Mg_{0.1}O:Al_2O_3 2mol\%$. The magnetron sputtering sources with 76 mm diameter had a balanced magnetic field, leading to a low flux of highly energetic electrons to the substrate [7]. As substrates a- and c-plane sapphire and single crystalline ZnO (001 and 110) as well as borosilicate glass were used. Prior to deposition, the substrates were ultrasonically cleaned in isopropanol and heated in the vacuum chamber at 400 °C for 30 min.

The substrate temperature was varied from 370 to 970 K, while the total sputtering pressure (≈ 0.3 Pa) and the sputtering power (50 W) were held constant. The deposition rate was about 10 nm/min at low substrate temperatures and decreased at higher temperatures due to zinc evaporation [13].

The films were structurally characterized by X-ray diffraction with a conventional Bragg-Brentano diffractometer using filtered CuK_α ($\lambda=0.15418$ nm) radiation. Transmission electron microscopy on film cross sections (XTEM) was used to investigate the microscopic structure of some of the films. The film composition and the epitaxial quality was analyzed by Rutherford backscattering (RBS) combined with channeling [14].

The carrier concentration in these films was adjusted by small additions of oxygen or hydrogen to the Ar sputtering gas. X-ray diffraction was used to characterize the phases and the structural perfection (strain, grain sizes). The optical characterization has been performed by reflection and transmission measurements in the spectral range from 200 to 3300 nm. By room-temperature and temperature-dependent (50-300 K) Hall and conductivity measurements, the electrical transport properties were measured. The work function was measured by X-ray and ultra-violet photoelectron spectroscopy (XPS, UPS). To that purpose the samples were transferred without breaking the vacuum from the deposition chamber to the XPS system. The XPS and UPS analysis were been perfomed by MgK_α (1253.6 eV) and He I (21.22 eV) radiation, respectively.

RESULTS AND DISCUSSION

Structural analysis

The phase analysis has been performed by X-ray powder diffraction, which is displayed in Fig.1. The ZnO:Ga films grow in the zincite (wurtzite) crystallographic structure on all substrates. While the c-axis of the zincite unit cell is oriented perpendicular on glass, a- and c-sapphire substrates, it is oriented parallel to the surface on r-plane sapphire (not shown). The lattice mismatch between ZnO and sapphire leads to tensile stress in the grown films, which can be inferred from the shift of the (002) diffraction peak of ZnO towards higher 2θ values compared to the position of the (002) peak of the JCPDS powder diffraction file. The films grown on borosilicate glass do not grow epitaxially and are obviously less strained, visible from a better fit of the (002) peak position to that of the powder diffraction file.

By cross-sectional transmission electron microscopy the atomar structure of some heteroepitaxial films was analyzed, displayed in Fig.2. While the films grown on a-plane sapphire (110) exhibit a very homogeneous structure without extended crystallographic defects, the films on (001) sapphire display a much higher defect density and a clear columnar structure, which is also known from ZnO growth on glass substrates. Obviously, the better lattice alignment of ZnO on a-plane sapphire leads to a better heteroepitaxial growth, which can also be inferred from the XRD patterns (see Fig.1). The XTEM analysis was performed for Al-doped ZnO films. But it is expected that the XTEM results are comparable to ZnO:Ga films, since the XRD patterns of ZnO:Ga films and ZnO:Al films, which were reported recently [15], are quite similar. Though the structural quality is clearly different between ZnO:Ga films on a- and c-plane sapphire, the XRD analysis is not suited to differentiate between films of different electrical parameters, for instance by variation of deposition temperature and/or oxygen partial pressure. This can be inferred from annealing experiments in vacuum at a temperature of 500 °C up to 8 hours, which lead to an increase of the Hall mobility, but which did not change the XRD patterns, both with respect to (002) diffraction peak position and halfwidth as well as to the rocking curve width [16], see the electrical characterization in the next paragraph.

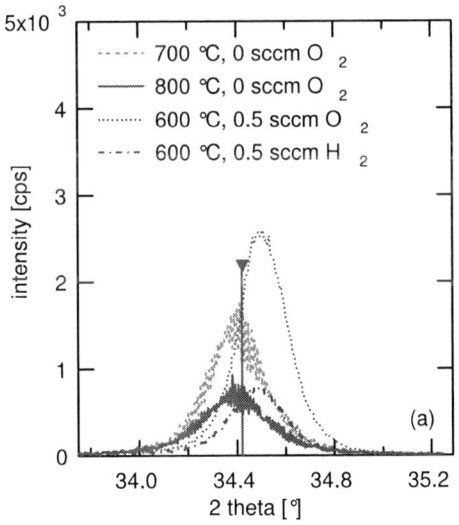
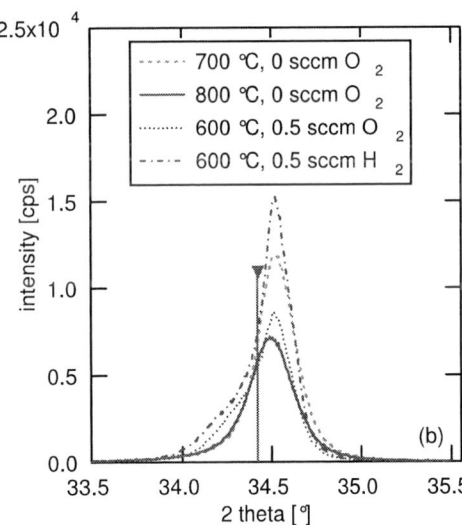

Fig.1: X-ray diffraction patterns of ZnO:Ga films on (a) borosilicate glass and (b) c-plane sapphire substrates at different deposition conditions (see the legends). The bar with the triangle displays the position of the (002) diffraction peak from the JCPDS powder diffraction file (card 1-47).

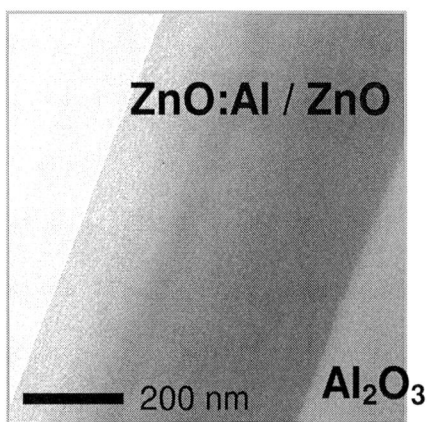 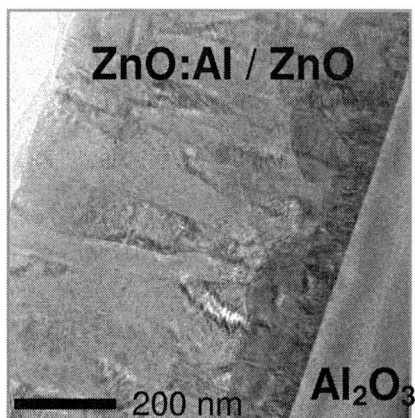

Fig.2: XTEM micrographs of two ZnO:Al films deposited on a-plane sapphire (left) and c-plane sapphire (right). The difference in the density of crystallograhic defects is clearly visible.

Rutherford backscattering

By Rutherford backscattering (RBS) analysis the crystallinity of the grown ZnO film has been checked over the film depth by comparing RBS spectra for random and channeling incidence (along the 110 direction) of the He ion beam (1.7 MeV). It was found that a film grown on (110) sapphire at 573 K exhibits a pronounced channeling minimum yield χ_{min} of about 6% at the film surface. But, towards the film-sapphire interface, a strong dechanneling of the ions is observed, which can be explained by an increasing defect density from the the ZnO:Al surface towards the film-substrate interface.

The dopant (Al) concentration was 3.5at%, which matches well with the aluminium content of the target (2.9 at%).

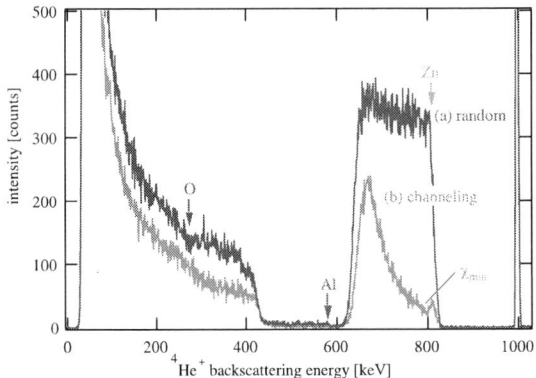

Fig.3: Rutherford backscattering spectra of a ZnO:Al film (250 nm) on a (110) sapphire substrate in random (a) and channeling (b) direction (1.7 MeV ^{4}He$^+$, scattering angle 170°). The energies for scattering at surface atoms of Zn, Al and O are marked by arrows.

Electrical measurements

The ZnMeO:Al films (Me – metal) were contacted by Ni/Au contacts in the van der Pauw geometry and measured at room temperature. The Hall mobilities and resistivities of the different films are displayed in Fig.4 as a function of the carrier concentration. It can be seen that the ZnO:Ga films exhibit much higher carrier concentrations compared to the ZnCaO:Al films ($N>10^{20}$ cm^{-3}). The Hall mobilities of the ZnO:Ga films are independent on the substrate (ZnO or sapphire) and it increase with increasing carrier concentration, a behaviour, which is typical for electrical grain barrier limited transport [17,18]. At $N \approx 4\cdot10^{20}$ cm^{-3} the Hall mobilities of the thin films approach the data reported of single crystalline ZnO, displayed as continuous line in Fig.4 (left). The ZnCaO:Al films, deposited at comparable sputtering conditions, exhibit much lower carrier concentrations and quite different mobilities for films on sapphire and homoepitaxial films on ZnO.

The trend of the Hall mobilities for the ZnCaO:Al and the ZnO:Ga films (Fig.4, left) and especially the very low mobilities at $2\cdot10^{18}$ cm^{-3} $<N<5\cdot10^{19}$ cm^{-3} is comparable to Hall mobility measurements reported by Makino et al. in 2005 [19]. This trend of the Makino data was explained by electrical grain barrier limited transport and points to a high grain boundary trap density in the range of $2\cdot10^{13}$ cm^{-2} [18,20]. This is quite surprising taking into account the well ordered epitaxial growth on sapphire and ZnO substrates in our own experiments and on lattice matched ScAlMgO$_4$ substrates in the case of Makino et al. Obviously, the lattice defects not easily detected by the typical structural analysis methods (XRD, rocking curve measurements),

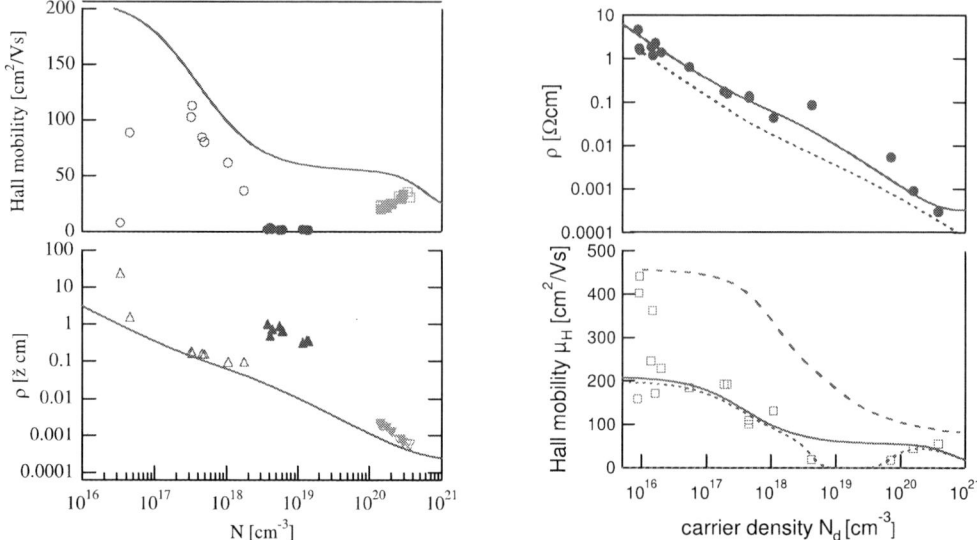

Fig.4: Left: Resistivity and Hall mobility as a function of the carrier concentration of ZnO:Ga (open and filled squares and down triangles) and ZnCaO:Al (open and filled circles and open and filled upward triangles) films on sapphire (filled symbols) and ZnO (open symbols) single crystalline substrates. The continuous lines in all figures are semiempirical fit curves to the literature data (mobility and resistivity) of single crystalline ZnO [18]. Right: Hall mobility and resistivity data of ZnO:Ga films epitaxially grown on lattice matched ScMgAlO$_4$ substrates by pulsed laser deposition (PLD), which displays a similar "mobility gap" as in the left figure[18,19]. The dotted lines are the mobility and resistivity fit curves of Makino et al. [19], which are obviously too high (mobility) or too low (resistivity).

lead to high defect densities at grain boundaries, inducing a charge carrier depletion at the grain boundaries by electron trapping. Only at very low carrier concentrations N<2·10^{18} cm^{-3} mobilities start to increase again, which is now due to a complete carrier depletion of the grains, making the potential barriers between the grains to vanish. At N≈5·10^{17} cm^{-3} the ZnCaO:Al mobilities approach the ZnO single crystal mobilities, which is a hint that the grain interiors are nearly defect free. The main scattering process at these carrier concentrations is ionized impurity scattering.

The alloying of ZnO with group II elements can be used to modify (extend) the band gap of zinc oxide. This has been analyzed by photoelectron spectroscopy (XPS and UPS) and optical band gap analysis. Fig.5 shows the position of the Fermi level relative to the valence band maximum E_F-E_{VBM}, optical band gap E_g, work function Φ and resistivity ρ for undoped ZnCaO and Al-doped ZnMgO:Al films as a function of the relative oxygen (positive) or hydrogen (negative) partial flow during sputtering. It can be seen that the work function increases with increasing oxygen partial pressure. Furthermore, the work functions of the ZnCaO and ZnMgO:Al films are significantly lower than that of pure ZnO, which are in the range of 4.5 eV. This is expected, since the addition of group II elements to ZnO should shift especially the conduction band edge upwards, as was found by density functional theory calculations, see for instance Robertson et al. [21]. This property of ZnMe(II)O alloys can be used for band gap tailoring and band alignment engineering of heterojunctions. The resistivities of the undoped ZnCaO films are very high compared to the Al-doped ZnMgO:Al films. Obviously, these films can be, due to the low electronegativity of Ca not be doped efficiently by zinc interstitials, i.e., under oxygen-deficient deposition conditions. If such films should be used as transparent electrodes, an extrinsic dopand has to be added, too(Al or Ga, for instance).

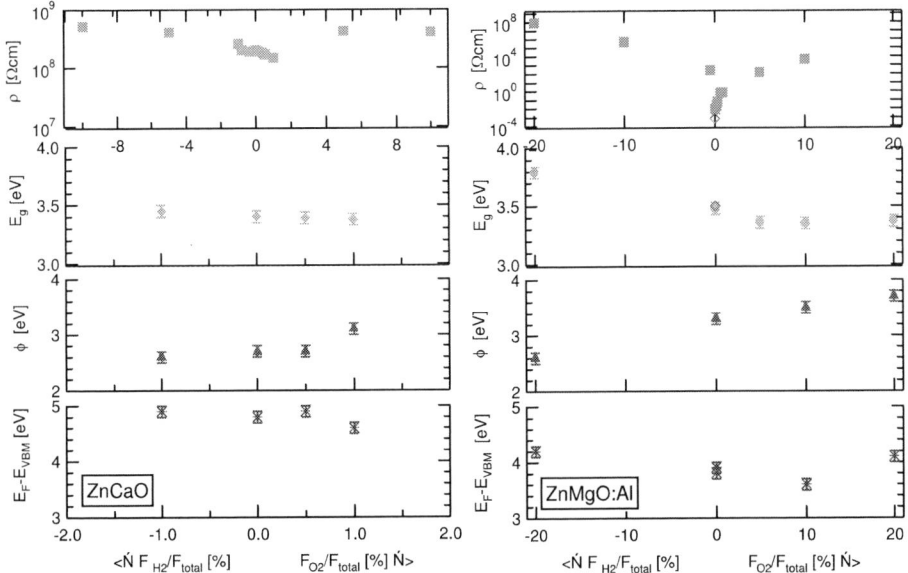

Fig.5: Position of the Fermi level relative to the valence band maximum E_F-E_{VBM}, optical band gap E_g, work function Φ and resistivity ρ for undoped ZnCaO films (left) and Al-doped ZnMgO:Al (right) as a function of the oxygen or hydrogen partial flow (in %) during magnetron sputtering. Positive F_{gas}/F_{total} values correspond to oxygen, negative ones to hydrogen addition.

CONCLUSIONS

ZnO:Ga and ZnMeO:Al alloy films have been prepared epitaxially on ZnO and sapphire single crystals by reactive RF magnetron sputtering from oxidic targets. Though the structural characterization by X-ray diffraction shows an in-plane and out-of plane orientation of the films relative to the substrate orientation, the films exhibit a high density of defects and, most probably, small angle grain boundaries, which significantly influence the electrical transport properties. The ZnCaO:Al and the ZnO:Ga films show significantly different carrier concentrations for comparable deposition conditions, which points to the different dopant activation (Ga or Al) when ZnO is alloyed with Ca. Also, the transport properties of ZnCa:Al films on sapphire and ZnO substrates are very different. While the homoepitaxial films on ZnO exhibit Hall mobilities approaching that of single crystalline ZnO, the ZnCaO:Al films on sapphire show very low Hall mobilities, which can tentatively explained by electrical grain boundary limited transport. These results point to the decisive role of crystallographic defects and electrical grain boundaries on the Hall mobility. The work function measurements by XPS and UPS yielded lower work functions for ZnCaO and ZnMgO films, which can be used for band gap and band alignment engineering of zinc oxide.

ACKNOWLEDGMENTS

The help of Matthias Resch, Peter Völz, Martin Roloff and Steven Franke in film preparation and electrical characterization is greatly acknowledged. I thank Rainer Grötzschel for the opportunity to perform RBS measurements at the institute of ion beam physics at the Forschungszentrum Dresden-Rossendorf (Germany).

REFERENCES

[1] U. Özgür, Y. I. Alivov, C. Liu, A. Teke, M. A. Reshchikov, S. Dogan, V. Avrutin, S.-J. Cho, and H. Morkoc, J. Appl. Phys. **98,** 041301-1...103 (2005).

[2] *Transparent Conductive Zinc Oxide: Basics and Applications in Thin Film Solar Cells*; *Vol.*, edited by K. Ellmer, A. Klein, and B. Rech (Springer, Berlin, 2008).

[3] K. Ellmer, *Electrical Properties*, in *Transparent Conductive Zinc Oxide: Basics and Application in Thin Film Solar Cells*, edited by K. Ellmer, A. Klein, and B. Rech (Springer, Berlin, 2008), p. 44.

[4] G. Galli and J. E. Coker, Appl. Phys. Lett. **16,** 439-441 (1970).

[5] J. Ma, F. Li, H. Ma, and S. Li, Thin Solid Films **279,** 213-215 (1996).

[6] Y. Igasaki and H. Saito, J. Appl. Phys. **69,** 2190-2195 (1991).

[7] R. Cebulla, R. Wendt, and K. Ellmer, J. Appl. Phys. **83,** 1087-1095 (1998).

[8] T. Minami, MRS Bull. **25,** 38-44 (2000).

[9] Y. R. Ryu, S. Zhu, J. M. Wrobel, H. M. Jeong, P. F. Miceli, and H. W. White, J. Cryst. Growth **216,** 326-329 (2000).

[10] I. Ohkubo, A. Ohtomo, T. Onishi, Y. Matsumoto, H. Koinuma, and M. Kawasaki, Surf. Sci. **443,** L1043-L1048 (1999).

[11] K. Nakahara, T. Tanabe, H. Takasu, P. Fons, K. Iwata, A. Yamada, K. Matsubara, R. Hunger, and S. Niki, Jap. J. Appl. Phys. **40,** 250-254 (2001).

[12] K. Ellmer, J. Phys. D: Appl. Phys. **34,** 3097-3108 (2001).

[13] R. Wendt and K. Ellmer, Surf. Coat. Techn. **93,** 27-31 (1997).

[14] *Handbook of Modern Ion Beam Materials Analysis*; *Vol.*, edited by J. R. Tesmer and M. Nastasi (MRS, Pittsburgh, 1995).

[15] P. Kuppusami, G. Vollweiler, D. Rafaja, and K. Ellmer, Appl. Phys. A **80,** 183-186 (2005).

[16] K. Ellmer and G. Vollweiler, Thin Solid Films **496,** 104-111 (2006).

[17] J. Y. Seto, J. Appl. Phys. **46,** 5247-5254 (1975).

[18] K. Ellmer and R. Mientus, Thin Solid Films **516,** 4620-4627 (2008).

[19] T. Makino, Y. Segawa, A. Tsukazaki, A. Ohtomo, and M. Kawasaki, Appl. Phys. Lett. **87,** 022101-1...3 (2005).

[20] T. Makino, Y. Segawa, A. Tsukazaki, A. Ohtomo, and M. Kawasaki, phys. stat. sol (c) **3,** 956-959 (2006).

[21] J. Robertson, K. Xiong, and S. J. Clark, Thin Solid Films **496,** 1-7 (2006).

Optical and Electrical Characteristics of Amorphous InGaZnO after Thermal Annealing

Satoshi Taniguchi, Norihiko Yamaguchi, Takao Miyajima and Masao Ikeda
Photonic Devices Laboratory, Advanced Materials Laboratories, Sony Corporation, Atsugi-Tec,
4-14-1, Asahi-Cho, Atsugi-Shi, Kanagawa, 243-0014 Japan
e-mail:Satoshi.Taniguchi@jp.sony.com

ABSTRACT

We studied electrical characteristics of an amorphous InGaZnO (a-IGZO) layer thermally annealed. To clarify the influence of a substrate and annealing conditions, the annealing time dependence and the depth profiles of carrier and impurity concentration were estimated using Hall effect-measurement and secondary ion-microprobe mass spectroscopy. We observed photoluminescence (PL) spectra from the a-IGZO layer as-deposited and annealed under various conditions. Intense broad deep-level emission (DLE) was detected at liquid nitrogen temperature. The PL intensity of DLE from the annealed layers was studied in relation to carrier concentration and was found to depend solely on carrier concentration despite a range of deposition and annealing conditions. To explain this result, we propose the interaction of atomic oxygen with intrinsic donor.

INTRODUCTION

Recently, transparent oxide semiconductors based on ZnO-materials have been attracting much attention in the field of transparent electronics and flexible electronics. Since the first demonstration of high-performance thin-film transistors (TFTs) fabricated on a plastic substrate [1], amorphous InGaZnO$_4$ (a-IGZO) has been intensively studied due to its unique properties such as transparency derived from its large band-gap, high electron mobility (~10 cm^2/V-s) even in the amorphous phase, low growth temperature down to room temperature and low carrier concentration controlled with oxygen (O$_2$) partial pressure during deposition [1-3]. Given its high mobility and reliability, a-IGZO may be a favorable channel material for TFTs instead of amorphous Si, since these properties are required for TFTs adopted to giant flat-panel displays of organic light emitting diodes or liquid crystals.

For a-IGZO-based TFTs, thermal annealing is an indispensable process for TFT fabrication [4, 5]. Nonetheless, there have been few investigations of the electrical properties of annealed layer. We studied the carrier and impurity distribution in the depth profile using Hall-effect measurements and secondary ion-microprobe mass spectroscopy (SIMS). We found that extrinsic impurity diffusion from the glass substrate into the a-IGZO layer influences the carrier distribution and confirmed that the behavior of the intrinsic donor such as oxygen vacancy (V_O) is evident on the a-IGZO layer on the sapphire substrate.

While photoluminescence (PL) is a useful technique to investigate layer quality, there are few reports on PL observation of a-IGZO layer. We observed PL spectrum from the a-IGZO layer with various carrier concentrations [6]. The intensity of deep-level emission (DLE) observed around 700 nm fell as the carrier concentration was increased by controlling O_2 partial pressure during deposition. We consider that this behavior can be attributed to the non-radiative deep center accompanying the intrinsic donor. In this paper, the annealing influence on the PL spectrum and carrier concentration was investigated.

EXPREIMENTAL DETAILS

An a-IGZO layer was deposited on a SiO_2 glass substrate (Corning #1737) or a sapphire substrate with the substrate temperature at room temperature or 80 °C by faced target sputtering using a polycrystalline IGZO target. The carrier concentration was varied from 10^{16} cm^{-3} to 10^{18} cm^{-3} by varying O_2 partial pressure from 0.003 Pa to 0.005 Pa. The layer thickness was 700-1000 nm.

The carrier concentration was evaluated by Hall-effect measurement using the van der Pauw configuration. The depth profile of the carrier concentration was estimated using this measurement and repeatedly HCl wet etching the a-IGZO layer. Thermal annealing in air was carried out using a hot-plate at an annealing temperature (T_a) ranging from 150 °C to 350 °C without capsulation. The annealing time (t_a) dependence of the carrier concentration was estimated using repeated Hall-effect measurement and accumulative annealing at constant T_a. We adopted Au/Ni ohmic contact metal deposited by electron beam evaporation as the stable electrode against HCl etchant and annealing. SIMS measurements were done to identify the depth profiles of impurities.

The Optical properties of the a-IGZO layer were characterized by PL spectral measurements at 77 K by using 325 nm excitation line of a He-Cd laser and a power density of 1.3 W/cm^2 as the light source. For the PL measurement of a-IGZO layer, the sapphire substrate was adopted to avoid unintentional luminescence and extrinsic impurity incorporation from the glass substrate during annealing. Samples were annealed under an O_2 or nitrogen (N_2) ambient using an infrared furnace system to study the relationship between carrier concentration and the PL spectrum.

RESULTS AND DISCUSSION

Electrical properties of a-IGZO layer on glass substrate

To study the carrier distribution in the a-IGZO layer, the depth profiles of carrier and impurity concentration were evaluated using Hall-effect measurements with repeated-step etching and SIMS measurements. Figure 1a shows the depth profile of carrier concentration from the a-IGZO layer as-deposited and annealed at 250 °C for 600 s in air. Although the as-deposited sample shows uniform distribution of carrier concentration through layer thickness, the annealed sample shows the gradual increase of carrier concentration toward the layer surface except for the interface between the a-IGZO layer and the glass substrate. In figure 1b, the hydrogen (H), the boron (B), and the carbon (C) atom distributions are shown from the layer as-deposited and annealed at 250 °C for 3600 s in air. The depth profiles indicate that the annealed layer includes B in the order of 10^{18} cm^{-3}, while there is little trace of B in the as-deposited layer. The B depth profile plotted in figure 1a (solid line) agrees well with the carrier distribution from the annealed layer in spite of the t_a difference.

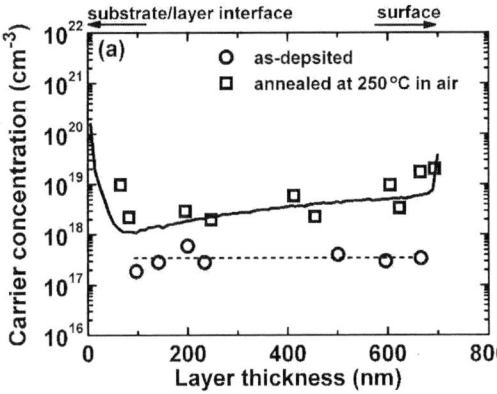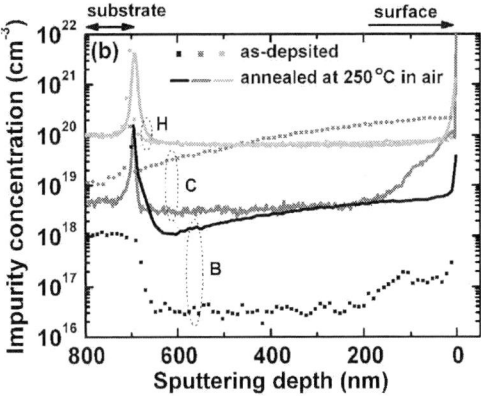

Figure 1. Depth profiles of carrier concentration from Hall-effect measurements using repeated-step etching (a) and impurity concentration of H, B, and C atom from SIMS measurement (b) from the a-IGZO layer as-deposited and annealed at 250 °C in air. Annealing time was 600 s (a) and 3600 s (b), respectively.

Thus, we consider that the carrier distribution in figure 1a is a result of B thermally diffusing into the layer from **the glass substrate** or **the interface between the substrate and the a-IGZO layer.** It would seem, therefore that introducing impurity-blocking layer such as SiO_2 or SiN between the a-IGZO layer and the glass substrate can block extrinsic impurity incorporation from the substrate or **the interface.** In order to exclude the influence of the extrinsic impurity, we simply adopted sapphire substrate which is thermally stable and high quality, for a further experiment.

Electrical properties of a-IGZO layer on sapphire substrate

In figure 2, t_a dependence of carrier concentrations is shown using the a-IGZO layer deposited on a sapphire substrate and annealed at 250 °C and 350 °C in air, respectively. In this figure, two distinguishable stages are observed from each t_a dependence. In stage I, a rapid increase of carrier concentration is seen from the each dependence, and in stage II, there shows the carrier saturation at T_a of 250 °C and the gradual decrease at T_a of 350 °C after stage I. The reason of the gradual decrease in stage II was studied using Hall-effect measurements with repeated-step etching. Figure 3 shows the distribution of carrier concentration from the a-IGZO layers annealed at 350 °C for 30 s and 4200 s. While the carrier concentration is uniform in thickness from the layer annealed for 30 s, the carrier gradually decreases toward the surface from the layer annealed for 4200 s. These distributions are different from that of the annealed layer in figure 1a. Since we can exclude the influence of extrinsic donors from the sapphire substrate, as we found no trace of impurities in SIMS measurements, the behavior in figure 3 can be attributed to the intrinsic donor. Since the depth profile for t_a of 4200 s in figure 3 was obtained from the sample in stage II in figure 2, the carrier decrease in stage II can be attributed to the carrier deficiency starting from the surface in figure 3. To elucidate the origin of this behavior, we investigated the optical properties of the layer annealed under different conditions.

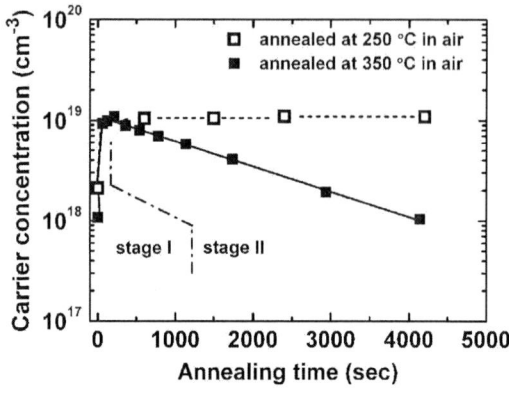

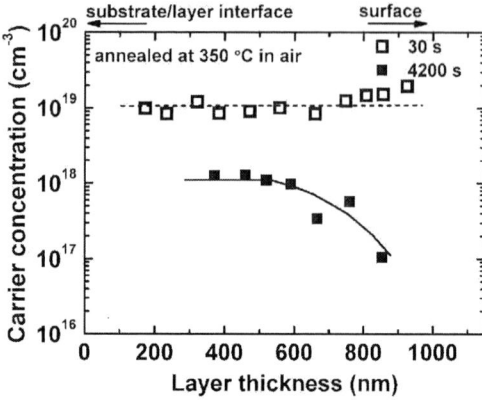

Figure 2. Annealing time dependence of carrier concentration of the a-IGZO layer deposited on sapphire substrate. Annealing temperatures were 250 °C and 350 °C.

Figure 3. Depth profile of carrier concentration from Hall-effect measurements using a step-etched a-IGZO layer deposited on a sapphire substrate. Carrier concentration of as-deposited the a-IGZO is about 2×10^{18} cm^{-3}. Annealing was done at 350 °C for 30 s and 4200 s in air.

Optical properties of a-IGZO layer on sapphire substrate

We used a sapphire substrate to avoid both unintentional luminescence and impurity incorporation from the glass substrate. The layer was annealed at T_a ranging from 150 °C to 350 °C for 1.5 h in air, under an O_2, or under a N_2 ambient respectively. Figure 4 shows the relationship between carrier concentration and DLE peak intensity and the inset in figure 4 depicts the PL spectrum from the layers as-deposited or annealed under an O_2 ambient.

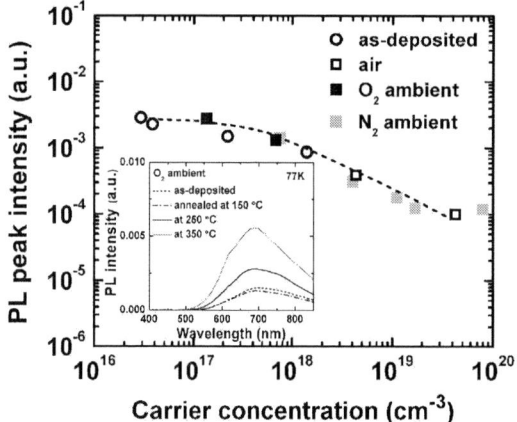

Figure 4. Carrier concentration dependence of DLE peak intensity from the a-IGZO layer as-deposited, annealed in air, under an O_2 ambient, or under a N_2 ambient. The inset shows the PL spectra from the a-IGZO layer annealed at various temperatures for 1.5 h under an O_2 ambient.

As the inset shows, PL spectra from a-IGZO layer consists of a broad DLE peaking around 700 nm and DLE exhibits enhancement in intensity with higher T_a.

In a previous report [6], we have shown that DLE intensity is reduced with raised carrier concentration by the decreased O_2 partial pressure during deposition, and we noted that V_O could be the origin of the carrier and non-radiative center. In the present work, carrier concentration and DLE intensity shows similar tendency despite different deposition and annealing conditions. This may be because, as seen in figure 5, an interaction of atomic oxygen (O) and V_O plays an important role during annealing. Here, the V_O is again supposed to act as intrinsic donor as well as non-radiative center. During annealing under an O_2 ambient, O is supplied through the O-stabilized surface and diffuses into the layer to inactivate V_O. Thus, carrier concentration decreases and DLE is enhanced together with decrease of non-radiative center. This will be supported by the depth profile in figure 3, where carrier deficiency occurs especially toward the surface during annealing in air. The O_2 in air is speculated to inactivate V_O in the same way of annealing under an O_2 ambient. On the other hand, annealing under a N_2 ambient enhances V_O generation by the release and out-diffusion of O toward the surface. Therefore, carrier concentration increases accompanied by DLE reduction. While we believe that O incorporation into the a-IGZO layer and the inactivation of V_O is the dominant process in stage II in figure 2, the rapid carrier increase in stage I, probably can be attributed to the different donor activation process. We would offer that carrier stabilization and reproducible TFT characteristics are influenced by capping layer such as SiO_2 or SiN for the purpose of suppressing O incorporation or removal from the a-IGZO surface.

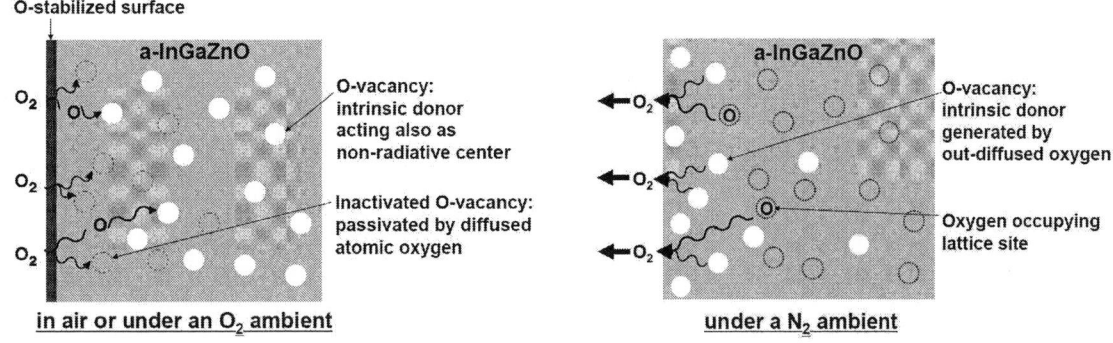

Figure 5. Proposed process during annealing in air, under an O_2 or under a N_2 ambient.

SUMMARY

The agreement of the depth profiles of carrier and impurity concentration indicates that boron in the glass substrate thermally diffuses into the a-IGZO layer and acts as donor. From the annealing time dependence of carrier concentration, the a-IGZO layer deposited on a sapphire substrate shows carrier decrease for a long-time annealing at 350 °C. This carrier decrease is found to be due to the carrier deficiency occurring in the layer, especially near the surface.

PL spectra of the a-IGZO layer under various annealing conditions are estimated. The relationship between the intensity of deep-level emission and carrier concentration is constant in varying annealing conditions. From these results, we consider that interaction between atomic oxygen and O vacancy acting as the intrinsic donor and non-radiative center plays an important role during annealing. Introducing both an impurity blocking layer and a capping layer is necessary to avoid the influence from the substrate and the ambient conditions.

ACKNOWLEDGEMENTS

The authors would like to thank Ichiro Hase for fruitful discussions.

REFERENCES

1. K. Nomura, H. Ohata, A. Takagi, T. Kamiya, M. Hirano, and H. Hosono, *Natuer* **432**, 488-492 (2004).
2. H. Yabuta, M. Sano, K. Abe, T. Akiba, T. Den, H. Kumomi, K. Nomura, T. Kamiya, and H. Hosono, *Appl. Phys. Lett.* **89**, 112123 (2006).
3. D. Kang, H. Lim, C. Kim, I. Song, J. Park, Y. Park, and J. G. Chung, *Appl. Phys. Lett.* **90**, 192101 (2007).
4. H. Hosono, K. Nomura, Y. Ogo, T. Uruga, and T. Kamiya, *J. Non-Cryst. Solids*, **354**, 2796-2800 (2008).
5. H. Q. Chiang, B. R. McFarlane, D. Hong, R. E. Presley, and J. F. Wager, *J. Non-Cryst. Solids*, **354**, 2826-2830 (2008).
6. N. Yamaguchi, S. Taniguchi, T. Miyajima, and M. Ikeda, *abst. of The 5ht International Workshop on ZnO and Related Materials,* Michigan, D4 (2008).

AUTHOR INDEX

Akita, Y. 132
Al-Amoody, F. 104
Aoi, T. 92
Avrutin, V. 126
Banerji, P. 172
Barnett, A. 198
Baski, A. A. 153
Bhattacharya, A. 108
Brillson, L. J. 49
Cantwell, G. 49
Chattopadhyay, S. 172
Cheng, T. 146
Chiu, K. 159
Cho, S. 114
Den, T. 43
Dong, Y. 49
Douglas, J. F. 177
Doutt, D. 49
Drzaic, P. 69
Ellmer, K. 221
Fagan, J. 177
Fang, Z. Q. 49
Feng, P. X. 204
Fenwick, W. 7, 55
Ferguson, I. 7, 55
Foussekis, M. 153
Fried, M. 31
Fukuda, K. 98
Funaki, M. 187
Gassenbauer, Y. 75
Ghosh, K. 108
Goyal, A. 43
Gruner, G. 69
Gupta, R. K. 108

Han, G. 63
Hasegawa, T. 15, 25, 86
Hatalis, M. 216
Hayashi, R. 92
Hecht, D. 69
Hirano, M. 120
Hirohata, K. 37
Hirose, Y. 15, 86
Hitosugi, T. 15, 25, 86
Hoang, N. L. H. 15, 25
Horvath, Z. 31
Hosaka, M. 132
Hosono, H. 120
Hsu, W. 146
Hu, L. 69
Huang, J. 146
Huang, W. 104
Ikeda, M. 229
Inoue, K. 187
Irvin, G. 69
Itagaki, N. 43
Iwasaki, T. 43, 120
Izyumskaya, N. 126
Jain, F. 104
Jamil, M. 7, 55
Jean, R. 159
Jin, C. S. 63
Jo, J. 167
Juhasz, G. 31
Kahol, P. K. 108
Kamiya, T. 120
Kamiyama, T. 1
Kao, Y. 159
Kasai, J. 15, 25

AUTHOR INDEX

Kasami, M. 187
Kato, H. 98
Kato, Y. 132
Khan, S. A. 216
Kim, H. 167
Klein, A. 75
Konuma, S. 86
Korber, C. 75
Kumomi, H. 43, 92, 120
Labadi, Z. 31
Ladous, C. 69
Li, N. 7, 55
Lin, C. 146
Lin, G. 146
Liu, C. 146
Liu, K. 114, 146
Look, D. C. 49
Major, C. 31
Majumdar, S. 172
Makise, K. 187
Melton, A. 7, 55
Meng, F. 146
Miao, W. 63
Migler, K. 177
Misra, V. 210
Mita, M. 132
Miyajima, T. 229
Morell, G. 204
Morkoc, H. 126
Mosbacker, H. L. 49
Muth, J. F. 210
Myers, M. 49
Nakao, S. 15, 25, 86
Nause, J. 55

Nemeth, A. 31
Nishi, Y. 37
Nomura, K. 120
Novak, S. 126, 210
Obrzut, J. 177
O'Connell, M. 69
Oi, H. 132
Oka, N. 1, 37, 92
Omura, H. 120
Ono, Y. 132
Papadimitrakopoulos, F. 104
Park, Y. 69
Petrik, P. 31
Rao, G. V. 75
Reshchikov, M. A. 126, 153
Rodriguez, A. 104
Sakata, O. 132
Sanno, Y. 1
Sato, Y. 1, 37, 92
Sauberlich, F. 75
Schafranek, R. 75
Shibuya, T. 187
Shigesato, Y. 1, 37, 92
Shimada, R. 126
Shimada, T. 15, 25, 86
Shinozaki, B. 187
Shur, M. 114
Siahmakoun, A. 198
Simien, D. 177
Song, J. J. 49
Suarez, E. 104
Sugimoto, W. 98
Sugimoto, Y. 132
Sun, S. 138

AUTHOR INDEX

Suresh, A. 210

Syed, M. 198

Takasu, Y. 98

Tamulailtis, G. 114

Tan, L. K. 63

Taniguchi, S. 229

Tayim, D. 49

Thomas, D. 69

Thompson, C. V. 63

Tompa, G. S. 138

Uppireddi, K. 204

Utsuno, F. 187

Wachau, A. 75

Wang, S. 55

Wellenius, P. 210

Wilkerson, J. 198

Willner, B. I. 138

Xu, T. 55

Yamada, N. 15, 25, 86

Yamaguchi, N. 229

Yamamoto, I. 37

Yang, B. 204

Yano, K. 187

Yoshimoto, M. 132

Yu, H. 55

Yun, J. 167

Zaidi, T. 7

Zgrabik, C. 49

Zhang, J. 49